FOOD MICROBIOLOGY RESEARCH TRENDS

FOOD MICROBIOLOGY RESEARCH TRENDS

MARTA V. PALINO
EDITOR

Nova Science Publishers, Inc.
New York

For permission to use material from this book please contact us:
Telephone 631-231-7269; Fax 631-231-8175
Web Site: http://www.novapublishers.com

LIBRARY OF CONGRESS CATALOGING-IN-PUBLICATION DATA

Food microbiology research trends / Marta V. Palino (editor).
p. ; cm.
Includes bibliographical references and index.
ISBN-13: 978-1-60021-723-4 (hardcover)
ISBN-10: 1-60021-723-0 (hardcover)
1. Food--Microbiology. 2. Food contamination. I. Whitnall, Lyndal S.
[DNLM: 1. Food Microbiology. 2. Food Contamination. 3. Research. QW 85 F6845 2007]
QR115.F6623 2007
664.001'579--dc22

2007015418

Published by Nova Science Publishers, Inc. ✦ New York

Contents

Preface		**vii**
Expert Commentary	Future Trends on Applying Predictive Models in a Quantitative Microbiological Risk Assessment Framework *A. Valero and G. Zurera*	**1**
Chapter 1	The Applications of DNA and Protein Arrays in Microbial Food Safety and Processing *Christine M. Carey and Magdalena Kostrzynska*	**3**
Chapter 2	Research Trends in Lactic Acid Fermentation *E. H. Drosinos and S. Paramithiotis*	**39**
Chapter 3	Conceptual and Methodological Foundations for Developing Microbial Risk Assessment Models *F. Pérez-Rodríguez, R. M. García-Gimeno and G. Zurera*	**93**
Chapter 4	Current Status on the Etiology, Epidemiology, Food Safety Implications and Control Measures in Escherichia coli O157:H7 Infections *G. Normanno*	**131**
Chapter 5	Lactic Acid Fermentation of Fruits and Vegetables - An Overview *Ramesh C. Ray and Smita H. Panda*	**155**
Chapter 6	Control of Harmful Food-Borne Gram-Negative Bacteria by Weakening Cell Membranes *Hanna-Leena Alakomi and Maria Saarela*	**189**
Chapter 7	Real-Time Detection of Foodborne Pathogen Contamination in Food Using GC/MS and Electronic Sensors Integrated with Pattern Recognition Techniques *Ubonrat Siripatrawan*	**217**
Index		**245**

PREFACE

This book presents new and important research from aound the world in the field of food microbiology. Included in the scope are: Physiology, genetics, biochemistry, and behavior of microorganisms that are either used to make foods or that represent safety or quality problems; Effects of preservatives, processes, and packaging systems on the microbiology of foods; Methods for detection, identification and enumeration of foodborne microorganisms or microbial toxins; Microbiology of food fermentations; Predictive microbiology; Microbial ecology of foods; Microbiological aspects of food safety and Microbiological aspects of food spoilage and quality.

Chapter 1 - With the continued emergence of foodborne pathogens and strain diversification, control and analysis of our food supply has become increasingly reliant on molecular-based methods. DNA and protein array analyses are emerging molecular techniques, with potential to become leading tools in food microbiology research associated with bacterial, protozoan and viral pathogens. Microarray has varied applications in microbial food safety, including bacterial identification and screening, as well as gene expression profiling in response to stressors and various microbial systems containing mixed bacterial populations. Oligonucleotide microarrays can detect virulence genes and simultaneously discriminate among foodborne human pathogens, such as *Bacillus, E. coli*, *Salmonella* and *Shigella* strains. Foodborne pathogens may also be used as bioweapons against humans, livestock and agricultural products. As part of a successful biodefence strategy and emergency preparedness, DNA and protein arrays represent broad-spectrum techniques capable of simultaneously identifying multiple foodborne pathogens, their associated toxins and virulence genes. Microarray can correlate phenotypic antibiotic resistance of starter cultures and probiotics used in food to the specific genetic determinants, as well as examine the production and utilization of fermented products by bacteria. Microarray can also be used as a biosensor, whereby bacterial proteins and analytes can be rapidly detected and characterized. Native toxins associated with cholera, diphtheria, anthrax, tetanus and staphylococcal enterotoxins have been detected and quantified using a multiplex antibody-based microarray. As well, protein arrays are beneficial for serological typing. As complete genomic sequences of microbial pathogens and hosts become available, microarray technology will enable a more detailed investigation of the interaction and establishment of infection of food pathogens in human and animal hosts. This chapter will discuss the widespread application and recent investigations of DNA and protein array in food microbiology, focusing on microarray potential in food safety research and processing.

Chapter 2 - Spontaneous fermentation has been applied for thousands of years mainly as a method for preservation of excess foodstuff and secondarily as a way to obtain spicy and flavorful foods. Numerous fermented foods are consumed worldwide, owing their diversity to the variety of the raw materials. The microorganisms responsible for fermentation constitute a part of the microbiota indigenously present in the raw materials. Lactic acid fermentation is one of the oldest and most widespread types of fermentation. It has been successfully applied in cereals, fruits, vegetables, meat and milk yielding an extensive range of foods and beverages. During the last decades, a great amount of effort has been made to study these ecosystems and the biodiversity of the microorganisms that form their microbiota. Furthermore, research has focused on the microbial properties that affect the physicochemical and organoleptic properties of the final products, their nutritional value as well as their safety. The outcome of years of research is quite promising but further investigation is still essential.

Chapter 3 - According to the definition provided by the FAO (Food Agriculture Organization), Risk Assessment (RA) is a process consisting of four stages: Hazard Identification (HI), Hazard Characterisation (HC), Exposure Assessment (EA) and Risk Characterisation (RC); the result of each stage is the starting point for the next stage (FAO/WHO, 1995). Lamerding and Fazil (2000) defined RA for microbial hazards (MRA) as the process that provides an estimation of the probability and impact of an adverse health effect attributable to potentially contaminated food.

RA techniques have been developed over the years by disciplines such as Civil Engineering, Economics and Investment, and Toxicology. This experience has been used as a foundation to establish an approach to food safety. RA was first applied to the field of food microbiology just a few years ago, since it needed to be developed to a greater extent for this purpose. The dynamic and complex nature of biological systems (natural variability, growth, death, survival, etc) that are subjected to MRA adds an extra layer of difficulty not found in other types of RA (e.g., chemical and environmental hazards) where the object studied is an inert system, which does not undergo modifications (Lammerding and Fazil, 2000). Hence, models must be developed in order to simulate or predict the behaviour of microorganisms as accurately as possible.

Chapter 4 - *Escherichia coli* O157:H7 is an important foodborne pathogen. It was recognized as a cause of severe human illness only in the early 1980's. Since then it has been implicated in foodborne disease outbreaks in many countries throughout the world. Cattle are the main reservoir of this microrganism, and transmission of the infection to humans occurs primarily through the consumption of contaminated food. The pathogenicity of *E. coli* O157:H7 is associated to genes encoding for a number of virulence factors, especially the verotoxin-encoding genes. The infective dose is extremely low - in the order of a few bacterial units. *E. coli* O157:H7 is acid-tolerant and survives in acidic environments such as the gastric barrier and acidic foods. The most severe illnesses induced by *E. coli* O157:H7 are Hemorrhagic Colitis and Hemolytic Uremic Syndrome. Both these clinical forms are characterized by severe morbidity and might sometimes be lethal. Children are the most commonly affected age group and they are affected by the highest death rate. *E. coli* O157:H7 is one of the main food safety hazards having important implications for human health worldwide, since outbreaks lead sometimes to cases of mortality. This review examines the etiological and epidemiological aspects of *E. coli* O157:H7 infection and focuses on the food safety concerns raised by *E. coli* O157:H7 and on control methods for the prevention of food poisoning.

Chapter 5 - Lactic acid (LA) fermentation has many benefits including feasible in small scale, inexpensive, does not require additives and confers organoleptic characteristics to food stuffs. It not only produces LA but also imparts taste, aroma and flavour to the fermented products and preserve nutrients, vitamins such as ascorbic acid and coloured pigments (β-carotene and anthocyanin) of fruits and vegetables, which are considered as antioxidants. Fermented fruits and vegetables contribute significantly to the diets of rural populations in a number of developing countries. The main reasons of LA fermentation are nutritional, physiological and hygienic aspects of the process. Fermentation of fruits and vegetables can occur 'spontaneously' by the natural lactic bacterial surface microflora, i.e. *Lactobacillus* spp., *Leuconostoc* spp., *Pediococcus* spp., etc.; however, the use of starter culture such as *Lactobacillus plantarum* and *Lb. acidophilus* provides consistency and reliability of performance. LA bacteria are considered to be powerful probiotics that improves human intestinal microbial balance and enhance health by inhibiting the growth of pathogens such as *Escherichia coli, Salmonella* and *Staphylococcus*. The probiotic species include *Lb. plantarum*, *Lb. acidophilus*, *Lb. rhamnosus* and *Lb. gasseri*.

Many products like *kimchi*, *sauerkraut* and pickles are produced by LA fermentation of fruits and vegetables, but most economically relevant are fermentations of olives, cucumbers and cabbage. Fermented vegetables commonly produced in Asia are *gundruk* (fermentation of mustard, radish, cauliflower and other leafy vegetables in Nepal and Bhutan), *kimchi* (fermented Korean cabbage and radish), *sinki* (fermented radish tap roots in Nepal and Bhutan), *kanji* (fermented carrots in India and Pakistan), *dhamuoi* (fermented cabbage and other vegetables in Vietnam), *dak-gua-dong* (fermented mustard leaves in Thailand), *hua-chai po* (fermentation of carrots and turnips in Thailand) and *khalpi* (fermented cucumbers in Nepal). The other examples of LA bacterial and yeast fermentations are *gari*, *fufu*, *lafun*, *dawa dawa* and *agbelima* which are produced in African continent from cassava, a tropical root crop. Likewise, sweet potatoes are fermented by LA bacteria into lacto-pickle and lacto-juice.

Chapter 6 - Among common food spoilage microbes there are a number of Gram-negative species, some of which are responsible for many serious infections such as various serovars of *Salmonella enterica* and *Escherichia coli*, and others which create various hygienic problems and spoilage of foods by their metabolic products. Gram-negative bacteria are typically inherently resistant to many external agents intended for antimicrobial action, and for this reason they pose a specifically difficult problem in the food protection. This resistance is largely due to the membrane structures of the cell, i.e. outer membrane (OM) present as the outermost layer in all Gram-negative bacteria. However, it is possible to specifically weaken the OM by various agents that disintegrate the lipopolysaccharide layer. Such agents are collectively termed permeabilizers.

Many parameters effect the survival and growth of microbes in food, one of which is the pH. Weak organic acids, like citric and lactic acid, in addition to their antimicrobial property due to the lowering of the pH, also function as permeabilizers of the Gram-negative bacterial outer membrane. Permeabilization of Gram-negative bacteria by substances suitable for food protection, e.g. lactic acid, would offer an optimal mean to prevent the growth of these microbes in food materials and during different stages of food production.

This chapter focuses on the role Gram-negative bacteria as harmful bacteria and gives an overview of the resistance mechanism of Gram-negative bacteria OM as well as clarifies the mechanisms of antimicrobial agents intended to weaken these protective structures.

Chapter 7 - Among common food spoilage microbes there are a number of Gram-negative species, some of which are responsible for many serious infections such as various serovars of *Salmonella enterica* and *Escherichia coli*, and others which create various hygienic problems and spoilage of foods by their metabolic products. Gram-negative bacteria are typically inherently resistant to many external agents intended for antimicrobial action, and for this reason they pose a specifically difficult problem in the food protection. This resistance is largely due to the membrane structures of the cell, i.e. outer membrane (OM) present as the outermost layer in all Gram-negative bacteria. However, it is possible to specifically weaken the OM by various agents that disintegrate the lipopolysaccharide layer. Such agents are collectively termed permeabilizers.

Many parameters effect the survival and growth of microbes in food, one of which is the pH. Weak organic acids, like citric and lactic acid, in addition to their antimicrobial property due to the lowering of the pH, also function as permeabilizers of the Gram-negative bacterial outer membrane. Permeabilization of Gram-negative bacteria by substances suitable for food protection, e.g. lactic acid, would offer an optimal mean to prevent the growth of these microbes in food materials and during different stages of food production.

This chapter focuses on the role Gram-negative bacteria as harmful bacteria and gives an overview of the resistance mechanism of Gram-negative bacteria OM as well as clarifies the mechanisms of antimicrobial agents intended to weaken these protective structures.

In: Food Microbiology Research Trends
Editor: Marta V. Palino, pp. 1-2
ISBN 978-1-60021-723-4

Expert Commentary

FUTURE TRENDS ON APPLYING PREDICTIVE MODELS IN A QUANTITATIVE MICROBIOLOGICAL RISK ASSESSMENT FRAMEWORK

A. Valero and G. Zurera
Departament of Food Science and Technology
University of Córdoba (Spain)

Currently there are several choices available into mathematical modelling framework to microbial risk assessors, from descriptive evaluations until more complex models and detailed analysis. In fact, in a Quantitative Microbiological Risk Assessment (QMRA) scheme, a combination of both methodologies is commonly incorporated in order to provide qualitative or quantitative information regarding a specific foodborne hazard. In many times methodology to be applied will depend on the severity of the problem to deal with, as well as the necessary sources and management options to face this problem.

Predictive models usually differ in their complexity, but not always an application of a complex model is useful in a QMRA scheme. The generalization capacity of a predictive model should be compared to its ease of use, its robustness and accuracy, or its validation performance in foods. In any case, estimations generated from a predictive model must be as closer as possible to a real case.

A QMRA process presents some limitations, such as the lack of data availability that leads to an establishment of assumptions when estimating the microbial risk in a target population. Further research should be focused on validating these risk estimations on independent data to obtain a more reliable QMRA. Another remarkable point lies on integrating developed predictive models inside a QMRA study. Nowadays several models exist in scientific literature, but they are not applicable in some cases to food industries or they can not be used properly by risk managers and risk assessors mainly due to an inadequate integration with expert knowledge and epidemiological or food consumption data. Therefore, predictive models should be designed in order to facilitate this integration in a microbial risk estimation process. This estimation has to be dynamic and updating in accordance to food industries necessities and evolution of population patterns. Future studies have to follow these guidelines to ensure an adequate control of microbial risk in foods.

Food safety measures are focused on generating data and covering some aspects about lack of information from the scientific community. The main points to be addressed are:

Data requirement: A QMRA process needs both qualitative and quantitative data to determine the exposure level to a foodborne pathogen in foods at the moment of consumption, and to characterize this exposition to estimate the probability of illness/population/year. However, there are relatively few available data at this respect, and a lack of specificity to identify the strains implicated in food outbreaks. Likewise, some data sources regarding epidemiology and consumption levels are scarce. As an example, dose-response data to estimate the probability of illness to different exposure levels of the pathogen are limited for some specific food commodities and microbiological hazards.

QMRA framework: It should be flexible enough to adopt different risk management necessities. However, foodstuff sectors rarely have strict control on the overall food chain, from the raw material to the final food product. This is the case of certain foods destined to international trade in which the raw material proceed from one country, and the final food product is elaborated in another one. In this situation it is quite difficult to apply a QMRA approach and critical control points in a HACCP plan. However, critical levels of a particular foodborne hazard could be quantified through the use of Food Safety Objectives (FSO). Once FSOs are established, food industries should transfer them to their own criteria in order to be applied in a food product/process.

Risk management: The QMRA process starts by setting the different requirements of a risk manager to take decisions concerning risk. There is a priority to carry out a consistent risk management process, determining the appropriate levels of health protection. It should be developed an international agreement inside the risk management framework in order to compare different food hygiene programs.

Currently there are a group of scientists and risk assessors that do not expect effective QMRA approaches as a short term and, on the other hand, another group of them criticise that is very difficult to obtain available data to determine a suitable level of food safety. However, the establishment of measures leading to a performance of QMRA at an international level is possible, once an agreement among the different countries involved will be reached.

In: Food Microbiology Research Trends
Editor: Marta V. Palino, pp. 3-38
ISBN 978-1-60021-723-4

Chapter 1

THE APPLICATIONS OF DNA AND PROTEIN ARRAYS IN MICROBIAL FOOD SAFETY AND PROCESSING

Christine M. Carey and Magdalena Kostrzynska
Food Research Program, Agriculture and Agri-Food Canada,
93 Stone Rd. W, Guelph, Ontario, Canada

ABSTRACT

With the continued emergence of foodborne pathogens and strain diversification, control and analysis of our food supply has become increasingly reliant on molecular-based methods. DNA and protein array analyses are emerging molecular techniques, with potential to become leading tools in food microbiology research associated with bacterial, protozoan and viral pathogens. Microarray has varied applications in microbial food safety, including bacterial identification and screening, as well as gene expression profiling in response to stressors and various microbial systems containing mixed bacterial populations. Oligonucleotide microarrays can detect virulence genes and simultaneously discriminate among foodborne human pathogens, such as *Bacillus, E. coli, Salmonella* and *Shigella* strains. Foodborne pathogens may also be used as bioweapons against humans, livestock and agricultural products. As part of a successful biodefence strategy and emergency preparedness, DNA and protein arrays represent broad-spectrum techniques capable of simultaneously identifying multiple foodborne pathogens, their associated toxins and virulence genes. Microarray can correlate phenotypic antibiotic resistance of starter cultures and probiotics used in food to the specific genetic determinants, as well as examine the production and utilization of fermented products by bacteria. Microarray can also be used as a biosensor, whereby bacterial proteins and analytes can be rapidly detected and characterized. Native toxins associated with cholera, diphtheria, anthrax, tetanus and staphylococcal enterotoxins have been detected and quantified using a multiplex antibody-based microarray. As well, protein arrays are beneficial for serological typing. As complete genomic sequences of microbial pathogens and hosts become available, microarray technology will enable a more detailed investigation of the interaction and establishment of infection of food pathogens in human and animal hosts. This chapter will discuss the widespread application and recent investigations of DNA and protein array in food microbiology, focusing on microarray potential in food safety research and processing.

1. Introduction

Microbial pathogens can contaminate food and water supplies, with the associated illness occurring after the pathogen or its toxin has been consumed. Although some foodborne pathogens can be controlled, the prevalance of foodborne disease is substantial, with *Salmonella* being the most frequent pathogen (Lee et al. 2001). While a major cause of morbidity in less developed countries, foodborne pathogens are responsible for numerous cases of gastrointestinal illness in developed countries, resulting in substantial economic losses due to lost productivity and health care costs. It has been estimated that there are 76 million cases of foodborne illness in the United States, of which, 325,000 resulted in hospitalizations and 5000 deaths every year (Mead et al. 1999). As a result, the economic burden of foodborne illnesses has been estimated to be 9-32 billion (US$)(Mead et al. 1999). While many foodborne illnesses are mild and associated with acute gastrointestinal symptoms, some can be serious and life-threatening, primarily affecting children in developing countries, the elderly and immunocompromised individuals. Due to the rapid expansion of food trade and increase in population mobility, foodborne infection has become an international issue. Food contamination can occur during food harvesting, processing, storage and preparation. Despite the introduction of the Hazard Analysis and Critical Control Points (HACCP) concept and a variety of food safety regulations, the occurrence of foodborne illnesses has continued to increase in the industrialized world (Maurice 1994).

Foodborne pathogens include enteric bacteria, viral pathogens, and protozoans. While not a comprehensive list, common foodborne pathogens include *Listeria monocytogenes*, *Staphylococcus aureus*, *Clostridium botulinum*, *C. perfringens*, *Bacillus cereus*, *Escherichia coli* O157:H7, *Salmonella* spp., *Shigella* spp., *Yersinia enterocolitica, Vibrio*, *Campylobacter jejuni, Brucella* spp., hepatitis A virus (HAV), noroviruses (NV), *Cryptosporidium parvum*, *Giardia lamblia*, and *Toxoplasma gondii*. Some of the aforementioned pathogens also pose a threat as a bioterrorism agent. New foodborne diseases are likely to emerge as a result of pathogen evolution, changes in agricultural and manufacturing processes, as well as changes in human hosts. Consequently, reliable and comprehensive methods are required to safeguard the public food supply from current, emerging and potential bioterrorism foodborne pathogens.

A primary factor limiting pathogen detection in food science is the complexity of food samples. Food products are exposed to various microorganisms, often at high concentrations, and thus, sensitive and specific detection methods are needed. While molecular-based methods have improved detection limits and specificity, a common shortfall of current microbial detection, genotyping and serotyping methods is the limited ability for multi-pathogen screening. Ideally, detection and genotyping strategies should provide detailed information on the species or strain, in a timely and cost efficient manner. DNA microarray is a recent technology capable of screening samples for the presence of multiple pathogens, with the ability to elicit information about strain diversification.

The immense flexibility in microarray applications is in part due to different types of probes and compatibility with different targets (cDNAs, total genomic DNA, total cellular RNA and crude nucleic acid extracts from samples). The technology is becoming increasingly important in various biological disciplines, as a large fraction of a microorganism's genome or proteome can be investigated in a single assay. DNA and protein arrays have immense potential in food microbiology, particularly in the disciplines of food safety research and processing. Of particular importance to food safety and bioterrorism, DNA and protein arrays are amenable to multiple-pathogen detection platforms. Recent progress in the field of microarray technology can be attributed to previously developed techniques, such as multiplex PCR, which can easily be adapted to a microarray platform, offering improved specificity and speed of hybridization analysis. With respect to detection of foodborne pathogens, protein arrays have the distinct advantage of automation, small sample and reagent volumes, and possibility of parallel sample analysis.

2. DNA Microarray

2.1. DNA Microarray Principles and Applications

As the spectrum of foodborne pathogens is very broad and oftentimes the etiological agent is unknown, multiple microorganism screening is needed. DNA microarray is conducive to multiple pathogen screening for the presence of multiple markers, enabling the determination of a complete genetic profile of a single strain or distinguish one strain from a very large collection of possible alternatives. Therefore, this approach is potentially useful for the screening of multiple microbial isolates in a diagnostic assay.

DNA microarray represents a powerful technology for foodborne pathogen detection. In a microarray, DNA fragments (primarily PCR products) or oligonucleotides are immobilized onto a solid support (Al-Khaldi et al. 2002). Fluorescently labelled DNA isolated from samples, as well as RNA or cDNA prepared from mRNA, are hybridized to DNA or oligonucleotides on the array and detected by laser scanning. Various support materials include nitrocellulose and nylon, which support hundreds of probes to form a low-density array. Glass microscope slides are conducive for the deposition of more probes (in the order of tens of thousands) and are referred to as a high density microarray. Genomic microarrays and oligonucleotide microarrays are two major types of arrays, both of which can simultaneously detect foodborne pathogens or measure the expression of thousands of genes.

In a genomic microarray, customized DNA probes are spotted onto a solid surface using robotic-mediated technology. The spotted DNA can be PCR fragments, genomic fragments from DNA library or open reading frames from a microorganism strain, thus, the spotted DNA can vary in length. Expression analysis using DNA microarray requires extraction of total RNA from the microbiological samples (experimental and reference samples) and synthesis of cDNA by reverse transcription. The experimental and reference cDNA samples are labelled with different fluorescent dyes, typically Cy3 and Cy5, and are hybridized onto a glass slide containing DNA sequences. Fluorescent signals are collected from both samples, statistically analyzed by specialized software and the gene expression is computed as a fold increase or decrease (up-regulation or down-regulation). As DNA microarrays can be designed to include genomic DNA from several related microbial strains (mixed genomic microarray), the technique is useful for foodborne pathogen detection, identification based on fingerprinting genomic isolates, phylogenetic classification and studying genetic diversity (Al-Khaldi et al. 2002; Bekal et al. 2003; Taboada et al. 2004).

In oligonucleotide microarrays, the probes are designed to correspond to parts of a known or predicted sequence. Oligonucleotides can be synthesized and printed onto the array or synthesized *in situ* directly onto the microarray support. In the latter approach, which is an example of an oligo-chip, oligonucleotide probes (usually 25 nucleotides long) are synthesized directly on a water surface using a photolithographic manufacturing process (Wilson et al. 2002). Key variables associated with *in situ* oligonucleotide synthesis involve the type of surface and immobilization chemistry used to retain the oligonucleotides to the substrate. As with other hybridization-based techniques, oligonucleotide microarrays depend on the oligoprobes for hybridization specificity and detection sensitivity. Therefore, oligoprobes must be carefully selected to confer high gene-specificity, as well as high sensitivity to detect low-level targets. Commercially available high-density oligonucleotide arrays can include thousands of genes, covering an entire genome. Some companies which currently offer oligonucleotide chips include Affymetrix, Agilent, GE Healthcare, and Ocimum Biosolutions. High density oligonucleotide arrays enable parallel hybridization to occur on the same surface, thereby increasing probe redundancy and enabling multiple independent detections of a diagnostic DNA sequence (Lipshutz et al. 1999). Oligonucleotide microarrays can be used for pathogen detection and genomic analyses. As oligonucleotide microarrays consist of multiple sequence variants of a target gene, the method provides more sensitive detection of polymorphisms and differentiation among strains. Detection of single-point mutations are important for epidemiological and phylogenetic studies, whereby mutational evolutionary changes can be detected in response to novel environments.

2.1.1. Advantages of DNA Microarray

The wealth of information generated, benefit of multi-sample analysis, and assay flexibility are primary advantages of DNA microarrays. As commercial kits and multiplex PCR applications only cover a small range of pathogens, DNA microarray supports broad-spectrum foodborne pathogen detection. This is of particular importance during foodborne outbreaks, when the etiological agent is unknown or strain level required. Of particular relevance to food quality and control, Bavykin et al. (2001) developed a portable system for microbial sample preparation and oligonucleotide microarray for rapid detection of *Escherichia coli, Bacillus subtilis,* and *Bacillus thuringiensis*. As well, DNA microarray can be used for screening samples for suspected presence of biowarfare agents, such as botulinum toxin and staphylococcal enterotoxin B (SEB) (Ligler et al. 2003). In addition to foodborne pathogen detection, DNA microarray can also be used for gene discovery, sequence identification, mapping and polymorphism. Oftentimes, these applications are performed in a more time-efficient manner then conventional molecular-based techniques.

Foodborne microbial pathogens and their associated toxins can be identified using culture-based, biochemical, immunological and PCR/DNA hybridization-based methods. Antibody, gene sequence and functional signature detection-based tests primarily exist for common foodborne pathogens that routinely impact commercial products. However, some foodborne pathogens have fewer detection methods available or methods are highly specialized and time-consuming. The use of a PCR-microarray offers several advantages over the aforementioned techniques. Firstly, due to the PCR amplification process, PCR-microarray offers increased sensitivity, which can be a critical issue when dealing with foodborne pathogens, such as *Escherichia coli* O157:H7, that typically require few cells to initiate infection. Secondly, microarray hybridization improves assay specificity, by not detecting non-specific PCR products. Lastly, with the use of multiple probes, the microarray format is conducive to parallel analysis of multiple genetic markers in a single experiment. As foodborne pathogen identification is based on the presence of multiple genetic characteristics, a PCR-microarray can reduce the occurrence of both false negative and positive results. Furthermore, when compared to detection with gel electrophoresis, DNA microarray was 32 times more sensitive (Call et al. 2001).

Microarrays are useful for investigations which analyze a large number of genes or when the sample volume is small. As well, DNA microarrays can be used to assess gene expression within a single sample, or compare gene expression between a treatment and control. For instance, Moen et al. (2005) used a DNA microarray to study nongrowth survival conditions of *Campylobacter jejuni* under temperature and oxygen tension, as compared to a control.

In terms of DNA microarray experimental criteria, design of short oligonucleotides improves hybridization specificity, as a single mismatched base will prevent pairing (Sergeev et al. 2004). Conversely, use of longer oligonucleotides is amenable to detection of gene families, where significant variation has occurred (Sergeev et al. 2004). As oligonucleotides can be designed based on available sequences, there are numerous oligonucleotide design possibilities, thereby eliminating the need for reference DNA samples, which are required for amplicon-based technologies.

2.1.2. Disadvantages of DNA Microarray

Although DNA microarray can be highly specific, reproducible and sensitive, the presence of inhibitors and extensive sample preparation procedures remains a limiting factor. Polysaccharides and humic acids are particularly troublesome inhibitors present in food products. Moreover, food processing may degrade DNA and introduce PCR inhibitors, leading to ambiguous results. Consequently, the complexity of food samples requires stringent and possibly lengthy extraction and purification techniques to ensure efficient recovery of nucleic acids and removal of inhibitors (Lipp et al. 2001; Meyer 1999).

Target foodborne pathogens are often present on food and food surfaces at low levels, as compared to non-target microbes. At present, microarrays alone are not amenable to low detection levels and therefore, are not suitable for detection of less abundant foodborne pathogens. However, PCR, which is amenable for low level detection, can be coupled with microarray for additional confirmation of amplicon identification. Microarray research is not ideal for quantitative assays, particularly when coupled with PCR.

Oligonucleotides, particularly shorter ones, exhibit variations in signal intensity between perfectly matched oligonucleotides of the same length but different sequences. This problem often requires a high redundancy design, where the signal of several oligonucleotides is combined and statistically normalized for analysis (Lemarchand et al. 2004). Additionally, short oligonucleotide microarrays require a large number of probes, contributing to high experimental costs.

In terms of microarray setup, there are often difficulties associated with visual control of the oligonucleotide spotting procedure. As such, an adequate quality control system is imperative to ensuring spotting uniformity. As well, the hybridization step oftentimes requires extensive optimization, thus greatly increasing experimental costs associated with DNA microarray. While able to provide extensive information on microbial gene expression and pathogen detection, the high costs of DNA microarray limit widespread application of this technique, particularly within smaller laboratory settings. Of particular relevance to microarray design, using the same sample, Li et al. (2002) compared datasets between long cDNA-based and oligonucleotide probes and observed a significant difference in gene expression results. Li et al. (2002) attributed the discrepancy, at least in part, due to cross-hybridization. Oligonucleotide microarrays results were found to be more reliable for evaluating changes in gene expression than data from long cDNA microarrays, indicating that the microarray platform must be carefully selected according to experimental objectives.

Although capable of high-throughput detection and quantification of nucleic acids, DNA microarray presently has low sensitivity and poor resolution at the species and subspecies levels (Wang et al. 2006). Because species-level analyses do not necessarily provide sufficient information for clinical and food microbiology applications, subspecies microarray probes are needed. Progress in microarray-based research is limited to organisms whose genomes have been fully characterized and requires construction of reliable databases for detection of gene and protein expression. Other technical challenges include, obtaining reproducible results between experiments and platforms, high costs associated with duplicate oligonucleotide assays to compare two conditions, reducing or eliminating false positive results and obtaining high quality DNA and RNA. As well, synthesis of gene-specific primers required to amplify each gene in a genome, for spotting and labelling with fluorophores contribute to high costs of DNA microarray. Statistical analysis of vast amounts of data and informatics has also confounded microarray research.

2.2. DNA Microarray Applications in Detection, Characterization and Genotyping of Foodborne Bacteria

2.2.1. Detection of Bacterial Pathogens

Detection and enumeration of bacteria on food or food contact surfaces represents an integral component of quality control and assurance within the food industry. While simple, convenient, adaptable and generally inexpensive, culturing and confirmatory techniques do not provide information regarding potential bacterial pathogenicity or virulence. Although the complexity of food samples has complicated conventional bacterial detection, genetic-based methods offer improved detection sensitivity and specificity. However, these methods are commonly pathogen-specific and can only detect a single pathogen at a time. Due to methodological limitation of direct culturing, numerous waterborne pathogens are difficult to detect and accurately enumerate (Lemarchand et al. 2005). Consequently, microarray has recently been standardized for monitoring pathogens in wastewater, thereby illustrating potential application of microarray in screening for foodborne pathogens (Maynard et al. 2005). DNA microarray utilizes molecular signatures specific to particular forms of contamination, reducing much of the guesswork associated with suspect pathogen identification.

In combination with PCR, oligonucleotide microchips are powerful tools for the detection and simultaneous discrimination among foodborne bacterial pathogens, with the ability to detect several virulence factors of *Salmonella, Shigella*, and *E. coli* strains (Chizhikov et al. 2001). In fact, Chizhikov et al. (2001)demonstrated that unexpected and uncharacterized genes led to ambiguous gel electrophoresis results of multiplex PCR products, whereas microchip analysis of microbial virulence factors was useful for the automated identification and characterization of *E. coli*, *Shigella dysenteriae*, *S. flexneri* and *Salmonella enterica* serovar Enteritidis. Thus, hybridization of fluorescently-labelled PCR products was a more specific identifier then band size on an agarose gel.

A critical component of using DNA microarray for detection relies on the selection of suitable gene targets to design probes. Development of a high specificity assay reflects the amount of information available in the sequence database. The small subunit ribosomal RNA (16S rRNA for prokaryotic microorganisms and 18S rRNA for eukaryotic microorganisms), intergenic spacer region (between 16S and 23S genes) and virulence factors have often been used as targets to detect closely related microorganisms. Although widely used for phylogenetic classification and bacterial identification, the 16S rRNA gene is only 1500bp long and contains highly conserved regions. Thus, it cannot yield signature sequences for closely related strains and microorganisms cannot be differentiated at the strain level using the 16S rRNA gene. Furthermore, the presence of multiple 16S rRNA copies in various organisms can complicate quantification of abundance (Klappenbach et al. 2001). As an alternative, 23S rRNA may also be used, however, available databases are smaller and there has been limited experience using 23S rRNA as a detection target.

Enterohemorrhagic *Escherichia coli* (EHEC) is capable of causing serious illness. Pathogenic and non-pathogenic strains of *E. coli* differ based on the presence or absence of numerous DNA elements that contribute to the specific virulent behaviour. Call et al. (2001) constructed a low-density microarray for the efficient detection and genotyping of *E. coli* O157:H7 following single or multiplex reactions. Based on four virulence loci (intimin, Shiga-like toxins I and II, and hemolysin A), the microarray was a sensitive detection system for genotyping O157:H7 strains and other EHEC. Results indicated that the microarray was 32-fold more sensitive than gel electrophoresis, with the ability to detect amplicons from less than one cell equivalent of genomic DNA (1 fg).

Also considered a potential bioweapon, food contamination with staphylococcal enterotoxins (SEs) is one of the leading causes of gastroenteritis (Bean et al. 1996). Presently, there are 17 major serological types of heat-stable enterotoxins, and multiple SEs are commonly found in *S. aureus* strains (Jarraud et al.). Serological typing, which is commonly used for detecting SEs, is impractical due to antigenic similarities among enterotoxins, as well as it is a relatively complex and time-consuming method. Conversely, Sergeev et al. (2004) developed a rapid and reliable one-tube oligonucleotide microarray for the simultaneous detection and identification (genetic typing) of almost all known *ent* genes. Using previously characterized staphylococcal isolates containing 16 *ent* genes, the microarray was able to detect each toxin gene, with the ability to identify as many as nine different *ent* genes in one *S. aureus* strain. Additionally, the oligonucleotide microarray revealed that some of the isolates contained previously undetected *ent* genes, thereby representing a more sensitive method for SE detection.

Another important foodborne bacterial pathogen, *Campylobacter* is associated contamination of untreated water, unpasteurized milk, contact with pets and handling of undercooked meat products. As poultry and poultry products are considered a major source of *Campylobacter* and play a significant role in transmission, Keramas et al. (2003) developed a sensitive DNA microarray suitable for the rapid detection of *Campylobacter* spp. directly from faecal cloacal swabs. Universal bacterial and *Campylobacter* specific sequences were amplified by a multiplex PCR targeting the 16S rRNA, 16S rRNA intergenic region and specific *Campylobacter* genes. Amplicons were labelled with Cy5 and hybridized to probes immobilized onto the microarray to enhance assay sensitivity and specificity. The described assay was able to differentiate between the closely related *Campylobacter* species, *C. jejuni* and *C. coli*, both of which account for the majority of human infections (Oberhelman and Taylor 2000). The detection limit of the various capture probes corresponded to three *Campylobacter* genome equivalents. Suitable for mass screening of *Campylobacter* in food industry production, animal and clinical samples, DNA microarray represents an alternative to the laborious isolation and culture-based detection method currently employed.

DNA microarray is an important technique used to examine the microbiological quality of minimally processed vegetables and pre-handled food, which can be contaminated with multiple bacterial pathogens. As minimally processed vegetables are typically consumed without heat treatment or other sanitation process, Ikeda et al. (2006) developed a DNA microarray targeting 16S rRNA sequences for the detection of *Bacillus cereus, Salmonella enterica* serovar Enteritidis, and *Yersinia enterocolitica* in fresh vegetables (bean sprouts and lettuce). Diluted bacterial cultures inoculated into fresh vegetable samples were specifically detected and provided faster results, as compared to conventional culture methods used for these bacteria.

Multipathogen detection, with the use of DNA microarray, is essential to ensure microbiological food quality, thus improving food safety and reducing health costs associated with foodborne illnesses and poisoning. Wilson et al. (2002) developed a Multi-Pathogen Identification (MPID) microarray for the identification of 18 pathogens, including 11 bacteria, 5 RNA viruses and 2 eukaryotes. On average, 91% of the 53 660 pathogen probes used in the MPID performed as predicted. Using six diagnostic primer-pairs, the microarray detection limit was 10 fg of *B. anthracis* DNA, while agarose gel electrophoresis was unable to visualize samples at this concentration. By examining the specific diagnostic regions and creating specific fingerprints, the MPID was used for high throughput screening of bacterial and viral pathogens.

In a similar effort to apply DNA microarray for multi-pathogen detection, Hong et al. (2004) developed an oligonucleotide microarray for the rapid detection of common foodborne bacteria (*Bacillus cereus, Clostridium botulinum, C. perfringens, Listeria monocytogenes, Staphylococcus aureus, Streptococcus pyogenes, Campylobacter jejuni, Escherichia coli, Proteus vulgaris, Salmonella enterica, Shigella dysenteriae, Vibrio cholerae, V. parahaemolyticus, and Yersinia enterocolitica,*). Although *S. enterica* and *Y. enterocolitica* gave a weak cross-reaction with *E. coli*, detection was not affected. With the exception of *C. perfringens* and *S. pyogenes*, the other pathogenic bacteria showed high sensitivity and specificity for the oligonucleotide array. As well, Lin et al. (2005) developed an oligonucleotide array for the identification of six foodborne pathogens and *Pseudomonas aeruginosa* grown on selective media. The sensitivity and specificity of the oligonucleotide array were as follows: *Bacillus cereus* (100 and 77%), *E. coli* (100 and 100%), *Listeria monocytogenes* (100 and 90%), *Pseudomonas aeruginosa* (100 and 100%), *Salmonella* (100 and 100%), *Staphylococcus aureus* (100 and 100%), and *Vibrio parahaemolyticus* (100 and 94.2%), respectively. Although other species of *B. cereus* group cross-hybridized with *B. cereus* probes, further confirmation was obtained through morphological examination. Subsequent hemolysis test was needed to differentiate *Listeria innocua*, which cross-reacted with *L. monocytogenes* probes.

With varied applications, DNA microarray has a high potential for automation, portable analysis of food products and incorporation into a mass screening microsystem (Bavykin et al. 2001). While conventional and multiplex PCR offer a significant advantage over costly and time-consuming conventional culture-based techniques, these methods are limited to the detection of a few pathogenic bacteria. Conversely, amplifying conserved regions of bacterial DNA with universal primers and subsequent hybridization with species specific oligonucleotide probes enables the sensitive and specific detection of multiple foodborne pathogens.

2.2.2. Characterization and Genotyping of Foodborne Pathogen Isolates

Microbial genotyping can be used to track the spread of infectious diseases in human and animal populations (vanBelkum 2003). While whole genome sequencing enables detection of differences between microorganisms at the single nucleotide level, the technique is too costly for multiple sample comparison of numerous species isolates, as is the case with global infectious disease surveillance programs. Pulse field gel electrophoresis (PFGE) has commonly been used to discriminate between related and non-related isolates, however, the technique requires a combination of costly restriction enzymes to increase the discriminatory capacity of the assay. Uncertainty of the genetic identity of two bands of equal size requires Southern blotting, thus prolonging the time required to obtain analytical results. Conversely, as thousands of probes can be placed on an array, offering high discriminatory resolution, DNA microarray represents an attractive alternative to present genotyping methods. With the emergence of genomic data, DNA microarray offers an alternative means to characterize pathogenic isolates. DNA microarray applications in genotyping include phylogenetic analysis and species identification of *Campylobacter* (Dorrell et al. 2001; Volokhov et al. 2003), EHEC (Call et al. 2001), *Listeria* (Volokhov et al. 2002) and *Staphylococcus* (Fitzgerald et al. 2001) and species identification among mixed bacterial communities (Volokhov et al. 2003; Volokhov et al. 2002).

Genomic hybridization on a whole genome array can detect the presence or absence of DNA regions of sufficient similarity to the sequenced isolate in other microorganisms, enabling a genome-wide comparison of their genetic contents. The presence of open reading frames (ORF) of the sequenced genome in sample isolates are examined in a single hybridization experiment. Intergenic spacer regions represent an alternative array format, with the potential to efficiently discriminate between different bacterial strains of a species. Short oligonucleotide arrays can be used to identify the single nucleotide polymorphisms (SNPs) that are thought to be responsible for genetic variation. With respect to identifying foodborne bacteria, microarrays involve labelling a specific DNA sequence from a target gene. With the use of multiple oligonucleotides for each target gene, microarray increases detection and discrimination between closely related species. Thus, species can be identified based on the hybridization pattern of bacterial DNA with species-specific microarray probes. Because microarray technology has the potential to rapidly determine the complete genetic profile of a microorganism in one hybridization experiment, the technique has valuable applications for genetic screening and identification of microorganisms.

The abundance of sequence databases has enabled the opportunity to identify 'genomic signatures' or 'DNA fingerprints', which are short DNA sequences that uniquely ascertain the presence or absence of bacteria or virulence genes. Additionally, DNA signatures can be used to detect the presence of one or more virulence genes, such as *Bacillus* genes which code for important virulence factors, enterotoxins and exotoxins (Sergeev et al. 2006), as well as differentiation between closely related microorganisms (Willse et al. 2004). Tembe et al. (2007) used an integrated high-performance computing (HPC) software tool for oligonucleotide fingerprint identification (TOFI) for several bacteria and plasmid sequences. TOFI uses a multi-step approach, in which DNA sequences common to target sequence and biological near-neighbor sequences are discarded. From the remaining sequences, DNA microarray probes are designed. Each DNA probe is subsequently aligned with sequences in the reference nucleotide database to predict cross-hybridizations. Tembe et al. (2007) used TOFI to identify DNA fingerprints for the plague-causing *Yersinia pestis* strain CO92. Based on TOFI integrated analysis, approximately 150 *in silico* DNA fingerprints, of which, further screening based on additional experimental constraints (ie. presence of restriction enzyme cleavage sites) identified 99 *in silico* fingerprints. Using a customized DNA microarray chip, probes were further analyzed for discriminating power and cross-hybridizations using the near-neighbour genome *Y. pseudotuberculosis*. Quantifiable threshold values for target and non-target responses were assessed and a sufficiently large number of probes would allow for detection redundancy. While additional experiments are needed to test the *Y. pestis* probes against a standard panel of non-target genomes, continued improvement in TOFI parameter selection will greatly improve the use of oligonucleotide fingerprint identification for microarray-based analysis of foodborne pathogens.

Using an alternative design principle, Pannucci et al. (2004) used multi-locus sequence typing (MLST) to identify candidate virulence factors and analysis of genetic polymorphisms that are coupled to *Bacillus anthracis* virulence factors, thereby enabling the identified polymorphic sites to be incorporated into a low cost, rapid microarray-based genotyping assay. Using microarray-based mini sequencing, detection primers were synthesized for each of the six identified single nucleotide polymorphisms (SNPs) within the *pag*A locus of pX01. Based on seven *B. anthracis* strains which exhibited genotype diversity at one or more of the *pag*A sites, microarray-based sequencing provided accurate sequence data for six SNP sites investigated, as compared to traditional automated sequencing. The assay can be applied to investigate and identify virulence-linked nucleic acid signatures for biothreat agents. Of course, the success of this method depends on the availability of sequence data for the pathogen of interest.

Oftentimes not all species of a genus pose a human threat. Therefore, genotyping-based analyses are important in assessing potential health implications of contaminated food products. For instance, Wilson (1995) detected *Listeria* species in 5% of examined ready-to-eat foods, of which, *L. monocytogenes* (49%) and *L. innocua* (36%) were the prominent species identified. *L. monocytogenes* is the causative agent of listeriosis, while *L. innocua* is non-pathogenic. Because *Listeria* species are identified by few biochemical traits, Volokhov et al. (2002) developed a rapid oligonucleotide microarray-based assay for the detection and discrimination of *L. monocytogenes*, *L. ivanovii*, *L. innocua*, *L. welshimeri*, *L. seeligeri* and *L. grayi*. To simplify microarray identification, six target bacterial virulence factor genes (*iap*, *hly*, *inlB*, *plcA*, *plca*, and *clpE*) were amplified in a single-tube multiplex PCR assay, the ssDNA was fluorescently labelled and hybridized on a "universal" *Listeria* microchip containing oligoprobes for all six virulence factor genes. The authors noted that in order to maximize assay sensitivity, the labelling and PCR step should be separate. Volokhov et al. (2002) was able to reduce bacterial misidentification associated with genomic instability by using more than one gene to characterize pathogenic *Listeria* species. Based on multiple species-specific sequences within the *iap* gene, ten individual oligoprobes were designed per species, enabling discrimination between all six *Listeria* species.

Of the six recognized *Listeria* species listed above, *L. monocytogenes* is most commonly attributed to human disease and represents a potentially fatal foodborne disease associated with the consumption of contaminated dairy products. Although greater then 14 serotypes have been designated, the majority of human and animal cases are associated with three different *L. monocytogenes* serotypes (1/2a, 1/2b and 4b) (Graves and Swaminathan 2001). While commonly used for subtyping, pulse-field gel electrophoresis (PFGE) provides limited information on genetic characterization of isolates and does not identify the presence or absence of specific genes. Consequently, Borucki et al. (2003) developed a mixed-genome array based on ten different *L. monocytogenes* strains containing 585 probes. The microarray was able to discriminate between closely related *L. monocytogenes* isolates and identified genetic markers that can be used for epidemiological pathogenesis studies. However, microarray results identified 92 polymorphic probes, thereby identifying a need for further analysis of additional strains of known phenotype (virulent versus attenuated) on a microarray containing a greater number and diversity of polymorphic probes. Further use of a mixed-genome array for the identification of specific genetic differences among isolates with defined phenotypes will provide insight into the epidemiology and potential pathogenesis of *L. monocytogenes*.

Commonly associated with contamination of raw chicken, the *Campylobacter* genus consists of 16 closely related species, of which, *C. jejuni* is recognized as a leading cause of bacterial foodborne diarrheal disease worldwide (Allos 2001). As conventional *Campylobacter* genotyping methods are tedious and time-consuming, Volokhov et al. (2003) developed a composite microchip for microarray detection and discrimnation of thermophilic *Campylobacter jejuni, C. coli, C. lari* and *C. upsaliensis* by using species-specific oligoprobes against *fur, glyA, cdt, ceuB-C* and *fliY* genes. Alleles of *fur* and *glyA* genes were used for identification and discrimination of the four *Campylobacter* species investigated. *C. jejuni* and *C. coli* were discriminated based on the *ceuB-C* and the *fliY*, whereas *cdt* genes were used as additional genetic markers specific for C. *lari* and *C. upsaliensis* or *C. jejuni*. Although some cross-hybridization was observed for the *glyA* probes of *C. jejuni* and *C. lari* with *C. upsaliensis*, the use of six independent oligoprobes for each gene improved assay specificity and the cross-hybridization of one or two spots did not affect species identification. Using six independent oligoprobes for each gene, redundancy of sequences within genes and of genes within species assured detection of the four *Campylobacter* species despite sequence divergence. The microarray was also able to efficiently and specifically detect each *Campylobacter* species present in the mixed bacterial population samples, as well as unambiguously identify target species in the presence of DNA from various non-*Campylobacter* species. Additionally, Dorrell et al. (2001)developed a microarray consisting of all 1731 coding sequences of *C. jejuni* strain NCTC 11168. The microarray analyzed 11 *C. jejnui* strains of different origins and the results demonstrated that 1300 of the sequences (78.6%) were common to all strains.

A contaminant of prepared poultry and egg products, *Staphylococcus aureus* is a major human pathogen which causes life-threatening septicaemia, endocarditis and toxic shock syndrome (TSS). As well, *S. aureus* is also responsible for economically detrimental mastitis infections in cows and sheep. Fitzgerald et al. (2001) developed a microarray which represented greater than 90% of the *S. aureus* COL strain genome. The microarray was used to characterize the genomic diversity, evolutionary relationships and virulence gene distribution between 36 strains of divergent clonal lineages (detected by multilocus enzyme electrophoresis), including those strains resistant to methicillin (MRSA) and organisms which cause TSS. Results identified eighteen zones of difference, ten of which had genes which encode putative virulence factors or proteins mediating antibiotic resistance. Horizontal transfer of the *mec* gene into distinct *S. aureus* chromosomal backgrounds occurred at least five times, illustrating that MRSA evolved multiple independent times, rather then from a single ancestral strain. In this study, microarray provided information relevant to host and disease specificity, population genetics and evolution of *S. aureus*.

While whole genome sequencing is too expensive for multiple sample comparison of a large number of species isolates, whole genome microarray can detect the presence or absence of similar DNA regions in sufficiently related microorganism, thus enabling genome-wide comparison of their genetic contents. *In situ* oligonucleotide arrays, which increase the number of probes per array, can further enhance the discriminatory capabilities of microarray, enabling the detection of smaller genetic differences on a whole-genome scale. Furthermore, by incorporating sample amplification prior to hybridization, microarray sensitivity can be greatly enhanced. As several fingerprints can be simultaneously placed on a microarray, the method supports multiple foodborne bacterial analysis in a single assay, as well as provides increased detection redundancy. With the continued emergence of genomic data, development of more cost-effective protocols, more robust and simplified formats, DNA microarray offers an alternative means to characterize pathogenic isolates and will likely to become a prominent technique in foodborne microbial genotyping studies.

2.2.3. Virulence Factors

In addition to foodborne-associated pathogen detection and antibiotic resistance genes, DNA microarray may also be used for the determination of virulence factors. Numerous genes associated with virulence are regulated by specific conditions. Investigating gemone-wide gene expression patterns under relevant conditions, such as a physiological change during interaction with the host, can elicit candidate virulence factors. Conversely, comparative genomics may also be used to identify virulence genes. For instance, using a whole genome microarray, Salama et al. (2000) identified a class of candidate virulence genes by their coinheritance with a pathogenicity island.

Various *E. coli* strains have acquired genetic determinants (virulence factors) which render the microorganism pathogenic towards both humans and animals. Adhesins, host cell surface-modifying factors, invasins, toxins, and secretion systems represent virulence factors associated with *E. coli* pathogenic mechanisms. As strains of the same pathotype are genetically similar and carry the same virulence determinants involved in infection, detection of these virulence genes represents an ideal target for the determination of pathogenicity of any given *E. coli* isolate. Consequently, Bekal et al. (2003) developed a DNA microarray to differentiate *E. coli* pathotypes. Based on virulence factors, the microarray classified *E. coli* isolates into various pathogenic groups and identified the relatedness of the groups. Furthermore, Bekal et al. (2003) suggested that the microarray had the potential to detect emerging pathotypes arising from horizontal gene transfer of virulence genes. While based on genomic DNA previously characterized from clinical samples, PCR amplification of bacterial genomes will likely be necessary for identification of small quantities of pathogens from complex food samples.

An important component of safeguarding the public food supply also entails a comprehensive understanding of the protective nature of microorganisms under environmental stresses. For instance, bacterial species can comfortably tolerate mild acidic conditions by mounting a cellular response to protect the intracellular pH. Essentially, microorganisms are able to modify gene expression to support optimal growth and infection under slightly acidic conditions. As such, Weinrick et al. (2004) investigated the cellular response of *Staphylococcus aureus* to mild acid conditions in order to identify the affected gene set as a mild stimulon. Using transcriptome profiling by microarray, Weinrick et al. (2004) revealed that the transcript levels of over 400 *S. aurues* genes were affected. In particular, extracellular virulence factors were affected, as were genes associated with the regulation of virulence gene expression. It was also observed that glucose-induced down-regulation of certain staphylococcal virulence genes was largely the result of pH reduction due to glucose fermentation.

While many genes responsible for pathogen virulence and pathogenesis have not been identified, the rapid accumulation of information regarding microbial genome sequencing will contribute immensely to identifying the genetic elements which are likely involved with pathogenesis. As such, accumulation of this data will further our knowledge on the fundamental molecular basis of pathogenesis of foodborne diseases, as well as elicit the disease-causing potential of isolated microorganisms. The varied applications of DNA microarray enable investigations on virulent gene expression response to environmental stressors, as well as differentiating pathogens based on the presence of virulence factors.

2.2.4. Antibiotic Resistance Genes

Increased antimicrobial resistance is a major concern associated with foodborne pathogens. A microorganism can be naturally or inherently resistant to an antibiotic. Acquired resistance occurs when a strain of bacterial species commonly susceptible to an antibiotic becomes insensitive. Resistance can be acquired through mutations in the relevant genes or horizontal transfer. Also referred to as vertical transmission, resistance acquired through mutation will be confined to the mutant clone, thus, emergence and spread will depend on the clone's ability to multiply and infect new hosts. A more common method, horizontal transfer involves the uptake of resistant genes from another species.

Use of antibiotics in livestock species may accelerate the development of antibiotic-resistance strains of microbial pathogens, potentially complicating treatment for both animals and humans (Roy and Sen 2006). Antibiotic abuse associated with feeding animals and improper treatment of patients has resulted in increased drug resistance characteristics detected in foodborne pathogens (Barza 2002; Wallace et al. 2000; Woese 1987). With increasingly widespread occurrence of antibiotic resistance genes appearing in microbial foodborne pathogens, there is an increased need for detection and identification of antibiotic resistance genes.

Multi-drug resistance has emerged among pathogenic and beneficial indigenous bacteria. Starter cultures and probiotic bacteria represent a reservoir for gene transfer to human and animal pathogens, and therefore, should be confirmed free of antibiotic resistant genes. Although PCR and hybridization analysis are commonly used to detect antibiotic resistance genes in bacteria, these methods are time-consuming when multiple resistance genes are investigated. DNA microarray represents a sensitive and specific method to screen for the presence of antibiotic resistance genes in pathogenic and commensal bacteria, as well as in potential probiotics and starter cultures (Perreten et al. 2005). Additionally, microarrays can detect antibiotic resistance genes that are not phenotypically expressed in vitro (Perreten et al. 2005).

A hybridization microarray was developed to detect 90 of the most prevalent and transferable antibiotic resistance genes in Gram positive bacteria (Perreten et al. 2005). Resistance genes were detected using two different oligonucleotides, thereby providing increased specificity and sensitivity. Hybridization analysis of genomic DNA obtained from *Bacillus*, *Clostridium*, *Enterococcus*, *Lactococcus*, *Lactobacillus*, *Listeria*, *Staphylococcus*, and *Streptococcus* verified the sensitivity of 125 of 137 oligonucleotide targets and identified 71 resistance genes. Oligonucleotides provided higher hybridization specificity then PCR products, with a shorter hybridization time. With improved rapid screening of bacteria harbouring antibiotic resistance genes, microarray could have future applications in food safety and surveillance programs.

Kastner et al. (2006) examined probiotics and starter cultures for the presence of antibiotic resistance genes. Based on results from the disk diffusion method, twenty-seven isolates exhibiting phenotypic resistance were screened for 90 known antibiotic resistance genes using microarray hybridization. Results were confirmed by PCR amplification or Southern hybridization. Tetracycline resistance gene was identified in five *Staphylococcus* isolates used as meat starter cultures, *Bifidobacterium lactis* DSM 10140, and *Lactobacillus reuteri* SD 2112 probiotic cultures. The study highlights the presence of antibiotic resistance genes in food bacteria and identifies the need to assess commercial starters and probiotic cultures for the presence of gene transferability. Genetic screening by microarray provides a more comprehensive analysis of antibiotic resistance genes, as compared to disc diffusion tests or methods focusing on single genes. As well, (Call et al. 2003) used DNA microarray probes specific for 17 tetracycline-resistance genes in a variety of bacterial species, including the foodborne pathogens *E. coli*, *Acinetobacter*, *Pseudomonas*, *Enterobacter*, and *Citrobacter*. The researchers noted that DNA microarray has the potential to be an efficient and cost-effective tool for the screening of several antibiotic-resistance genes.

2.2.5. Food Processing, Fementation and Starter Cultures

Quality control and ensuring food safety are important in the food production process. With various methods and techniques, food processing removes pathogenic microorganisms and toxins, enhances flavour, increases food consistency and preservation or enhances food products through the addition of beneficial microorganisms (starter cultures or probiotics). Food supplies can become externally contaminated during food harvesting, processing, storage and preparation. As part of ensuring a safe food supply and optimizing food processing practices, DNA microarray has been used to evaluate the presence of virulence factors in probiotics and starter cultures, as well as monitor lactic acid bacteria community dynamics during fermentation. Additionally, DNA microarray has potential applications in microbial source tracking during food processing.

Although few DNA microarray fermentation studies currently exist, the method has increasingly been applied to the exploration of genome-wide transcriptional profiles, whereby transcriptional profiles have generated patterns specific to a growth condition, developmental stage or drug treatment (Ye et al. 2001). Thus, the signature profiles present a potential valuable diagnostic tool to optimize fermentation process and identify which genes are co-regulated in the production process. Enabling proper strain selection or improvement, DNA microarray can enhance the quality, taste and improve processing during the industrial food fermentation.

Lactic acid bacteria (LAB) exhibit a critical role in manufacturing of fermented milk products, vegetables, meat and wine. For instance, to ensure proper ripening and preservation, kimchi (a type of Korean pickled vegetable) is produced through lactic acid fermentation of Chinese cabbage. Depending on fermentation conditions and ingredients, fermented kimchi can have a different bacterial composition. Consequently, Bae et al. (2005) developed a genome-probing microarray (GPM) to quantitatively monitor LAB community dynamics during fermentation of kimchi. Deposited as probes on glass slides, each GPM contained genomic DNA from 149 different strains, including 138 type strains of LAB. When compared to a 50-mer oligonucleotide microarray, the specificity of the GPM was about 10 to 100-fold higher. With the use of the GPM assay, approximately 100 diverse LAB species were quantitatively analyzed to be active in kimchi fermentation, illustrating the applicability of GPM in LAB community dynamics involved with fermentation.

DNA microarray has also been used to investigate anaerobic gene expression patterns of *Bacillus subtilis* on a genomic scale (Ye et al. 2000). A facultative anaerobe, *B. subtilis* can grow under anaerobic conditions, either with nitrate or nitrite as the electron acceptor or by fermentation. *B. subtilis* is a key microorganism in the production of soya-based traditional fermentation. The ability of *Bacillus* species to ferment in acidic, neutral, and alkaline pH ranges has lead to the development of new commercial enzyme products (Schallmey et al. 2004). In order to optimize the fermentation process, it is imperative to understand gene expression of relevant genes under anaerobic and fermentation conditions. In the described DNA microarray, certain *B. subtilis* genes were observed to be upregulated or downregulated under anaerobic conditions. Affected genes were associated with a variety of cell functions, including carbon metabolism, electron transport, iron uptake, antibiotic production, and stress response. Some genes were specifically regulated during anaerobic growth in nitrite, while others were primarily affected during fermentative growth, indicating a complex regulatory circuitry of anaerobic metabolism. Therefore, in better understanding the influence of anaerobic and aerobic culture conditions on gene expression, *B. subtilis* fermentation can be optimized.

Changes in extracellular parameters during food fermentation may also impact final product quality. Consequently, Schoondermark-Stolk et al. (2006) examined the effect of pH and salt concentration on the formation of fusel alcohols in *Saccharomyces cerevisiae*. The fusel alcohol, 3-methyl-1-butanol, was used as an indicator substance relevant for the taste of fermented foods. Using optimal fermentation conditions for 3-methyl-1-butanol (pH 3.0 and 0% NaCl) and yeast cultivation standard conditions (pH 5.0), the genome-wide expression profile of *S. cerevisiae* identified seven genes which were related to the 3-methyl-1-butanol pathway. Thus, cDNA microarrays can provide insight into molecular processes during specific fermentative conditions, thereby identifying target genes for potential improvement of yeast strains used in industrial processes.

Consistent and predictable rates of acid production are important for the success of fermented milk products. Therefore, as pasteurization kills most natural LAB present in raw milk, starter cultures are added to initiate fermentation and control the rate and extent of acid development. Although of immense value to the food industry, use of starter cultures and probiotics must be carefully evaluated. These beneficial bacteria should be verified to be free of antibiotic resistance genes, virulence factors and other potential disease-causing traits.

Consequently, Kastner et al. (2006) used a hybridization microarray to examine probiotics and starter cultures in the food industry for the presence of antibiotic resistance genes. Tetracycline resistance gene was identified in five *Staphylococcus* isolates used as meat starter cultures.

Enterococci are associated with food spoilage, particularly in meats. Although certain enterococcal strains can cause human disease, enterococci are also used as human probiotics and are important for ripening and the aroma of certain traditional cheeses and sausages. Consequently, proper use of enterococcal strains as probiotics or starter cultures in food warrants a comprehensive investigation to determine which factors are important in enterococcal virulence. Virulence factors such as adhesins, invasins and haemolysin have been described (Franz et al. 2003). As such, DNA microarray represents an ideal platform to screen enterococcal strains and other potential starter cultures or probiotics for the presence of virulence factor genes, as well as antibiotic resistance genes.

2.3. DNA Microarray Applications in Detection of Foodborne Viruses

Viruses, such as rotaviruses, astroviruses, hepatitis A virus (HAV), and human enteric caliciviruses, including noroviruses, have been associated with food and waterborne illnesses in humans, with an increasing incidence of foodborne transmission (Motarjemi et al. 1995). Human infection by foodborne viruses occurs either during preparation by infected food handlers or contact with sewage sludge or polluted water. Strict intracellular parasites, viruses cannot replicate in food or water. Consequently, although a source of contamination, food processing and manufacturing will not result in an increase in viral contamination of food. Therefore, food contamination by viruses relies on viral stability, quantity of virus shed by infected individuals, method of food or water processing, infection dose and host immune susceptibility. Salads, cold meats, shellfish, confectionary and infected food handlers are common sources of virus-contaminated foods. Noroviruses (NV) and HAV are most commonly transmitted by food. Although viral infections were not commonly diagnosed until recently, epidemiological studies suggest that caliciviruses may be as frequent a cause of foodborne illness as *Salmonella* (Djuretic et al. 1996; Mead et al. 1999). Consequently, the costs of viral illnesses can be high due to frequent occurrence and high transmissibility (Vinjé et al. 1997; Vinjé and Koopmans 1996).

Quality control based on detecting fecal indicator organisms in food is an unreliable predictor of viral contamination (Croci et al. 2000; LeGuyader et al. 1998). The high detection limit of electron microscopy (EM) and presence of unculturable viruses limit their use for detection of food- and waterborne viruses. Complexities of viral flora and large number of constantly evolving viral serotypes render antibody-based viral detection techniques nearly impossible. While PCR detection increases detection sensitivity of viruses in food, methods are unable to assess viability and therefore provide limited information to manufacturing and processing sectors. Conversely, DNA microarray provides simultaneous monitoring and broad-spectrum detection of foodborne enteric viruses and associated subtypes.

Recent epidemiologic studies indicate that noroviruses are one of the most frequent causes of acute non-bacterial gastroenteritis (Goodgame 2006). NV can be genetically subdivided into five genogroups (GI-GV), of which, genogroups I, II and IV occur in humans.

Pagotto et al. (2004) used an oligonucleotide microarray to detect NV in fecal samples. Stool samples were PCR-amplified prior to labelling and microarray hybridization. The microarray hybridized to specific sequences of all three open reading frames and simultaneously characterized noroviruses genogroups I and II. Pagotto et al. (2004) were able to detect and partially genotype NV, illustrating potential microarray applications for identification in complex food samples. Due to the lack of universal primers for PCR amplification and sequence variability in viruses, Pagotto et al. (2004) are currently investigating a non-gene specific amplification method to alleviate challenges associated with relying on specific primers to amplify target nucleic acid. Jääskeläinen and Maunula (2006) also investigated the applicability of microarray detection of noroviruses and astroviruses. The specificity, precision and accuracy of microarray detection was assessed by analyzing 202 previously collected clinical stool specimens (19 genotyped, 74 RT-PCR tested and 109 archived samples analyzed by EM). The slide contained 50 arrays, each of which consisted of seven detection primers for noroviruses, eight for astroviruses and five random, unspecific primers as a negative control. Microarray detected all but one of the 19 genotyped samples (which consisted of 9 NV-GI and 10 NV-GII). Additionally, the detection primers successfully differentiated between the GI and GII samples. Initially, conventional RT-PCR was able to detect 35 positive samples for noroviruses, whereas microarray detected 45 samples of the 74 tested stool samples. EM of the archived samples visualized small round viruses which likely contained sapo-, picorna- and parvoviruses, in addition to noroviruses and astroviruses, however, further identification of these samples was not conducted. From the 109 archived stool samples, microarray detected 36 samples containing noroviruses GII, 25 samples contained astroviurses, double infection occurred in five samples and 43 samples were negative. Repeated freeze-thaw of the archived stool samples may explain the relatively low yield of noroviruses and astroviruses. While microarray successfully identified NV, the genetic diversity of this virus groups represents a challenge in the design and construction of a broadly reacting molecular assay. In conclusion, detecting primers for the genogroups were able to detect all but one NV, GII-7, in which the primer failed to anneal to the RNA template.

As food samples can be contaminated with various pathogenic microorganisms, DNA microarray represents an ideal platform to discriminate among pathogens or species, as well as specifically identify all organisms. Wilson et al. (2002) developed a Multi-Pathogen Identification (MPID) microarray for the identification of 18 pathogens, including 11 bacteria, 5 RNA viruses and 2 eukaryotes. On average, 91% of the 53 660 pathogen probes used in the MPID performed as predicted. By examining the specific diagnostic regions and creating specific fingerprints, the MPID was used for high throughput screening of bacterial and viral pathogens.

Use of DNA microarrays for detecting viral pathogens in clinical samples can also have prospects if applied to agriculture and food science. For instance, Wang et al. (2002) used a combined random PCR-microarray to screen virally-infected tissue culture cells and clinical specimens isolated from human respiratory tracts. Although the study focused on respiratory tract viruses, the principle of random PCR microarray can be used for similar complex food specimens to detect known and characterized food-associated viruses. In using the most conserved sequences within a viral family as the microarray oligonucleotide probes, unsequenced or uncharacterized members of a viral family can be detected, as well as unique hybridization patterns could be used to discern related viral serotypes.

Sequence variability in viruses and lack of universal primers for PCR amplification remain limiting factors for microarray detection. As viruses of the same species or serotype can differ by more than 30% at the nucleotide level, identification by whole-DNA hybridization is inefficient (Chizhikov et al. 2002). Since the degree of evolutionary conservation varies widely among different parts of viral genomes and within viral genes, highly conserved regions can be used for efficient and broadly specific PCR amplification of DNA from all members of a group or for finding regions unique for members of a particular genotype of interest. Thus, short genotype-specific oligonucleotides represent a promising approach for classifying viruses with varying degrees of relatedness.

2.4. DNA Microarray Applications in Detection of Foodborne Protozoa

Toxoplasma, *Cryptosporidium*, *Giardia* and *Cyclospora* are intestinal protozoan parasites which have become increasingly associated with foodborne illnesses. These protzoans are present in the environment as (oo)cysts, which are environmentally robust. Foodborne outbreaks of cryptosporidiosis, giardiasis and cyclosporiasis have been associated with (oo)cyst contamination of fresh produce (particularly fruit, vegetables and salads) and other lightly cooked or raw foods (Carey et al. 2004; Mansfield and Gajadhar 2004; Wang et al. 2002). DNA microarray has increasingly been employed for the detection of foodborne bacterial and viral pathogens (Chizhikov et al. 2001; Volokhov et al. 2002; Wang et al. 2002), however, no studies to date have investigated protozoa on food and food products. An oligonucleotide microarray has been developed for the parallel environmental detection and discrimination of *Entamoeba histolytica*, *E.* dispar, *Giardia lamblia*, and *C. parvum* (Wang et al. 2004). Transmitted by the fecal-oral route and commonly associated with outbreaks through the consumption of untreated water (Johnston et al. 2003), outbreaks attributed to these protozoan pathogens have also occurred through contaminated food (Chaidez et al. 2005). In this microarray, genus-level specificity was achieved, with the ability to differentiate between members of the genera *Entamoeba*, *Giardia*, and *Cryptosporidium*. Furthermore, the hybridization of diagnostic targets to species- and genotype-specific probes generated unique profiles enabling the identification and differentiation of *E. dispar* from *E. histolytica*, *G. lamblia* assemblages A and B, and *C. parvum* genotypes 1 and 2 in a single assay. The oligonucleotide microarray demonstrated a lower detection limit of five trophozoites of *G. lamblia*. As protozoan infections linked to food are not commonly detected (Nichols 2000), there is an increasing need to develop a cost efficient multiple-protozoan pathogen detection method, such as DNA microarray.

Based on a single nucleotide polymorphism of the *hsp70* gene, a microarray was developed to detect *Cryptosporidium* species, differentiate between closely related species and isolates of *C. parvum*, as well as identify common *Cryptosporidium* genotypes associated with human disease (Straub et al. 2002). Despite perfect mismatch discrimination, the microarray unambiguously distinguished between *C. parvum* genotype I and genotype II isolates (now recognized as *C. hominis* and *C. parvum*, respectively) and demonstrated the potential to differentiate between other isolates and species on a single microarray. While designed as a potential method for water utilities and public health officials for assessing contamination of water supplies, the described method also has potential applications in food safety.

Globalization of food production has created new routes of transmission for protozoa. Consequently, advances in diagnostic detection methods and surveillance systems have increased the number of protozoa that may be linked to food. While measures are required to prevent protozoan-causing illnesses and increase efficiency of inactivation procedures, protozoan detection remains a critical control point in safe-guarding the public food supply, particularly with respect to imported foods. DNA microarray represents a method conducive to multi-pathogen detection and as such, warrants further investigation with respect to protozoa detection strategies.

2.5. DNA Microarray Applications in Biodefense

In addition to foodborne pathogens, microbial toxins also have the potential to be used as bioweapons. Important components of a biodefence strategy are intelligence, detection, surveillance and diagnosis. As such, rapid, sensitive, and accurate detection of bioterrorism agents, as well as human and agricultural pathogens, require a broad-spectrum assay capable of discriminating between closely related bacterial, protozoan or viral pathogens. Use of microarray as a parallel detection tool (PDT) can simultaneously analyze thousands of DNA sequences, including pathogen virulence genes, in a limited amount of time. With lower costs and reduced time of testing, as compared to traditional identification methods, DNA microarray represents a broad-spectrum detection approach able to distinguish between genetic and functional signatures (ie. virulence factors).

Given the importance of virulence factors for pathogenesis, it is important to screen samples for virulence factors as part of a biodefence detection strategy. Molecular genetic analysis have identified important virulence factors and key regulatory proteins associated with toxin production and capsule biosynthesis of *Bacillus anthracis*, which is responsible for the disease anthrax. *B. anthracis* must carry pX01 and pX02, two virulence factor encoding plasmids, which are responsible for pathogenic characteristics (Pannucci et al. 2004). *B. anthracis* specific toxins *lef* (lethal gene), *cyaA* (edema factor) and *pagA* (protective antigen) are carried on pX01, while pX02 encodes proteins which attribute to *B. anthracis* virulence (Green et al. 1985; Makino et al. 1989; Uchida et al. 1993; Vietre et al. 1995).

The most severe form of intraocular bacterial infections (endophthalmitis) is caused by *Bacillus* spp. Members of *Bacillus cereus* group, which cause exotoxin-mediated diarrheal food poisoning, also carry genes coding for several important virulence factors. Due to the genetic and phenotypic similarity among *B. cereus* group members and the importance of virulence factors towards pathogenesis, Sergeev et al. (2006) developed an oligonucleotide microarray for the simultaneous detection and identification of nineteen *Bacillus* virulence factors relevant to food safety and human health. Three types of *Bacillus* virulence factors were investigated; enterotoxins (*nheA, nheB, nheC, hblA, hblC, hblD, entFM, cytK* and *bceT*), phospholipases (*spH, plc* and *piplc*) and *B. anthracis* specific toxins (*lef, cyaA* and *pagA*). The chromosomal marker BA-5449 and three capsule genes (*capA, capB* and *capC*) were used to identify *B. anthracis* (Volokhov et al. 2004), which causes anthrax. Initially, samples were subjected to a multiplex PCR amplification step and subsequent amplicon identification by hybridization to an oligonucleotide microarray. Gene specific primers were used for multiplex PCR assays and microarray probes were designed to identify *Bacillus* virulence factors. Each target gene was represented by five independent sequences (or spots) in triplicate on the microarray. Use of multiple probes increased the confidence of gene identification, while multiple short probes increased the probability of identifying a genetic variation of the gene of interest. The combined PCR-microarray represents a sensitive and specific assay, with the ability to screen for numerous *Bacillus* virulence factors in a single assay. In contrast, other methods such as PCR, are unable to provide definitive identification of amplified sequences or differentiate between *Bacillus* species. For instance, *B. cereus* and *B. thuringiensis* cannot be differentiated on the basis of biochemical characteristics (McKillip 2000) and their genomes have a high level of similarity (Ivanova et al. 2003; Rasko et al. 2004). Results revealed that *Bacillus* virulence factors, which are relevant for food safety and human health, were present in several combinations in the strains analyzed. With the exception of *B. subtilis* strains 1267 and 1268, each *Bacillus* strain investigated had a unique "virulence factor signature", suggesting that microarray can be used for strain typing, in addition to identifying virulence factors present in an isolate.

During a potential bioterrorism situation, when time is of the essence, multi-pathogen screen by DNA microarray is critical to protecting the food supply. Consequently, Sergeev et al. (2004) developed a single-chip multi-pathogen oligonucleotide microarray (FDA-1) for the simultaneous analysis of sixteen *Staphylococcus aureus* enterotoxins, *Listeria* spp. (*iap* gene), *Campylobacter* spp (*glyA* and *fur* genes) and *Clostridium perfringens* toxins (*iA*, *cpa*, *cpe*, *etxD*, *cpb1*, *cpb2*). The *fur* (ferric uptake regulator) and *glyA* (encodes serine hydroxymethyltransferase) genes were used to discriminate between *C. jejuni*, *C. coli*, *C. lari* and *C. upsaliensis*. The iap gene was used to discriminate between *L. monocytogenes*, *L. ivanovii*, *L. innocua*, *L. welshimeri*, *L. seeligeri* and *L. grayi*. Many of the mentioned pathogens are relevant for food safety, while some are also considered potential bioweapons. Following PCR, amplicons were fluorescently- and chemically-labelled, and were hybridized to a microarray containing species-specific oligoprobes that correspond to variable DNA regions. Because the amplified variable region of the fragment will only hybridize to the corresponding species-specific oligoprobe on the FDA-1 microarray, bacterial species can be identified based on the pattern of hybridization. Sergeev et al. (2004) used short oligonucleotides (17-35nt) to reduce the cost of the array chip, effectively cover the target sequence and ensure detection of minor genetic variants in target genes in the bacterial population. Each bacterial gene was identified by hybridization with four to six independent oligonucleotides. As identification was based on multiple genetic characteristics, the FDA-1 assay limited the probability of false positive and false negative results. Additionally, the FDA-1 assay analyzed a minimum of six different samples on one slide, thereby reducing assay costs.

2.6. Detection and Microbial Gene Expression under Environmental Stress and Food Storage Conditions

An important component of safeguarding the public food supply also entails a comprehensive understanding of the behaviour of microorganisms under environmental stresses and storage conditions. For instance, bacterial species can comfortably tolerate mild acidic conditions by mounting a cellular response to protect the intracellular pH (Weinrick et al. 2004). Using DNA microarray, gene expression profiling has been used to evaluate the microbial response to stressors and various microbial systems containing mixed bacterial populations on food surfaces. As well, the effect of food storage conditions on bacterial communities has also been evaluated using DNA microarray.

Although DNA microarrays have increasingly been applied to investigate microbial systems (Conway and Schoolnik 2003; Ye et al. 2001), most studies have focused on single-factor experiments (Conway and Schoolnik 2003). Consequently, Moen et al. (2005) used a DNA microarray to study non-growth survival conditions of *Campylobacter jejuni* under temperature and oxygen tension (anaerobic, microaerobic, and aerobic). As a large number of significant genes were down-regulated under certain temperature and oxygen conditions, as compared to the control, this information can improve intervention strategies to reduce *C. jejuni* in the food chain and in turn, reduce *C. jejuni*-associated disease. The authors suggest that a potential survival mechanism may include the down-regulation of most functions to save energy and produce polysaccharides and oligosaccharides for protection against harsh environments. As previously mentioned, Ye et al. (2000) investigated the survival and gene expression patterns of *Bacillus subtilis* using a DNA microarray, whereby certain *B. subtilis* genes were upregulated or downregulated under anaerobic conditions. As many genes with unknown function were induced under anaerobic conditions, the study demonstrates the use of DNA microarray to evaluate differential gene expression as a basis for future genetic and biochemical studies.

In addition to oxygen conditions, microbial communities on plant surfaces can also be affected by both climate and vegetable growth conditions. As part of maintaining quality control and minimizing food contamination, food storage conditions should be evaluated. Rudi et al. (2002) developed a DNA array-based method that targets16S rDNA, enabling the direct detection and quantification of microbial communities in ready-to-eat vegetable salads in a modified atmosphere. Comparisons were made with respect to the effect of storage at different temperatures for up to 12 days and the geographic origin of the crisphead lettuce (Spanish or Norwegian), the main salad component. Microarray results demonstrate that *Pseudomonas* spp. dominated both Norwegian- and Spanish-based lettuce salads, before storage and after storage at 4°C. While *Pseudomonas* dominated Norwegian lettuce stored at 10°C, *Enterobacteriaceae* and lactic acid bacteria dominated the microbial community in Spanish lettuce. This study illustrates the applicability of DNA microarray in assessing microbial communities on ready-to-eat vegetables with respect to food storage conditions and vegetable origin.

3. Protein Array

3.1. Protein Array Principles and Applications

Protein arrays could be used to analyze protein function on a whole genome level (MacBeath 2002; Templin et al. 2002). The arrays are used to identify protein-protein interaction, substrates of protein kinases, or targets of biologically active small molecules. In a protein array, different protein molecules are affixed at separate locations on a solid support (glass or synthetic material with a modified coated surface), in much the same manner as DNA microarray. The choice of surface chemistry determines the method of protein attachment to the functionalized support, which in turn influences whether the proteins require modification prior to arraying (Spielbauer and Stahl 2005). Proteins can be externally synthesized and purified or synthesized *in situ* and attached to the array. Common capture molecules in protein arrays include antigen-antibody, protein-protein, aptamers (synthetic molecules composed of nucleotides, ribonucleotides or peptides), enzyme-substrate and receptor-ligand. Bound target molecules can be detected by fluorescence, chemiluminescence or radioactivity. Fluoresence detection, which is safe, sensitive and has a high resolution, is commonly used for detection. Additionally, fluorescent detection is compatible with standard DNA microarray scanners, although some minor software adjustments may be required.

With the development of highly specific antibodies and low assay costs, ELISA is commonly used within the food industry. A wide variety of ELISA kits are commercially available for detection of foodborne pathogenic *E. coli, Salmonella* and *Listeria.* However, ELISA tests are typically pathogen-specific and oftentimes require 24-48 hours to obtain preliminary results. As an alternative, protein arrays have demonstrated an 80% agreement with ELISA results (Bacarese-Hamilton et al. 2004). Furthermore, protein array detection is amenable to high through-put analysis, reduces sample analysis time, is conducive to multi-pathogen and toxin detection, and represents a cost-effective alternative.

Antibody array, the most common type of protein array, involves spotting antibodies onto a protein chip. Antibodies are used as capture molecules to detect protein from cell lysate solution. The classical antibody sandwich assay involves binding a pair of antibodies to two discrete recognition surfaces on each protein. In this procedure, an array of antibodies are covalently bound to a solid surface and probed with various analytes. The second biotinylated antibody binds to the captured analyte, forming a unique complex. Using a universal antibiotin antibody conjugated to a fluorophore, the second antibody is detected by fluorescence.

3.1.1. Advantages of Protein Arrays

In recent years, protein arrays have become a method for screening and multiplex analysis (tens of thousands) of proteins in a single assay, enabling the analysis of protein function on a whole gemone level (MacBeath 2002; Templin et al. 2002). Because protein microarrays are conducive to multiple sample analysis, the resulting data can be subjected to rigorous statistical analyses and tested for accuracy and reproducibility. As well, the multiple sample analysis supports multi-pathogen and toxin detection of foodborne microbes, which has particular relevance to food safety and bioterrorism programs. Protein array platform can also be used to predict protein abundance or identify protein function. In addition to commercial diagnostics and assessment of environmental pollution, antibody microarrays also have great potential for quality control within the food industry. In particular, antibody microarrays have been used in biosensor detection of *E. coli* O157:H7 (Gehring et al. 2006), detection of foodborne pathogens and associated toxins (Ligler et al. 2003; Rucker et al. 2005), as well as serotyping (Anjum et al. 2006).

Detection of foodborne pathogens can be achieved through either immunological or molecular methods, such as PCR-based detection. As immunological detection techniques require specific antisera, the method is laborious and time consuming. In the case of PCR-based investigations, experiments are prone to failure and false positive results, as well as a limited ability to simultaneously detect multiple microbial pathogens. As protein arrays platforms can be adapted to existing automated processes previously developed for DNA microarray research, thus, there is potential cost benefits and reduced time requirements.

Furthermore, protein array detection is not restricted to the detection of DNA or RNA carrying microbes, spores and viruses. As antibodies are able to detect antigens present on the microbial surface, extraction and purification of RNA or DNA is not needed, thereby minimizing time requirements and influence of inhibitors affecting food sample analyses. With the ability to detect food-related toxins and pathogens, as well as serotypes, protein arrays can be used to elicit potential health threats within food supplies.

3.1.2. Disadvantages of Protein Arrays

Because not all genes will be concurrently expressed, monitoring protein interactions is an extremely complex process. Furthermore, protein synthesis and degradation can vary greatly under different metabolic or environmental conditions, thereby requiring repeated analysis of the proteome under strictly controlled conditions (Kusnezow and Hoheisel 2002). Although similar in detection principle to DNA microarrays, protein concentration in biological samples is greater then mRNAs, therefore, protein arrays require a wider detection range.

As the success of protein arrays is dependent on proteins maintaining their correctly folded conformations, array design can be particularly challenging. Because proteins are composed of many different amino acids, they have variable surface properties rendering it difficult to find a common surface suitable for different proteins with a broad molecular weight range and physio-chemical properties, advancements in protein array have been limited (Cheung et al. 1999).

Although antibodies are the best characterized and most effective protein array capture agent, they are subject to problems with specificity in commercial antibody preparations and antibodies do not exist for most proteins. Consequently, as antibodies cannot be manufactured with a known specificity and affinity, antibody-based microarrays require validation and optimization of individual antibodies, thereby lengthening the experimental design and optimization process. However, once optimized, protein arrays are easily incorporated as laboratory procedures, requiring minimal sample preparation time, as compare to DNA microarray.

3.2. Detection and Characterization of Food Isolates

Although various technologies exist for detection of foodborne pathogens, protein arrays have the distinct advantage of automation, small sample and reagent volumes, and possibility of parallel sample analysis. With respect to food safety and quality, protein arrays offer immense possibilities for multi-foodborne pathogen and toxin detection.

As conventional protein immobiliziation protocols can be time consuming, labour intensive and expensive, Lee et al. (2006) developed a sol-gel-derived bioactive platform as an alternative immobilization procedure required for the detection of pathogenic *E. coli* O157:H7. The nanoporous material provides a support matrix for analyte-sensitive reagents which are commonly added to the precursor solution. Biomolecules, such as antibodies, enzymes and receptors, are entrapped in the sol-gel derived matrix and have demonstrated improved stability which reflects the improved resistance to thermal and chemical denaturation, long term storage and operational stability (Kim et al. 2001). In this study, a mAb antibody against *E. coli* O157:H7 was immobilized on the sol-gel derived matrix by physical adsorption. According to the fluorescent intensity, which was proportional to the concentration of *E. coli* O157:H7, the detection limit was 10^2 CFU/ml. Based on these results, the authors suggest that a sol–gel-derived bioactive platform can be applied to and possibly improve protein chip based detection.

Using an array-based multiplex immumoassay, Thirumalapura et al. (2006) used antibodies to detect surface antigens of Gram-negative (*Escherichia coli* and *Salmonella enterica* serovar Typhimurium) and Gram-positive bacteria (*Staphylococcus aureus*, *Streptococcus pyogenes* and *Listeria monocytogenes*). The assay involved printing whole bacterial cells on nitrocellulose coated glass substrates and antibody binding was detected using fluorophore labelled secondary antibodies. Both Gram-positive and Gram-negative bacteria strongly adsorbed onto nitrocellulose coated slides, however, a statistically significant loss of fluorescence signal after washing was observed with *S. aureus*. The sensitivity of antibody detection for both *S.* Typhimurium and *L. monocytogenes* was 0.1 μg/ml, compared to 1 μg/ml using conventional immunofluorescence assays for antibody detection (Goldsby 2003). The described array supports the simultaneous detection of antibodies against multiple foodborne bacteria. Additionally, in using whole bacterial cells as the antigen source, there is no need for antigen purification.

Huelseweh et al. (2006) developed a protein chip for the simultaneous detection of potential biowarfare agents, including *Staphylococcus* enterotoxin B, *Yersinia pestis*, *Brucella melitensis*, and enterohemorrhagic *Escherichia coli* O157:H7. With a completion time of 1 to 1 ½ hours, the protein assay for each biological warfare agent was individually optimized and validated. Subsequent parallel analysis enabled the simultaneous detection of up to five biological warfare agents in a single microarray. Detection limits for the pathogens and toxins were comparable to ELISA results, however, the protein array was less time consuming. Furthermore, in the presence of irrelevant bacteria in 10 to 100-fold excess, specific target detection was not affected.

3.3. Serotyping Foodborne Pathogens

Serological classification based on antigenic differences is often used for typing (fingerprinting of foodborne pathogens). Current serotyping methods are only able to detect a single antibody-antigen reaction at a time, require well-experienced technologists to perform, are time consuming and require relatively high volumes of reagents. While DNA-based approaches, such as PCR, can be used to identify a particular serovar, the number of serovars analyzed in a single assay is limited and several genetic markers have not yet been developed or verified for identification of various serovars (McQuiston et al. 2004). Conversely, protein arrays can be used for the parallel analysis of multiple antigens for serotyping investigations.

Understanding the epidemiology of infections and differentiation between pathotypes is an important component of food safety. Commonly, O serotyping has been used to assess a pathogen threat. For instance, serological typing of *E. coli* O antigens has been used to identify and differentiate O serotypes commonly associated with disease. While a primary diagnostic tool, O serotyping is time-consuming, costly and may be subject to cross-reactions, particularly with the use of polyclonal antibodies. With increased applications, protein arrays have been used for parallel identification, quantification and functional analysis of different proteins.

Anjum et al. (2006) developed an antibody-based miniaturized array chip to detect bound whole-cell *E. coli* antigens and the corresponding immobilized antibody. The microarray was based on 17 O antisera of common *E. coli* pathogens associated with disease syndromes in humans and farm animals, as well as contaminated food products. Using 17 related control strains, results demonstrate a 100 % correlation of O types detected by the microarray chip and conventional typing. In a blind trial, the array correctly identified 88% of 100 previously serotyped *E. coli* isolates. Discrepancies with conventional serotyping data were attributed to the increased sensitivity of the protein array. Additionally, after storage, isolates became untypeable and resulted in poor signal intensity. However, array results were confirmed by latex agglutination, suggesting that protein array may have a higher accuracy than traditional O serotyping methods. With the ability to identify *E. coli* O serotypes in pure culture and mixed *E. coli* cultures, protein array is also less time-consuming and costly then conventional serotyping methods. The authors demonstrate the potential of protein array for serotyping as an alternative to traditional *E. coli* O serotyping with potential future applications with *Salmonella*.

Human *Salmonella* infections can lead to several clinical conditions, including enteric (typhoid) fever, uncomplicated enterocolitis and systemic infections by nontyphoid microorganisms. As the *Salmonella* genus consists of over 2,500 serovars (Cai et al. 2005), serological detection can be complicated. Serological testing procedures aim to determine the complete antigenic formulas of individual *Salmonella* isolates. Antigenic determinants include the somatic (O) lipopolysaccharides on the external surface of the bacterial outer membrane and flagellar (H) antigens which are associated the peritrichous flagella. In traditional *Salmonella* serotyping assays, the O serogroups and H phase 1 and 2 antigens need to be tested separately. However, Cai et al. (2005) constructed an antibody array to simultaneously detected both H phase antigens in a single assay. The microarray was constructed to include 35 polyclonal rat antiserum antibodies (11 O factor antisera, 9 H phase antisera, 13 H factor antisera, and 2 O multigroup antisera) for the identification of 20 commonly isolated and clinically important *Salmonella* serovars. The assay identified 86 of the 117 target *Salmonella* strains, partially identified 30 target *Salmonella* strains and excluded all 73 non-target strains. The described protein array represented a more rapid and potentially cost-effective alternative to the current slide agglutination method for *Salmonella* serotyping.

3.4. Detection of Food-Related Toxins

While many toxins are secreted from bacteria during infection, toxins can also be introduced by industrial and agricultural activities. As well, toxins, such as mycotoxins, can occur naturally in the food supply. Bacterial toxins secreted into the foodstuffs during growth (ie. Staphyloccus aureus enterotoxin and botulium toxins) can also cause foodborne illnesses. Therefore, as part of protecting the food supply, sensitive detection strategies must also be directed towards toxins. Traditional methods for detection of these toxins have employed laboratory techniques such as high pressure liquid chromatography (HPLC) and mass spectrometry are traditionally used for toxin detection, however, significant sample preparation steps are required to eliminate interference from other components of the sample matrix. While immunoassays require less extensive sample preparation, array technology supports the detection of multiple toxins in a single assay.

Ngundi et al. (2006) used a planar waveguide detection protein array to investigate the detection of bacterial toxins. With this technology, only fluorophores located at or near the surface of the waveguide are excited and signals from unbound molecules in the bulk solution are not detected. With improved sensitivity, the technology is beneficial for detecting low-abundance messengers of relevant biological information. In this study, derivatives of N-acetyl galactosamine (GalNAc) and N-acetylneuraminic acid (Neu5Ac) were used as receptors for protein toxins. While *Salmonella* Typhimurium, *Listeria monocytogenes*, *Escherichia coli* and staphylococcal enterotoxin B (SEB) did not bind to either carbohydrate, at a concentration of 100 ng/ml, the cholera and tetanus toxins bound to both GalNAc and Neu5Ac. The study indicated that toxin binding to the both carbohydrates was density dependent and semi-selective.

Using a sandwich antibody array, Ligler et al. (2003) measured toxins in clinical fluids, environmental and food samples. The method required minimal sample preparation, thereby reducing experiment time requirements. With toxin levels as low as 0.5 ng/ml, the assay was conducive for the rapid analysis of staphylococcal enterotoxin B (SEB), ricin, cholera toxin, botulinum toxoids, trinitrotoluene, and the mycotoxin fumonisin. Regardless of the food matrix tested, the limit of detection for SEB was identical to using buffer, however, the intensity of signal was lower for all food matrix tested (ham, milk and cantaloupe). The authors attribute this finding to increased sample viscosity and consequent reduction in the diffusion of toxin to the surface or sequestering of some of the toxin by sample matrix components. Therefore, for quantitative assays, a standard curve should be prepared for each food matrix and/or a positive control known toxin concentration in the corresponding food matrix should be incorporated in each assay.

Rucker et al. (2005) developed a multiplex antibody-based array for the detection of cholera toxin beta-subunit, diphtheria toxin, anthrax lethal factor and protective antigen, *Staphylococcus aureus* enterotoxin B, and tetanus toxin C fragment. Direct and competitive assays were evaluated. In the direct assay, fluorescently labelled toxins were directly captured by the antibody array. The concentration of the labelled toxin was determined by correlating binding information to the baseline amount. In the competitive assay, which provides information on the presence, identity and concentration of toxins, unlabeled toxins were used as reporters for the quantification of the native toxin in solution. While the direct assay was more sensitive, detection limits for unmodified toxins in the competitive assay were comparable to values previously reported for sandwich-format immunoassays of antibodies arrayed on planar substrates. Additionally, the competitive assay detects toxins in their native state in complex samples, which is paramount to food protection programs.

3.5. Protein Array Applications in Biosensing

Although laboratories offer complex instrumentation and possess highly qualified technical staff, there is also a need for portable technology for sample analysis. Consequently, biosensor technology represents a portable and rapid means of acquiring immediate "on-the-spot" results. The high specificity and sensitivity of biosensors support broad-spectrum detection of analytes in complex matrices, such as food, blood, serum, and urine. The technology has important applications in food quality and safety control, and is of particular importance for the rapid detection of total microbial contamination.

Biosensors involve a biological recognition component, such as receptors, nucleic acids, or antibodies which are in close contact with an appropriate transducer. Based on the method of signal transduction, biosensors can be divided into four groups: optical, mass, electrochemical and thermal sensors. Biosensors can rely on direct analyte detection or indirect detection, whereby the products of a biochemical reaction are detected by a sensor. Ideally, biosensors need to be efficient, rapid, specific, sensitive, and preferably quantitative. Biosensors should also be amenable to adaptations to allow testing of new analytes and support multi-analyte analysis. With the ability to contain hundreds or thousands of micrometer-diameter spots, microarrays pose enormous potential in biosensing. In a protein array biosensor, the target is detected by a single label-free reagent that does not affect biomolecular activity, thereby decreasing non-specific binding and increasing detection sensitivity of the biomolecular interaction.

As carbohydrate-cell interactions are ubiquitous in nature, such investigations have the potential to impact a variety of microbiolgical disciplines, including food safety. In particular, the carbohydrate microarray represents a method for rapid screening and testing of pathogens, which can potentially be used as a biosensor. As various cell types bind to carbohydrates, the carbohydrate "fingerprint" can be used to disseminate the type of bacteria present within a complex mixture (Mammen et al. 1998). Disney and Seeberger (2004) used a carbohydrate array to investigate the carbohydrate binding of specific bacteria and detect *E. coli*. Subsequently, *E. coli* captured by the arrays was further cultured for additional susceptibility tests. The study was able to distinguish between mutant and wildtype *E. coli*, the former having a reduced affinity for mannose. The ability to distinguish between different carbohydrate affinities of cells could have a beneficial impact on detection strategies, as virulence of many pathogens correlates with carbohydrate binding. Therefore, the carbohydrate binding profile could be used to assess pathogenicity.

Gehring et al. (2006) also investigated the use of an array sandwich immunoassay biosensor for detection of *E. coli* O157:H7. In this study, biotinylated capture antibodies were printed onto streptavidin-coated slides. Fluorescein-labeled reporter antibodies were used to detect the presence of captured *E. coli* O157:H7. The authors examined the ranges in which array biosensors could be used for detection of *E. coli* O157:H7 and explored the capture efficiency of antibody-mediated binding of *E. coli* to the array substrate. A sandwich fluorescent immunoassay in an array format was used to capture and detect *E. coli* O157:H7. Biotinylated capture antibodies generated a higher signal, compared to biotinylated protein G-bound capture antibodies. In terms of capture efficiency, as the number of bacteria available for capture decreased, capture efficiency increased to near 35%. The authors noted that the number of bacteria available for capture affected the optimal combination of capture and reporter antibodies. The findings suggest that antibody-based arrays, when used in sandwich assay format, may effectively be used to capture and detect *E. coli* O157:H7. Based on these finding, Gehring et al. (2006) plan to investigate antibody array detection in culture-enriched food samples (e.g., ground beef) which are artificially inoculated with pathogenic bacteria.

Delehanty and Lighler (2002) developed an antibody array biosensor for the characterization and rapid detection under flow conditions of proteins, bacterial and viral analytes both individually and in samples containing a mixture of analytes. The assay was able to detect cholera toxin, staphylococcal enterotoxin B, ricin, and *Bacillus globigii* at levels as low as 8 ng/mL, 4 ng/mL, 10 ng/mL, and 6.2 x 10^4 cfu/mL, respectively. These detection limits are comparable to those reported for other biosensor technologies. Apart from the ability to simultaneously detect proteins and bacteria, flow-based antibody array biosensor results were achieved within 15 minutes, which is a significant improvement over other antibody arrays that take hours to complete.

Contamination of food by microorganisms or their toxins can greatly affect human and animal health. Food processing and quality control remain a primary use of biosensor technology within the food industry. Using a microarray platform, biosensor detection within the food sector would be less expensive, more reliable, less time-consuming and amenable to multi-pathogen screening, compared to conventional biosensing methodologies.

4. Conclusion

DNA and protein array technologies offer immense flexibility in food-based applications, offering different types of probes and compatibility with numerous targets. Quality control and analysis of microbial pathogens is important in food safety. Microarray technology can be used to identify the etiological agent responsible for food contamination, track bacterial contamination sources throughout the manufacturing process and examine the production and utilization of fermented products by bacteria. As a large fraction of the microbial genome or proteome can be investigated in a single assay, DNA and protein arrays have varied applications in microbial food safety, including broad-spectrum pathogen and toxin detection, genotyping, serotyping, detection of antibiotic resistance genes, as well as gene expression profiling in response to stressors and various microbial systems containing mixed bacterial populations.

Because it is difficult to identify the etiological agent of concern prior to the onset of clinical symptoms, broad-spectrum techniques capable of simultaneously identifying multiple foodborne pathogens, virulence genes and associated toxins within food supplies remains a critical component of food safety programs and national defence strategies. Recent DNA microarray progress can be attributed to previously developed techniques, such as multiplex PCR, which can be easily adapted to a microarray platform, offering improved specificity and speed of hybridization analysis. With some modification, DNA microarray analysis could be used for field applications, which is of particular importance in food safety and bioterrosim programs. Oligonucleotide microarrays can detect virulence genes and simultaneously discriminate among foodborne human pathogens, such as *Bacillus, E. coli*, *Salmonella* and *Shigella* strains. Using DNA microarrays as a foodborne pathogen screening technique can also be combined with real-time PCR for further confirmation and quantitative analyses. Bacterial toxins, such as cholera toxin beta-subunit, diphtheria toxin, anthrax lethal factor and protective antigen, *Staphylococcus aureus* enterotoxin B, and tetanus toxin C fragment, have been detected using a multiplex antibody-based array for detection.

5. Future Perspectives

As a result of pathogen evolution, changes in agricultural and manufacturing processes, and changes in human hosts, new foodborne pathogens and strain diversification will likely emerge. Consequently, genotyping-based microarrays provide a means of detecting and characterizing divergent populations. The availability of complete genome sequences of several microorganisms has enabled DNA microarray technology to become a powerful tool to explore global gene expression profiles and measure genome-wide differences in genetic contents. Consequently, there is a definite need to continue microarray research in monitoring microbial gene expression in order to further our understanding of uncharacterized genes, probe microbial physiological adaptations in response to various environmental conditions, and identify virulence-associated genes. As well, microarrays can be used as screening tool in microbial gene expression studies, while real-time PCR can be used for more detailed analysis and confirmation of these microarray results.

Due to the inherent complexity of developing and manufacturing protein microarrays, available research using the technique lags behind DNA microarray. DNA microarray have become overwhelmingly dominant in microarray research, due to the relatively ease of synthesis and chemical robustness of DNA. However, both DNA and protein arrays have deficiencies in terms of routine method application within a laboratory setting. Although DNA microarray can be highly specific, reproducible and sensitive due to probe printing technologies and platform design, analytical limitations associated with efficient separation of target microorganism(s) from food sample matrices, presence of inhibitors and extensive sample preparation procedures limit this technique. To ensure efficient recovery of nucleic acids and removal of inhibitors from complex food samples stringent extraction and purification techniques are required, thus increasing overall sample processing times. Therefore, to realize the full potential of DNA microarray research, there is an increased need for continued improvement of sample preparation methods. Conversely, protein arrays rely on high-affinity, specific antibodies for detection for microbial identification, requiring careful screening and optimization of capture and detection antibodies. Because antibodies cannot be manufactured with a known specificity and affinity, antibody-based arrays require validation of individual antibodies. While this can be accomplished using Western Blot, antibody validation lengthens the experimental design and optimization process. However, once optimized, protein arrays are easily incorporated into the analytical and research laboratory setting, requiring minimal sample preparation time, as compared to DNA microarray. Furthermore, as immunoassays are commonly used within the food industry, the transition to protein arrays is relatively straightforward.

Protein arrays are highly informative tools used for high-throughput interaction studies of various biomolecules including proteins, DNA and small molecules as well as in biochemical investigations of protein activity for functional annotation and characterization. With increasing applications of DNA and protein arrays and to effectively interpret the enormous amounts of data generated, there is an increasing need for construction and maintenance of reliable databases. Consequently, continued progress and growth of internet-based data mining systems will further DNA and protein array research.

Although in its infancy, microarray-based technologies offer considerable opportunities and advancement in the field of food microbiology. With continued progress in method development, exploration of the microbial genome and proteome, and construction of reliable databases, DNA and protein arrays will become a prominent technology in food safety, quality control and assessment of pathogenecity.

References

Al-Khaldi, S.F., Martin, S.A., Rasooly, A., and Evans, J.D. (2002). DNA microarray technology used for studying foodborne pathogens and microbial habitats: minireview. *Journal AOAC International*, 85, 906-910.

Allos, B.M. (2001). *Campylobacter jejuni* infections: update on emerging issues and trends. *Journal Clinical Microbiology*, 32, 1201-1206.

Anjum, M.F., Tucker, J.D., and Sprigings, K.A. (2006). Use of miniaturized protein arrays for *Escherichia coli* O serotyping. *Clinical and Vaccine Immunology*, 13, 561-567.

Bacarese-Hamilton, T., Mezzasoma, L., Ardizzoni, A., Bistoni, F., and Crisanti, A. (2004). Serodiagnosis of infectious diseases with antigen microarrays. *Journal of Applied Microbiology*, 96, 10-17.

Bae, J.W., Rhee, S.K., Park, J.R., Chung, W.H., Nam, Y.D., Lee, I., Kim, H., and Park, Y.H. (2005). Development and evaluation of genome-probing microarrays for monitoring lactic acid bacteria. *Applied and Environmental Microbiology*, 71, 8825-8835.

Barza, M. (2002). Potential mechanisms of increased disease in humans from antimicrobial resistance in food animals. *Clinical and Infectious Diseases*, 34, 123-125.

Bavykin, S.G., Akowski, J.P., Zakhariev, V.M., Barsky, V.E., Perov, A.N., and Mirzabekov, A.D. (2001). Portable system for microbial sample preparation and oligonucleotide microarray analysis. *Applied and Environmental Microbiology*, 67, 922-928.

Bean, N.H., Goulding, J.S., Lao, C., and Angulo, F.J. (1996). Surveillance for foodborne-disease outbreaks- United States, 1988-1992. Morbidity and Mortality Weekly Report. *CDC Surveillance Summary*, 45, 1-66.

Bekal, S., Brousseau, R., Masson, L., Prefontaine, G., Fairbrother, J., and Harel, J. (2003). Rapid identification of *Escherichia coli* pathotypes by virulence gene detection with DNA microarrays. *Journal of Clinical Microbiology*, 41, 2113-2125.

Borucki, M.K., Krug, M.J., Muraoka, W.T., and Call, D.R. (2003). Discrimination among *Listeria monocytogenes* isolates using a mixed genome DNA microarray. *Veterinary Microbiology*, 4, 351-362.

Cai, H.Y., Lu, L., Muckle, C.A., Prescott, J.F., and Chen, S. (2005). Development of a novel protein microarray for serotyping *Salmonella enterica* strains. *Journal Clinical Microbiology*, 43, 3427-3430.

Call, D.R., Bakko, M.K., Krug, M.J., and Roberts, M.C. (2003). Identifying antimicrobial resistance genes with DNA microarrays. *Antimicrobial Agents and Chemotherapy*, 47, 3290-3295.

Call, D.R., Brockman, F.J., and Chandler, D.P. (2001). Detecting and genotyping *Escherichia coli* O157:H7 using multiplexed PCR and nucleic acid microarrays. *International Journal of Food Microbiology*, 67, 71-80.

Carey, C.M., Lee, J., and Trevors, J.T. (2004). Biology, persistence and detection of *Cryptosporidium parvum* and *Cryptosporidium hominis* oocyst. *Water Research*, 38, 818-862.

Chaidez, C., Soto, M., Gortares, P., and Mena, K. (2005). Occurrence of *Cryptosporidium* and *Giardia* in irrigation water and its impact on the fresh produce industry. *International Journal Environmental and Health Research*, 15, 339-345.

Cheung, V.G., Morley, M., Aguilar, G., Massimi, A., Kucherlapati, R., and Childs, G. (1999). Making and reading microarrays. *Nature Genetics*, 21, 15-19.

Chizhikov, V., Rasooly, A., Chumakov, K., and Levy, D.D. (2001). Microarray analysis of microbial virulence factors. *Applied and Environmental Microbiology*, 67, 3258-3263.

Chizhikov, V., Wagner, M., Ivshina, A., Hoshino, Y., Kapikian, A.Z., and Chumakov, K. (2002). Detection and genotyping of human Group A Rotavirus by oligonucleotide microarray hybridization. *Journal of Clinical Microbiology*, 40, 2398-2407.

Conway, T. and Schoolnik, G.K. (2003). Microarray expression profiling: capturing a genome-wide portrait of the transcriptome. *Molecular Microbiology*, 47, 879-889.

Croci, L., De Medici, D., Scalfaro, C., Fiore, A., Divizia, M., Donia, D., Consentino, A.M., Moretti, P., and Costantini, G. (2000). Determination of enterovirus, hepatitis A virus; bacteriophages and *E. coli* in Adriatic sea mussels. *Journal of Applied Microbiology*, 88, 293-296.

Delehanty, J.B. and Lighler, F.S. (2002). A microarray immunoassay for simultaneous detection of proteins and bacteria. *Analytical Chemistry*, 74, 5681-5687.

Disney, M.D. and Seeberger, P.H. (2004). The use of carbohydrate microarrays to study carbohydrate-cell interaction and to detect pathogens. *Chemistry and Biology*, 11, 1707-1707.

Djuretic, T., Wall, P.G., Ryan, M., Evans, H.S., Adak, G.K., and Cowden, J.M. (1996). General outbreaks of infectious intestinal disease in England and Wales 1992 to 1994. *CDR Reviews*, 6, R57-63.

Dorrell, N., Mangan, J.A., Laing, K.G., Hinds, J., Linton, D., Al-Ghusein, H., Barrell, B.G., Parkhill, J., Stoker, N.G., Karlyshev, A.V., Butcher, P.D., and Wren, B.W. (2001). Whole genome comparison of *Campylobacter jejuni* human isolates using a low-cost microarray reveals extensive genetic diversity. *Genome Research*, 11, 1706-1715.

Fitzgerald, J.R., Sturdevant, D.E., Mackie, S.M., Gill, S.R., and Musser, J.M. (2001). Evolutionary genomics of *Staphylococcus aureus*: insights into the origin of methicillin-resistant strains and the toxic shock syndrome epidemic. *Proceedings of the National Academy of Science of the Unites States of America*, 98, 8821-8826.

Franz, C.M.A.P., Stiles, M.E., Schleifer, K.H., and Holzapfel, W.H. (2003). Enterococci in foods- a conundrum for food safety. *International Journal Food Microbiology*, 88, 105-122.

Gehring, A.G., Albin, D.M., Bhunia, A.K., Reed, S.A., Tu, S.I., and Uknalis, J. (2006). Antibody microarray detection of *Escherichia coli* O157:H7: quantification, assay limitations and capture efficiency. *Analytical Chemistry*, 78, 6601-6607.

Goldsby, R.A. *Immunology*. 5th Edition. New York: W.H. Freeman and Company; 2003

Goodgame, R. (2006). Norovirus gastroenteritis. *Current Gastroenterology Reports*, 8, 401-408.

Graves, L.M. and Swaminathan, B. (2001). PulseNet standardized protocol for subtyping *Listeria monocytogenes* by macrorestriction and pulse-field gel electrophoresis. *International Journal of Food Microbiology*, 65, 55-62.

Green, B.D., Battisti, L., Koehler, T.M., Thorne, C.B., and Ivins, B.E. (1985). Demonstration of a capsule plasmid in *Bacillus anthracis*. *Infection and Immunology*, 49, 291-297.

Hong, B.X., Jiang, L.F., Hu, Y.S., Fang, D.Y., and Guo, H.Y. (2004). Application of oligonucleotide array technology for the rapid detection of pathogenic bacteria of foodborne infections. *Journal Microbiological Methods*, 58, 403-411.

Huelseweh, B., Ehricht, R., and Marschall, H.J. (2006). A simple and rapid protein array based method for the simultaneous detection of biowarfare agents. *Proteomics*, 6, 2972-2981.

Ikeda, M., Yamaguchi, N., Tani, K., and Nasu, M. (2006). Detection of food poisoning bacteria in fresh vegetables using DNA microarray. *Journal of Health Science*, 52, 36-42.

Ivanova, N., Sorokin, A., Anderson, I., Galleron, N., Candelon, B., Kapatral, V., Bhattacharyya, A., Reznik, G., Mikhailova, N., Lapidus, A., Chu, L., Mazur, M., Goltsman, E., Larsen, N., D'Souza, M., Walunas, T., Grechkin, Y., Pusch, G., Haselkorn, R., Fronstein, M., Ehrlich, S.D., Overbeek, R., and Kyrpides, N. (2003). Genome sequence of *Bacillus cereus* and comparative analysis with *Bacillus anthracis*. *Nature*, 423, 87-91.

Jarraud, S., Mougel, C., Thiolouse, J., Lina, G., Meugnier, H., Forey, F., Nesme, X., Etienne, J., and Vandenesch, F. (2002). Relationships between *Staphylococcus aureus* genetic background, virulence factors, *agr* groups (alleles), and human disease. *Infection and Immunity*, 70, 631-641.

Johnston, S.P., Ballard, M.M., Beach, M.J., Causer, L., and Wilkins, P.P. (2003). Evaluation of three commercial assays for the detection of *Giardia* and *Cryptosporidium* organisms in fecal specimens. *Journal of Clinical Microbiology*, 41, 623-626.

Jääskeläinen, A.J. and Maunula, L. (2006). Applicability of microarray technique for the detection of noro- and astroviruses. *Journal Virological Methods*, 136, 210-216.

Kastner, S., Perreten, V., Bleuler, H., Hugenschmidt, G., Lacroix, C., and Meile, L. (2006). Antibiotic susceptibility patterns and resistance genes of starter cultures and probiotic bacteria used in food. *Systematic and Applied Microbiology*, 29, 145-155.

Keramas, G., Bang, D.D., Lund, M., Madsen, M., Rasmussen, S.E., Bunkenborg, H., Telleman, P., and Christensen, C.B.V. (2003). Development of a sensitive DNA microarray suitable for rapid detection of *Campylobacter* spp. *Molecular and Cellular Probes*, 17, 187-196.

Kim, Y.D., Park, C.B., and Clark, D.S. (2001). Stable sol-gel microstructured and microfluidic networks for protein patterning. *Biotechnology and Bioengineering*, 73, 331-337.

Klappenbach, J.A., Saxman, P.R., Cole, J.R., and Schmidt, T.M. (2001). rrndb: the ribosomal RNA operon copy number database. *Nucleic Acids Research*, 29, 181-184.

Kusnezow, W. and Hoheisel, J.D. (2002). Antibody microarrays: promises and problems. *Biotechiques,* Suppl, 14-23.

Lee, W., Park, K.S., Kim, Y.W., Lee, W.H., and Chio, J.W. (2006). Protein array consisting of sol-gel bioactive platform for detection of *E. coli* O157:H7. *Biosensors and. Bioelectronics*, 20, 2292-2299.

Lee, W.C., Lee, M.J., Kim, J.S., and Park, S.Y. (2001). Foodborne illness outbreaks in Korea and Japan studied retrospectively. *Journal of Food Protection*, 64, 899-902.

LeGuyader, F., Miossec, L., Haugarreau, L., Dubois, E., Kopecka, H., and Pommepuy, M. (1998). RT-PCR evaluation of viral contamination in fibe shellfish beds over a 21-month period. *Water Science and Technology*, 2, 45-50.

Lemarchand, K., Berthiaume, F., Maynard, C., Harel, J., Payment, P., Bayardelle, P., Masson, L., and Brousseau, R. (2005). Optimization of microbial DNA extraction and purification from raw wastewater samples for downstream pathogen detection by microarrays. *Journal of Microbiological Methods*, 63, 115-126.

Lemarchand, K., Masson, L., and Brousseau, R. (2004). Molecular biology and DNA microarray technology for microbial quality monitoring of water. *Critical Reviews in Microbiology*, 30, 145-172.

Li, J., Pankratz, M., and Johnson, J.A. (2002). Differential gene expression patterns revealed by oligonucleotide versus long cDNA arrays. *Toxicological Sciences*, 69, 383-390.

Ligler, F.S., Taitt, C.R., Shriver-Lake, L.C., Sapsford, K.E., Shubin, Y., and Golden, J.P. (2003). Array biosensor for detection of toxins. *Analytical and Bioanalytical Chemistry*, 377, 469-477.

Lin, MC., Huang, A.H., Tsen, H.Y., Wong, H., and Chang, T.C. (2005). Use of oligonucleotide array for identification of six foodborne pathogens and *Pseudomonas aeruginosa* grown on selective media. *Journal of Food Protection*, 68, 2278-2286.

Lipp, M., Bluth, A., Eyquem, F., Kruse, L., Schimmel, H., and Van den Eed, G. (2001). Validation of a method based on polymerase chain reaction for the detection of genetically modified organism in various processed foodstuff. *European Food Research and Technology*, 212, 497-504.

Lipshutz, R.J., Fodor, S.P.A., Gingeras, T.R., and Lockhart, D.J. (1999). High density oligonucleotide arrays. *Nature Genetics*, 21, 20-24.

MacBeath, G. (2002). Protein microarrays and proteomics. *Nature Genetics*, 32, 526-532.

Makino, S., Uchida, I., Terakado, N., Sasakawa, C., and Yoshikawa, M. (1989). Molecular characterization and protein analysis of the cap region, which is essential for encapsulation in *Bacillus anthracis*. *Journal of Bacteriology*, 171, 722-730.

Mammen, M., Choi, S.K., and Whitesides, G.M. (1998). Polyvalent interactions in biological systems: implications for design and use of multivalent ligands and inhibitors. *Angew Chemistry*, 37, 2755-2794.

Mansfield, L.S. and Gajadhar, A.A. (2004). *Cyclospora cayetanensis*, a food- and waterborne coccidian parasite. *Veterinary Parasitology*, 126, 73-90.

Maurice, J. (1994). The rise and rise of food poisoning. *New Scientist*, 144, 28-33.

Maynard, C., Berthiaume, F., Lemarchand, K., Harel, J., Payment, P., Bayardelle, P., Masson, L., and Brousseau, R. (2005). Waterborne pathogen detection by use of oligonucleotide-based arrays. *Applied and Environmental Microbiology*, 71, 8548-8557.

McKillip, J.L. (2000). Prevalence and expression of enterotoxins in *Bacillus cereus* and other *Bacillus* spp. , a literature review. *Antoine Van Leeuwenhoek*, 77, 393-399.

McQuiston, J.R., Parrenas, R., Ortiz-Rivera, M., Cheesling, L., Brenner, F., and Fields, P.I. (2004). Sequencing and comparative analysis of flagellin genes *flicC*, *fljB*, and *flpA* from *Salmonella* spp. *Journal of Clinical Microbiology*, 42, 1923-1932.

Mead, P.S., Slutsker, L., Dietz, V., McCaig, L.F., Bresee, J.S., Saphiro, C., Griffin, P.M., and Tauxe, P.V. (1999). Food-related illness and death in the United States. *Emerging and Infectious Diseases*, 5, 607-625.

Meyer, R. (1999). Development and application of DNA analytical methods for the detection of GMOs in food. *Food Control*, 10, 391-399.

Moen, B., Oust, A., Langsrud, O., Dorrell, N., Marsden, G.L., Hinds, J., Kohler, A., Wren, B.W., and Rudi, K. (2005). Explorative multifactor approach for investigating global survival mechanisms of *Campylobacter jejuni* under environmental conditions. *Applied and Environmental Microbiology*, 71, 2086-2094.

Motarjemi, U., Käferstein, F., Moy, G., Miyagishima, K., Miyagawa, S., and Reilly, A. (1995). *Food technologies and public health*. WHO/FNU/FOS/95.12, Rome.

Ngundi, M.M., Taitt, C.R., McMurry, S.A., Kahne, D., and Ligler, F.S. (2006). Detection of bacterial toxins with monosaccharide arrays. *Biosensors and Bioelectronics*, 21, 1195-1201.

Nichols, G.L. (2000). Food-borne protozoa. *British Medical Bulletin*, 56, 209-235.

Oberhelman, R.A. and Taylor, D.N. *Campylobacter* infections in developing countries. In: I. Nachamkin and M. J. Blaser (ed.). *Campylobacter*. Washington: *ASM Press;* 2000; 139-153.

Pagotto, F., Corneau, N., Karthikeyan, K., and Bidawid, S. (2004). Current advances in norovirus: from foods to microarrays. Federal Food Safety and Nutrition Research Meeting. Kanata, Ontario, Canada, 3-5 October 2004, 80.

Pannucci, J., Cai, H., Pardington, P.E., Williams, E., Okinaka, R.T., Kuske, C.R., and Cary, R.B. (2004). Virulence signatures: microarray-based approaches to discovery and analysis. *Biosensors and Bioelectronics*, 20, 706-718.

Perreten, V., Vorlet-Fawer, L., Slickers, P., Ehricht, R., Kuhnert, P., and Frey, J. (2005). Microarray-based detection of 90 antibiotic resistance genes of gram-positive bacteria. *Journal of Clinical Microbiology*, 43, 2291-2302.

Rasko, D.A., Ravel, J., Okstad, O.A., Helgason, E., Cer, R.Z., Jiang, L., Shores, K.A., Fouts, D.E., Tourasse, N.J., Angiuoli, S.V., Kolonay, J., Nelson, W.C., Kolstro, A.B., Fraser, C.M., and Read, T.D. (2004). The genome sequence of *Bacillus cereus* ATCC 10987 reveals metabolic adaptations and a large plasmid related to *Bacillus anthracis* pXO1. *Nucleic Acid Research*, 32, 977-988.

Roy, S. and Sen, C.K. (2006). cDNA microarray screening in food safety. *Toxicology*, 221, 128-133.

Rucker, V.C., Havenstrite, K.L., and Herr, A.E. (2005). Antibody microarrays for native toxin detection. *Analytical Biochemistry*, 339, 262-270.

Rudi, K., Flateland, S.L., Hanssen, J.F., Bengtsson, G., and Nissen, H. (2002). Development and evaluation of a 16S ribosomal DNA array-based approach for describing complex microbial communities in ready-to-eat vegetable salads packed in a modified atmosphere. *Applied and Environmental Microbiology*, 68, 1146-1156.

Salama, N., Guillemin, K., McDaniel, T.K., Sherlock, G., Tompkins, L., and Falkow, S. (2000). A whole-genome microarray reveals genetic diversity among *Helicobacter pylori* strains. *Proceedings from the National Academy of Science in the Unites States of America*, 14668-14673.

Schallmey, M., Singh, A., and Ward, O.P. (2004). Developments in the use of *Bacillus* species for industrial production. *Canadian Journal of Microbiology*, 50, 1-17.

Schoondermark-Stolk, S.A., Jansen, M., Veurink, J.H., Verkleij, A.J., Verrips, C.T., Euverink, G.J.W., Boonstra, J., and Dijkhuizen, L. (2006). Rapid identification of target genes for 3-methyl-1-butanol production in *Saccharomyces cerevisiae*. *Applied Microbiology and Biotechnology*, 70, 237-246.

Sergeev, N., Distler, M., Courtney, S., Al-Khaldi, S.F., Volokhov, D., Chizhikov, V., and Rasooly, A. (2004). Multipathogen oligonucleotide micrarray for environmental and biodefense applications. *Biosensors and Bioelectronics*, 20, 684-698.

Sergeev, N., Distler, M., Vargas, M., Chizhikov, V., Herold, K.E., and Rasooly, A. (2006). Microarray analysis of *Bacillus cereus* group virulence factors. *Journal of Microbiological Methods*, 65, 488-502.

Sergeev, N., Volokhov, D., Chizhikov, V., and Rasooly, A. (2004). Simultaneous analysis of multiple staphylococcal enterotoxin genes by an oligonucleotide microarray assay. *Journal of Clinical Microbiology*, 42, 2134-2143.

Spielbauer, B. and Stahl, F. (2005). Impact of microarray technology in nutrition and food research.. *Molecular Nutrition and Food Research*. 49, 908-917.

Straub, T.M., Daly, D.S., Wunshel, S., Rochelle, P.A., Deleon, R., and Chandler, D.P. (2002). Genotyping *Cryptosporidium parvum* with an *hsp 70* single-nucleotide polymorphism microarray. *Applied and Environmental Microbiology*, 68, 1817-1826.

Taboada, E.N., Acedillo, R.R., Carrillo, C.D., Findlay, W.A., Medeiros, D.T., Mykytczuk, O.L., Roberts, M.J., Valencia, C.A., Farber, J.M., and Nash, J.H.E. (2004). Large-scale comparative genomics meta-analysis of *Campylobacter jejuni* isolates reveals low level of genome placticity. *Journal of Clinical Microbiology*, 42, 4566-4576.

Tembe, W., Zavaljevski, N., Bode, E., Chase, C., Geyer, J., Wasieloski, L., Benson, G., and Reifman, J. (2007). Oligonucleotide fingerprint identification for microarray-based pathogen diagnostic assays. *Bioinformatics*, 23, 5-13.

Templin, M.F., Stoll, D., Schrenk, M., Traub, P.C., Vohringer, C.F., and Joos, T.O. (2002). Protein microarray technology. *Trends in Biotechnology*, 20, 160-166.

Thirumalapura, N.R., Ramachandran, A., Morton, R.J., and Malayer, J.R. (2006). Bacterial cell micorarrays for the detection and characterization of antibodies against surface antigens. *Journal of Immunological Methods*, 309, 48-54.

Uchida, I., Hornung, J.M., Thorne, C.B., Klimpel, K.R., and Leppla, S.H. (1993). Cloning and characterization of a gene whose product is a trans-activator of anthrax toxin synthesis. *Journal of Bacteriology*, 175, 5329-5338.

vanBelkum, A. (2003). High-throughput epidemiologic typing in clinical microbiology. *Clinical and Microbiological Infections*, 9, 86-100.

Vietre, N.J., Marrero, R., Hoover, T.A., and Welkos, S.L. (1995). Identification and characterization of a trans-activator involved in the regulation of encapsulation by *Bacillus anthracis*. *Gene*, 152, 1-9.

Vinjé, J., Altena, S., and Koopmans, M. (1997). The incidence and genetic variability of small-round-structured viruses (SRSV) in outbreaks of gastroenteritis in The Netherlands. *Journal of Infectious Diseases*, 176, 1374-1378.

Vinjé, J. and Koopmans, M. (1996). Molecular detection and epidemiology of NLV in outbreaks of gastroenteritis in The Netherlands. *Journal of Infectious Diseases*, 174, 610-615.

Volokhov, D., Chizhikov, V., Chumakov, K., and Rasooly, A. (2003). Microarray-based identification of thermophillic *Campylobacter jejuni, C. coli, C. lari* and *C. upsaliensis*. *Journal of Clinical Microbiology*, 41, 4071-4080.

Volokhov, D., Pomerantsev, A., Kivovich, V., Rasooly, A., and Chizhikov, V. (2004). Identification of *Bacillus anthracis* by multiprobe microarray hybridization. *Diagnostic Microbiology Infectious Disease*, 49, 163-171.

Volokhov, D., Rasooly, A., Chumakov, K., and Chizhikov, V. (2002). Identification of *Listeria* species by microarray-based assay. *Journal of Clinical Microbiology*, 40, 4720-4728.

Wallace, D.J., Van Gilder, T., Shallow, S., Fiorentino, T., Segler, S.D., Smith, K.E., Shiferaw, B., Etzel, R., Garthright, W.E., and Angulo, F.J. (2000). Incidence of foodborne illnesses reported by the foodborne diseases active surveillance network (FoodNet)- 1997. FoodNet Working Group. *Journal of Food Protection*, 63, 807-809.

Wang, D., Coscoy, L., Zylberberg, M., Avila, P.C., Boushey, H.A., Ganem, D., and DeRisi, J. (2002). Microarray-based detection and genotyping of viral pathogens. *Proceedings from the National Academy of Science of the United States of America*, 99, 15687-15692.

Wang, Y., Barbacioru, C., Hyland, F., Xiao, W., Hunkapiller, K.L., Blake, J., Chan, F., Gonzalez, C., Zhang, L., and Samaha, R.R. (2006). Large scale real-time PCR validation on gene expression measurements from two commercial long-oligonucleotide microarrays. *BMC Genomics*, 7, 59-75.

Wang, Z., Vora, G.J., and Stenger, D.A. (2004). Detection and genotyping of *Entamoeba histolytica, Entamoeba dispar, Giardia lamblia*, and *Cryptosporidium parvum* by oligonucleotide microarray. *Journal of Clinical Microbiology*, 42, 3262-3271.

Weinrick, B., Dunman, P.M., McAleese, F., Murphy, E., Projan, S.J., Fang, Y., and Novick, R.P. (2004). Effect of mild acid on gene expression in *Staphylococcus aureus*. *Journal of Bacteriology*, 186, 8407-8423.

Willse, A., Straub, T.M., Wunschel, S.C., Small, J.A., Call, D.R., Daly, D.S., and Chandler, D.P. (2004). Quantitative oligonucleotide microarray fingerprinting of *Salmonella enterica* isolates. *Nucleic Acids Research*, 32, 1848-1856.

Wilson, I.G. (1995). Occurrence of *Listeria* species in ready to eat foods. *Epidemiological Infections*, 115, 519-526.

Wilson, W.J., Strout, T.Z., DeSantis, J.L., Stilwell, A.V., Carrano, A.V., and Andersen, G.L. (2002). Sequence-specific identification of 18 pathogenic microorganisms using microarray technology. *Molecular and Cellular Probes*, 16, 119-127.

Woese, C.R. (1987). Bacterial evolution. *Microbiology Reviews*, 51, 221-271.

Ye, R.W., Tao, W., Bedzyk, L., Young, T., Chen, M., and Li, L. (2000). Global gene expression profiles of *Bacillus subtilis* grown under anaerobic conditions. *Journal of Bacteriology*, 182, 4458-4465.

Ye, R.W., Wang, T., Bedzyk, L., and Croker, K.M. (2001). Applications of DNA microarrays in microbial systems. *Journal of Microbiological Methods*, 47, 257-272.

In: Food Microbiology Research Trends
Editor: Marta V. Palino, pp. 39-92

ISBN 978-1-60021-723-4

Chapter 2

RESEARCH TRENDS IN LACTIC ACID FERMENTATION

E. H. Drosinos and S. Paramithiotis
Laboratory of Food Quality Control and Hygiene,
Department of Food Science and Technology,
Agricultural University of Athens, Iera Odos 75,
11855, Athens Greece.

ABSTRACT

Spontaneous fermentation has been applied for thousands of years mainly as a method for preservation of excess foodstuff and secondarily as a way to obtain spicy and flavorful foods. Numerous fermented foods are consumed worldwide, owing their diversity to the variety of the raw materials. The microorganisms responsible for fermentation constitute a part of the microbiota indigenously present in the raw materials. Lactic acid fermentation is one of the oldest and most widespread types of fermentation. It has been successfully applied in cereals, fruits, vegetables, meat and milk yielding an extensive range of foods and beverages. During the last decades, a great amount of effort has been made to study these ecosystems and the biodiversity of the microorganisms that form their microbiota. Furthermore, research has focused on the microbial properties that affect the physicochemical and organoleptic properties of the final products, their nutritional value as well as their safety. The outcome of years of research is quite promising but further investigation is still essential.

1. INTRODUCTION

Fermentation has been applied for thousands of years mainly as a mean of preserving perishable foodstuff and secondarily as a way to obtain spicy and flavorful foods. Variables such as the geographical area, the availability of raw materials as well as the ambient temperatures, led to the production of a range of spontaneously fermented foods, which today are recognized as characteristic for each region. Although written historical references exist mainly for the mostly appreciated bread, cheese and wine, it is likely that other fermented products were also consumed as well.

The borderline between fermentation and spoilage depends upon the properties of the causative microorganism and the characteristics of the end product. Thus, fermented foods can be defined as food substrates invaded or overgrown by edible microorganisms whose metabolic activity results directly or indirectly in the production of non-toxic products with flavors, aromas and textures pleasant to human. Food fermentations can be classified in a number of ways (Steinkraus, 1997): by categories including (i) alcoholic beverages, (ii) vinegars, (iii) milks, (iv) fish or meat and (v) plant proteins; by classes incorporating (i) beverages, (ii) cereal products, (iii) dairy products, (iv) fish products, (v) fruit and vegetable products, (vi) legumes and (vii) meat products; by commodity to (i) fermented starchy roots, (ii) fermented cereals, (iii) alcoholic beverages, (iv) fermented vegetable proteins and (v) fermented animal protein.

In this chapter, the classification presented by Steinkraus (2002) has been preferred. According to this, fermented foods have been divided into eight categories:

i. fermentations producing textured vegetable protein meat substitutes from legume/cereal mixture, such as Indonesian tempe
ii. high salt/meat flavored amino acid/peptide sauce and paste fermentations, such as Chinese soy sauce and Philippine patis
iii. lactic acid fermentations, such as sauerkraut, cheeses and fermented olives
iv. alcoholic fermentations, including wines and Japanese sake
v. acetic acid fermentations such as apple cider and wine vinegars
vi. alkaline fermentations including African dawadawa, iru and ogiri
vii. leavened breads such as western yeast and sourdough breads
viii. flat unleavened breads.

In the next lines, an attempt has been made to present as concise as possible the world of lactic acid fermentation, a procedure carried out by lactic acid bacteria. In order to assist the presentation, a further subdivision has been made according to the raw materials that are subjected to lactic acid fermentation. For that purpose, a general subdivision to plant and animal-derived material has been made. Plant-derived materials were then further segregated to fruits, vegetables and cereals. The distinction between fruits and vegetables is rather difficult. In terms of food processing, fruits are nearly all acidic. This acidity naturally controls the type of microorganism that can grow and finally prevail in spontaneous fermentations. On the other hand, vegetables are characterised by their low acidity, which makes them susceptible to microbiological deterioration and can, in fact, provide an ideal substrate for food poisoning organisms when in a moist environment. Therefore, an acidification step is often unavoidable. Another useful aspect is that vegetables tend to be less sweet than fruits and often require some form of processing to increase their acceptance and edibility. As for the distinction between cereals and legumes, it was based on the high starch content of the former opposing the high protein content of the latter that also plays a role in the entailed processing. Regarding the fermentation of animal-derived material, it was further subdivided into meat and dairy products fermentation. Due to the huge amount of literature data currently available, an attempt is made to present an overview of each of the above reffered categories. Thus, aspects regarding the safety of lactic acid fermented products, the nutritional benefits that are obtained by their consumption along with production procedures, microbial ecosystems, the role of the lactic acid bacteria as well as research activities and future trends in the production of a wide range of well-known and highly appreciated products are introduced.

2. Lactic Acid Fermentation and Food Safety

Generally, many fermented foods are less likely a vehicle for food infection or intoxication than fresh food due to the antagonistic effect of the lactic acid bacteria. Three major risk factors can be identified during their production: quality of the raw materials, control of the fermentation process and level of sanitary and hygiene. The wholesomeness of the raw materials, accurate control of the fermentation process, as well as hygienic processing conditions are prerequisites for the production of safe and high quality final products. The application of Good Manufacturing Practises (GMP) and Good Hygiene Practises (GHP) as well as the implementation of HACCP system would ensure safety and quality of the final product. Precise knowledge of the potential agents for foodborne diseases is important for prolepsis reasons. The agents responsible for foodborne diseases can be categorized into physical, chemical and microbiological ones. The latter has drawn special attention by the scientific community since bacterial foodborne infections and intoxications constitute approximately 80% of all food related illnesses (Waites and Arbuthnott, 1990), whereas the formers have been studied to some extent depending on the scientific or commercial importance of the respective products. In the following paragraphs, the danger inflicted by the above mentioned agents as well as the effect of the fermentation process on them, are discussed.

Physical agents include all foreign bodies such as pieces of glass; metal, etc. that can be introduced into the product due to unawareness and upon consumption might trigger health problems. As fermentation can not contribute to the elimination of exogenously interjected physical materials, the implementation of GHP and GMP protocols is the only sufficient way and can provide with the desirable control.

Chemical agents are usually contained in the raw materials and through them they buck to the final product. They can be further subdivided into naturally occurring and environmentally polluting ones. The former ones include substances such as cyanogenic glycosides that are present in raw materials such as beans and tubers, while the latter include all chemical compounds that can be used during primary production of the raw materials and their presence in the resulting edible material has been documented, e.g. mycotoxins, pesticides, insecticides, hormones, etc. as well as compounds that infiltrate the production line due to environmental pollution, e.g. heavy metals and dioxins.

Cyanogenic glycosides are widely distributed in beans and tubers such as cassava. Their acid or enzymatic hydrolysis yields hydrocyanic acid, which is a potent respiratory inhibitor. It has been found that traditional methods of preparing lactic acid fermented foods from raw materials that contain cyanogenic glucosides, may reduce HCN content with fermentation being the most effective step (Reddy and Pierson, 1994).

Environmentally polluting chemical agents include chemicals that are used in the agricultural practice such as pesticides, insecticides, etc., in animal breeding such as hormones, antibiotics, etc. and chemicals due to environmental pollution such as heavy metals and dioxins. Depending on the dose, these chemicals may lead to acute intoxication or long-term health problems. Scarce data concerning the fate of the above mentioned chemical hazards during fermentation exists. Microbial degradation of organochlorine pesticide residues (OCPs) has been studied to some extend by soil and aquatic microorganisms (Faller et al., 1991; Haider et al., 1974; 1979; Quensen et al., 1990) and very few studies dealt with the effect of microorganisms used in food fermentation. Several *Lactobacillus* sp. and *Kocuria* sp. strains have been tested for their ability to degrade OCPs either in synthetic media or during sausage fermentation (Abou-Arab, 2002; Mirna and Coretti, 1979; Spiric et al., 1981; Bayari et al., 1997; 1998; Peric et al., 1980; Arino et al., 1992; 1993; 1995) and some *Streptococcus* sp., *Lactobacillus* sp. and yeast strains during various cheese production (Abou-Arab, 1997; 1999) revealing some of the factors that affect the degradation process.

Microbiological agents include microorganisms that can exert either infection or intoxication. For a bacterial foodborne infection to take place, contamination followed by survival or growth and consumption of sufficient quantity at adequate frequency by a susceptible consumer are prerequisite conditions (Nout, 1994). Bacterial foodborne infections can be caused by a wide range of pathogenic bacteria (e.g. *Escherichia coli, Salmonella* sp., *Listeria monocytogenes*). The efficacy of lactic acid fermentation for inhibiting these pathogens has been extensively studied. Although acidity may be the most important inhibitory agent, other antimicrobial factors such as the type of organic acid (acetic or lactic acid), production of carbon dioxide, hydrogen peroxide, ethanol, diacetyl and bacteriocins, as well as lowering of the redox potential, the water activity and nutrient depletion should not be neglected (Adams and Nicolaides, 1997). Only the synergistic effect of all the above might ensure the safety of the product. Food safety should not be relied exclusively upon acidity, since several enteropathogens, such as enterohaemorrhagic *E. coli* exhibit patterns of acid resistance and may survive certain fermentation processes, such as yoghurt and fermented meat that have been recognized as potential vehicles for infection by this microorganism (Motarjemi, 2002). All these factors that affect survival and growth of pathogenic bacteria in fermented products have been intensively studied and a great amount of data has accumulated over the past decades describing the response of different pathogens to adverse microenvironments. All this data has given birth to new approaches in food microbiology, namely microbiological risk assessment and predictive food microbiology. Both of them apply this knowledge in a logical, systematic and consistent way, the former to assess the chance that a certain food will cause illness and the latter to predict the responses of microorganisms to specified environmental variables. They are quite promising and rapidly developing tools due to the increased scientific attention that have received. Pathogenic bacteria are not the only source of foodborne infections; lactobacilli have also proved their pathogenic potential in isolated cases and in exceptional circumstances (Harty et al., 1994; Bernardeau et al., 2006). Among others, *Lactobacillus casei* and *Lb. rhamnosus* have been identified as the species most commonly responsible for lactobacillemia and infective endocarditis (Cannon et al., 2005), whereas even *Lb. rhamnosus* strain GG has been implicated for some bacteremia cases (Land et al., 2005; Salminen et al., 2002). On the other hand, food intoxication may also occur depending on the quantity and nature of the ingested toxin. Food intoxications may be subdivided to the ones caused by toxins produced during or after processing and toxic by-products of fermentation. Bacteria that can cause serious intoxications include *Clostridium botulinum*, *Cl. perfringens*, *Staphylococcus aureus*, *Bacillus cereus*, *Yersinia enterocolitica* and others. Processes that involve a cooking step have reduced risk of bacterial intoxication, whereas in fermentations involving non-cooked raw materials the combined effect of water activity, salt concentration, acidity, anaerobiosis, temperature and microbial competition should be optimized.

During raw material's production or storage, they might become contaminated with mycotoxins as a result of fungal growth. The last decade, several new and reliable techniques, for the quantitative determination of mycotoxins in foods, have been developed. As a result a plethora of reports regarding various food commodities infected by mycotoxins, have been published. Due to their toxic effects to humans and more accurately, to the long-term effects of chronic low level exposure, a great amount of research has taken place in order to establish decontamination strategies. These were initially achieved by physical methods such as oil or dry roasting (Marth and Doule, 1979; Lee, 1989), exposure to UV light (Yousef and Marth, 1989) or microwave irradiation (Farag et al., 1996), as well as chemical methods such as ammoniation (Park et al., 1988), treatment with sodium bisulfite (Hagler et al., 1982) and ozonisation (Dollear et al., 1968; Young et al., 1986; Zhang and Li, 1994) with varying level of success. Biological detoxification has also been applied and the results of the ongoing research are quite promising. Several bacteria, yeasts and moulds are able to degrade some mycotoxins (Bata and Lasztity, 1999; Karlovsky, 1999; Styriak and Concova, 2002; Shetty and Jespersen, 2006). Regarding to microorganisms used in lactic acid fermentations, the aflatoxin binding capacity of *Lb. rhamnosus* strains GG and LC 705 has been studied by El-Nezami et al. (1998; 2002a) revealing that these strains most effectively bound aflatoxin B_1 than aflatoxin B_2, G_1 and G_2. Moreover, these strains have been shown to bind effectively some of common *Fusarium* toxins such as deoxynivalenol, 3-acetyldeoxynivalenol, nivalenol, diacetoxyscirpenol, T-2 toxin and HT-2 toxin as well as zearalenone and its derivative α-zearalenol (El-Nezami et al., 2002b).

By products of fermentation may also cause toxic response to humans. Biogenic amine content of fermented products is usually characterised by great variability that occurs as a result of a complex equilibrium between the abiotic environment and the enzymatic activities of the microbial population. Several lactobacilli strains belonging to the species *Lb. buchneri*, *Lb. alimentarius*, *Lb. plantarum*, *Lb. curvatus*, *Lb. farciminis*, *Lb. bavaricus*, *Lb. homohiochii*, *Lb. reuteri* and *Lb. sakei* have been found amine positive either in vitro or in situ with tyramine being quantitatively the most important biogenic amine produced (Masson et al., 1996; Montel et al., 1999; Bover-Cid et al., 2001b; Pereira et al., 2001; Suzzi and Gardini, 2003). Another by product of fermentation is ethyl carbamate; a carcinogenic and mutagenic compound occurring in a variety of fermented foods from the esterification of ethanol with carbamic acid. The latter can be formed be several precursors including naturally occurring citruline, yeast metabolites from L-arginine and L-asparagine (Canas et al., 1989) as well as vicinal diketones and HCN liberated from cyanogenic glucosides, with heat and light enhancing its formation. Although ethyl carbamate formation mechanisms is not completely understood, research in wine and stone fruit (cherry, plum) indicated that a control over the level of precursors is possible to minimize ethyl carbamate levels in the final product (Nout, 1994).

Despite the huge amount of work that has taken place regarding the safety of fermented foods, limited information is available on the fate of parasites, viruses as well as prions during fermentation.

Only a few researchers have studied the survival of parasites during fermented sausage manufacture. Sommer et al. (1965) reported that *Toxoplasma* cysts are rapidly inactivated during manufacture of fermented sausage. Inactivation of *Trichinella* larvae has been studied in German-type sausages, Yugoslavian-type dry sausage as well as Genoa salami leading to the conclusion that a dry fermented sausage of water activity lower than 0.9 is very unlikely to contain them, even if prepared from pork already contaminated (Luecke, 1998).

Foods can serve as vectors for several viruses, among them poliovirus, hepatovirus and various gastroenteritis viruses like rotavirus, astrovirus and caliciviruses that provide Norwalk and Norwalk-like viruses (Untermann, 1998). Since viruses already present in livestock at the time of slaughter normally do not affect humans (Luecke, 1998), foods can be contaminated either directly by man or indirectly by contaminated water. Recent epidemiological evidence indicated that enteric viruses are the leading cause of foodborne disease in the U.S.A. (Sair et al., 2002). Given that only scarce scientific data exists, concerning the retention of infectivity during fermentation, effective preventive measures should be implemented to avoid contamination of the raw materials, intermediate of final product.

Prions are proteins that are characterised by an exceptional resistance to the extreme conditions involved in food processing. Prion diseases, or transmissible spongiform encephalopathies (TSE), are a group of fatal neurodegenerative brain diseases, which include among others scrapie and bovine spongiform encephalopathy (BSE) in animal and kuru and Creautzfeld-Jacob disease (CJD) in humans (Pedersen and Smith, 2002). Concerning the risk of human exposure through foods, TSE agents are capable of crossing the co-called 'species barrier', as has been demonstrated in both experimental and natural situations (Dormont, 2002), with offal as well as muscle tissues being indicated as carriers of significant levels of infectious prions (Bosque et al., 2002; Thomzig et al., 2003) . Although human TSEs are rare diseases, the fact that so far no applicable therapy exists, prophylaxis through the minimization of exposure should take place. That essentially involves the development of proper detecting strategies in order to identify infected tissues prior to their entry in the human food chain.

3. Nutritional Value of Lactic Acid Fermented Foods

Lactic acid fermentation can have multiple effects on the nutritional value of food either by modifying the level and bioavailability of nutrients, or by interacting with the gut microbiota and even the human immune system.

Microbial fermentation, generally leads to the decrease in the level of carbohydrates as well as some non-digestible poly- and oligosaccharides. The latter reduces side effects such as abdominal distention and flatulence. These side effects can occur due to various reasons, the most common of which is food intolerance; with lactose intolerance being the most frequent: it is caused by b-galactosidase deficiency that is rather widespread in adult populations. As a result, lactose reaches the colon where it is microbially degraded. The fermentation in the colon causes disturbances such as bloating, abdominal pain and sometimes diarrhoea. As lactic acid bacteria grow, they produce lactic acid that can be a mixture of the optical isomers L-(+) and D-(-). The latter cannot be metabolized by humans and excessive intake can result to acidosis, i.e. a disturbance of the acid-alkali balance in the blood (Nout and Motarjemi, 1997). On the other hand, the acidification of the substrate causes the reduction if its glycemic index. Fermentation by lactic acid bacteria have specific effects on various substances present in the raw material that they have an antinutritional or toxic effect such as oxalate, certain lectins, tannins and phytic acid or on the other hand are essential components of the human diet such as folates or possess interesting biochemical properties such as a number of lectins.

Very little is known about oxalate content of cereal grains, legumes and their products. Although oxalates are very poorly absorbed under normal non-fasting conditions in humans, excess ingestion may interfere with calcium metabolism and may result in chronic disease such as renal damage and stone formation. Antinutritional and chronic problems that can be linked to the above mentioned reason is minimal (Reddy and Pierson, 1994). Eka (1980) have studied the effect of the fermentation on the nutrient status of dawadawa, an African alkaline fermented food made by locust bean, and have reported a decrease up to 43% of oxalates during its production without, though, making clear whether this reduction is due to processing steps before fermentation such as soaking, dehulling and washing or to the fermentation itself.

Lectins are a group of bioactive proteins found in almost all organisms including plants, vertebrates, invertebrates, bacteria and viruses and thus are ubiquitous to human food. Some lectins originating from legumes and cereals have been found to be highly toxic to humans and animals after oral ingestion, most likely due to their binding to specific receptor sites on the surface of the intestinal epithelial cells, which seriously impair their ability to absorb nutrients from the gastrointestinal tract (GIT), thus causing serious growth retardation (Radberg et al., 2001). The removal of these antinutritional and toxic lectins has been studied to some extend. It has been found that lectins can be detoxified by traditional cooking since they are heat labile. As a result, fermented foods that are prepared using raw materials containing lectins, and a cooking step is incorporated, either as such or as steaming, before or after fermentation, can be lectin free. Moreover, limited hydrolysis of lectins has also been found to occur during lactic acid fermentation. A *Leuconostoc mesenteroides* strain, isolated from Indian fermented food batter (rice-soy idli batter), has been found to hydrolyze lectins from soy beans, navy beans, black beans and others, by excreting a mixture of protease, beta-*N*-acetylglucosaminidase and alpha-D-mannosidase that are involved in their hydrolysis (Reddy and Pierson, 1994). However, many lectins are non-toxic such as those of tomatoes, lentils, peas and other common foods. Furthermore several plant lectins are important tools in cell biology and immunology, with potential for clinical applications such as antitumor and anticarcinogenic activity (De Mejia and Prisecaru, 2005). Lectins of animal origin, such as galectins have also emerged as bioactive molecules that may possess immune-system modulation properties.

Condensed tannins are the predominant class of polyphenols that occur widely in food grains and legumes. They are considered to have antinutritional effect due to their interaction with proteins that lead to complex formation, resulting in decrease in protein digestibility and digestive enzyme inactivation. However this is of limited concern as they occur mainly in the outer layers or seed coats that are mostly removed from the substrate before fermentation (Reddy and Pierson, 1994).

Phytic acid occurs primarily as a salt of monovalent and divalent cations in discrete regions of cereal grains, legumes, some roots and tubers. The presence of phytate in foods causes concern because it decreases the bioavailability of minerals such as phosphorus, zinc, iron, calcium and magnesium and the solubility, functionality and digestibility of proteins by forming complexes. Phytate also interacts with enzymes, such as trypsin, pepsin, alpha-amylase, and β-galactosidase, resulting in a decrease of their activity (Reddy and Pierson, 1994). Lactic acid fermentation of cereals, legumes, and tubers provides the optimum pH for enzymatic degradation of phytates primarily as a result of the activity of endogenous phytases and secondarily due to microbial phytases, abrogating all the above mentioned negative effects.

Folate is a generic term used to describe the various forms of folic acid, one of the B vitamins. It is produced by various green leafy vegetables, cereals, legumes and by some microorganisms and is an essential component in the human diet. Vegetables and dairy products are the main source of folate for humans. Due to its recognized significance in human diet (Gregory, 2001; Kim, 1999; George et al., 2002; Ames, 1999; Brattstrom, 1996) intensive research has taken place regarding its fate during lactic acid fermentation. Some fermented milk products, especially yoghurt, buttermilk and different varieties of cheeses are already recognized as good dietary sources of folates (Forssen et al., 2000). Furthermore, it has been reported that yoghurt usually contains as much as the double quantity compared to milk or even higher (Hoppner and Lampi, 1990). However, folate production by lactic acid bacteria seems to be a strain depended property, e.g. *Lb. acidophilus* has been shown to produce folates in dairy products (Lin and Young, 2000) although Crittenden et al. (2003) reported its depletion by the strains of the same species that were studied. Folate production by *Lb. bulgaricus* has been reported (Lin and Young, 2000) while in some co-fermentations of *Streptococcus thermophilus* and *Lb. bulgaricus* have been shown that folates produced by *Str. thermophilus* were consumed by *Lb. bulgaricus* (Rao et al., 1984). In sourdough fermentation, Kariluoto et al. (2006) demonstrated that *Lb. acidophilus*, *Lb. brevis*, *Lb. plantarum* and *Lb. sanfranciscensis* consumed folates rather than produced, in both sterile and non-sterile rye flour. On the contrary, on the same study, *Lb. bulgaricus* strain as well *Str. thermophilus* seemed to have a positive effect on folate concentration. It is, though, possible to select folate producing starter cultures, as demonstrated by Jaegerstad et al. (2004) concerning vegetable fermentation.

Another very interesting effect of lactic acid fermentation is the hydrolysis of certain peptide sequences that were found to be toxic and inflict the so-called celiac disease or gluten-sensitive enteropathy. Although the exact mechanism is still more or less unknown, prolamins of wheat (α-, β-, γ- and ω-gliadin subgroups), rye (e.g. secalin) and barley (e.g. hordein) release a family of proline and glutamine rich polypeptides that are responsible for an inappropriate T-cell-mediated immune response (Silano and De Vincenzi, 1999). Some of these peptides, such as the 31-43 and 62-75 fragments of α-gliadin, and the 33-mer epitope, corresponding to the 57–89 fragment of a2-gliadin have been successfully hydrolysed by sourdough lactic acid bacteria under specific conditions (Di Cagno et al., 2002; 2004).

Several other compounds, with quite interesting properties, such as flavonoids, alkylresorcinols and glucosinolates, are also present in the raw materials and scarce information exists regarding their fate during lactic acid fermentation. Flavonoids are phenol derivatives widely distributed in plants. A series of attractive biochemical effects have been assigned to them, including action against cardiovascular diseases, cancer, inflammation and allergy (Middleton et al., 2000; Yao et al., 2004; Birt et al., 2001; Halliwell, 2007). Alkylresorcinols are amphiphillic 1,3-dihydroxybenzene derivatives, with an odd-numbered alkyl chain at position 5 of the benzene ring. Among others, they have been reported to be present in high levels in cereals, more accurately in the outer layers of the grains (Ross et al., 2003). They have been reported to have many biological effects, among them anticancer and antioxidant activities (Kozubek and Tyman, 1999). Glucosinolates are a group of sulphur-containing plant secondary metabolites found in plant families of the order Capparales, which includes the *Brassicaceae* family, widely used as raw material for vegetable fermentation. Their major hydrolysis products, isothiocyanates, have been found to act protectively against cancer, particularly in the bladder, colon and lung (Mithen et al., 2000; Song and Thornalley, 2007).

Human intestinal microbiota consists of a dynamic mixture of microorganisms, the composition of which is governed by the interactions between genetic and environmental factors (Isolauri et al., 2004). Due to its multiple functions, alteration would inevitably lead to disease. Thus, the maintenance of the normal intestinal flora and furthermore its protection or improval has been the research concept for decades. Currently, this is achieved mainly by the probiotic as well as the prebiotic concepts. The former is based on the belief that the support or the promotion of a beneficial balance among the autochthonous microbial population of the gastrointestinal tract may be achieved by the consumption of certain viable microorganisms; while the latter is based on the assumption that certain gut flora considered beneficial to human health, namely bifidobacteria and lactobacilli, may be selectively stimulated by indigestible by human enzymes, but fermentable dietary carbohydrates (Cummings et al., 2001). In order to select a strain as a probiotic culture, several continuously updated criteria should be fulfilled, which, in general, regard to their safety, functional and technological properties. The safety and non-pathogenicity of the strain is of major importance and aspects such as its potential virulence factors as well as the interactions with the host and the native microbiota of the GIT should be extensively studied. The functional properties refer to its survival of the extreme conditions in the GIT, and furthermore to its adherence to the epithelial surfaces and persistence in the GIT. Additionally, desired functional properties include antagonistic activity against potentially pathogenic microorganisms, particularly invasive Gram-negative pathogens, as well as a series of assets that are beneficial to the host. Finally, the technological properties relate to the desired characteristics of the final product and its ability to survive during processing, transport and storage. The main field for probiotic culture application has been the dairy products, mainly yoghurt since it has been recognized as a product with many desirable effects for consumers and it is considered by them as 'healthy' (Lourens-Hattingh and Viljoen, 2001b); with *Lb. casei* strain Shirota, *Lb. rhamnosus* strain GG and *Lb. johnsonii* strain La1 being the predominant ones (Reuter et al., 2002). However, a great variety of potentially probiotic strains have been studied, such as several *Lb. plantarum* strains (de Vries et al., 2006), *Lb. salivarius* CECT 5713 (Martin et al., 2006), *Lb. paracasei* A (Marzotto et al., 2006) and many more. Recently the idea of using probiotic bacteria as fermenting agents in meat products has developed. Many strains have been studied for their probiotic potential in fermented meat products, among them strains of *Lb. plantarum*, *Lb. paraplantarum*, *Lb. pentosus*, *Lb. brevis*, *Lb. paracasei* and *Lb. salivarius* (Pennachia et al., 2006; Klingberg and Budde, 2006). Moreover, similar efforts have been made with fruits, vegetables and their products. Betoret et al. (2003) have developed dried fruits with probiotic effect by vacuum impregnating *Saccharomyces cerevisiae* CECT 1347 and *Lb. casei* spp. *rhamnosus* CECT 245 into apple cylinders that were subsequently air dried. The development of vegetable juices with probiotic effect has been studied by Yoon et al. (2005; 2006). Cabbage and beet juice were fermented by *Lb. plantarum* C3, *Lb. casei* A4 and *Lb. delbrueckii* D7. The latter was also fermented by *Lb. acidophilus* LA39. By combining their fermentation performance as well as their ability to survive during storage at 4°C the most suitable strains were pointed out. The utilization of various cereals as a substrate for growth of various probiotic (Kedia et al., 2007) and potential probiotic strains (Charalampopoulos et al., 2002; Korakli et al., 2002) has also been studied. Furthermore, Palacios et al (2006) successfully applied several *Bifidobacterium* strains in breadmaking, while Martensson et al (2002) evaluated the ability of two lactobacilli strains and one *Bifidobacterium* strain of human origin to survive during refrigerated storage in three oat-based products with different carbohydrate composition and Angelov et al. (2006) managed to develop a new oat-based drink using the potentially probiotic strain *Lb. plantarum* B28.

Microencapsulation is a technique used to protect the probiotic cultures from the stringent conditions during manufacturing, such as high temperatures and during passage through the GIT, by retaining them within a protective polymer membrane or matrix. Several methods for microencapsulation have been reported so far, but none seems to be capable for industrial application (Kailasapathy, 2002). It has been successfully applied in both yoghurt and sausage manufacture (Adhikari et al., 2000; Kailasapathy, 2006; Muthukumarasamy and Holley, 2006). However, as far as sausage fermentation is concerned, it has been shown that microencapsulated *Lb. reuteri*, a reuterin producing strain, although increased its viability, had reduced inhibitory action against *E. coli* O157:H7 compared to the one of the free microorganism, during sausage fermentation (Muthukumarasamy and Holley, 2007). Application of this technique in industrial scale has certain disadvantages, among them the effect on the product organoleptic quality that needs to be addressed.

There is an extended list of claimed beneficial effects of probiotic bacteria to humans including, among others, reduction of lactose intolerance (Kim and Gilliland, 1983), reduction of the serum cholesterol levels (Gilliland et al., 1985), prevention of intestinal infections (Ushiyama et al., 2003), modulation of the immune response (Chen and Walker, 2005), anticarcinogenic activity (Commane et al., 2005), prevention of urogenital infection (Zarate and Nader-Macias, 2006), protection against traveler's (Koo and DuPond, 2006) infantile (Guandalini et al., 2000) and antibiotic-induced (Kim et al., 2003b) diarrhoea and allergic diseases (Tricon et al., 2006). This is a dynamically developing field and new data are continuously available.

The health benefits that are associated with the consumption of lactic acid fermented foods may be far more complex than the ones already described above. Several food constituents, either essential or non-essential to human nutrition, have been demonstrated to affect gene expression. Thus, a new, and quite promising, science, nutrigenomics, has emerged, aiming to study all these effects that bioactive compounds from food can have on gene expression. Taking into consideration that nutrients affect gene expression and heredity affects nutritional needs and responses, nutrigenomics and its subset nutrigenetics, aim to provide evidence-based dietary intervention, strategies for restoring health and fitness as well as preventing diet-related disease, based on an individual's genetic background (Evans et al., 2006; Trujillo et al., 2006; Afman and Mueller, 2006). Despite the scarcity in human clinical studies and well-established cause-and-effect relationships between the food affecting gene expression and the disease state, the use of new and innovative technologies such as microarrays and nanotechnologies can provide the vehicle through which, these high expectations can be accomplished.

4. Lactic Acid Fermentation of Plant-Derived Material

4.1. Fruit Fermentation

A wide variety of fruits have been traditionally utilized as a substrate for fermentation but only a few of them undergo lactic acid fermentation. Among them, olive fermentation has a worldwide commercial significance and therefore studied in greater detail. Other lactic acid fermented fruits of local economic importance, such as tempoyak and capper berries have only recently drawn the attention.

The main commercial table olive preparations are: (1) treated green olives in brine (Spanish-style), (2) untreated natural black olives in brine (Greek-style) and (3) ripe olives (California-style). Although lactic acid microbiota can be developed in all preparations, only the first one can be considered as lactic fermentation and thus only this will be discussed in this paragraph.

The preparation of treated green olives in brine is depicted in Figure 1 and essentially involves an initial treatment with a NaOH solution (lye treatment) that aims to facilitate the upcoming lactic fermentation. The lye solution used in this treatment, the immersion time as well as the penetration degree depends upon many factors with variety, maturity stage of the fruit and temperature being the most important ones. Generally, lye concentration varies between 1.3 and 2.6% (w/v) but it may as well be as high as 3.5%. During this stage, oleuropein, a phenolic glucoside that apart from being largely responsible for the bitterness of olive fruits, has been also shown to have a bactericidal effect, is hydrolyzed, and a proportion of the fruit carbohydrate content is leached into the solution. The washing step that follows is equally important as it aims to the removal of the excess lye solution. Inadequate washing would lead to the retention of residual lye that will prevent suitable final pH value. On the other hand, excess washing would remove fermentable carbohydrates or may even facilitate undesirable bacterial contamination. Lye-treated olives are then brined in approximately 10-13% NaCl, depending on the variety and the maturity degree. After the initial lowering of the salt concentration, salt is added and kept between 5-6% during fermentation, while at the end of the fermentation it is raised to 7% or higher. During this treatment, the aqueous solution, due to osmotic effects leads to leaching of carbohydrates, vitamins and minerals into the brine and the fermentation begins. The fermentation can be considered to take place in three stages (Fernandez et al., 1995). During the first stage, a complex microbiota coexists. Hermetic seals should be used, since growth of moulds should be avoided because of the pectinolytic enzymes that they might produce and can cause softening of the fruit. Non-sporulating Gram-negative bacteria characterize this stage. Their maximum growth is reached aroung 48th hour after brining and their population diminishes gradually to disappear by the end of the second stage. As fermentation progresses Gram-positive lactic cocci also begin to grow, but their growth is inhibited after a few days. Strains of the genus *Pediococcus* are most frequently found during the first stage and at the beginning of the second stage. Members of this genus together with Gram-negative bacteria contribute to the decrease in pH that facilitates growth of lactobacilli. The length of this stage varies from 2 or more days ending when pH reaches 6.0 and the lactic acid bacteria appear.

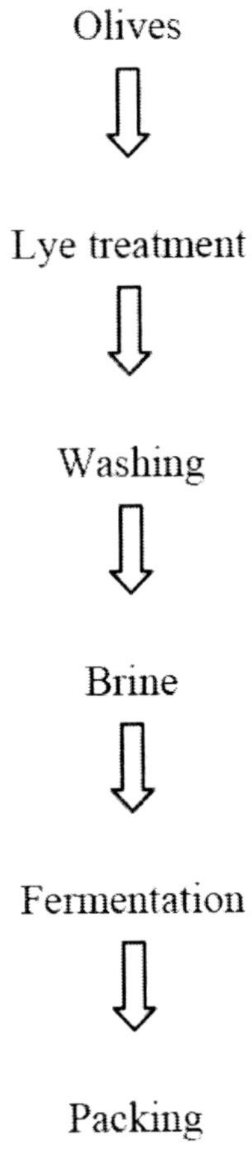

Figure 1. Preparation of treated green olives in brine.

The second stage starts at pH 6.0 and may last up to 10 to 15 days, after total disappearance of non-sporulating Gram-negative bacteria and pH around 4.5. This stage is characterized by rapid growth of lactobacilli and yeasts. The third stage lasts until fermentable substrates have been exhausted. Gram-negative and sporulating Gram-positive bacilli cannot be found; only *Lactobacillus* sp. coexisting with yeasts. Final pH value is typically 3.8-4.2. The dominant lactic acid bacteria species is *Lb. plantarum* and to a lesser extend *Lb. delbrueckii*; with some exceptions such as *Lb. casei* that was found to prevail in Sicilian fermented olives (Randazzo et al., 2004) Olives are kept in the same brine until sold. Usually stabilization is done by increasing the salt content to 8% or higher in order to prevent the fermentation of the only fermentable substrate left: lactic acid. Its consumption will lead to increase in the pH of the brine and undesirable changes might occur.

Aiming towards product standardization and reproducibility, several researchers have set about the development of starter cultures. Montano et al. (1993) have used *Lb. plantarum* as a starter culture for fermentation, whereas Duran Quintana et al. (1999) have used four *Lb. plantarum* strains as starters and have optimized the initial brine conditions for fermentation at low temperatures. The utilization of *Lb. pentosus* accompanied by *Enterococcus casseliflavus* as starters has been studied by De Castro et al. (2002). Recently Spyropoulou et al. (2001) and Caggia et al. (2004) studied the safety of green table olives fermented with several strains of lactic acid bacteria. In the first case, inhibition of *E. coli* O157:H7 was evident in all fermentation procedures regardless the treatment used, whereas in the latter, *L. monocytogenes* was able to grow irrespective of olive cultivar, lactic acid bacteria starter used, pH and titratable acidity of brine samples, or treatment applied. Thus, the production of bacteriocins seems to be a desirable property of starter cultures. Vega Leal-Sanchez et al. (2003) have used *Lb. plantarum* strain LPCO10 as a starter culture, a strain capable of producing plantaricins S and T; combination that results in a broad inhibitory spectrum including other lactic acid bacteria, propionibacteria and clostridia and can assist the strain to its predominance in this microenvironment. Rubia-Soria et al. (2006) have also screened several lactic acid bacteria strains for production of antimicrobial compounds. Alternatively, the addition of fermentable carbohydrate such as glucose or sucrose would lead to acidity production and pH value that would ensure the safety of the product (Chorianopoulos et al., 2005). The biogenic amine content of Spanish-style green olives has been assessed by Garcia et al. (2000; 2004). Putrescine was the amine found in the highest concentration and it seemed that it was produced during the active fermentation phase. Histamine, tryptamine, cadaverine and tyramine were also detected, with the latter two most probably being produced in during storage and related to the 'zapatera' spoilage. The polyphenolic compounds in olives have also drawn special attention, mainly in olive oil, due to their beneficial effects on human health. More accurately, hydroxytyrosol was reported to inhibit LDL oxidation (Auroma et al., 1998) and passive smoking-induced oxidative stress (Visioli et al., 2000); tyrosol was demonstrated by Giovannini et al. (1999) to protect against oxidized-LDL-induced injury in Caco-2 cells, while other phenolic compounds such as hydroxytyrosol acetate and lignans also possess antioxidant properties (Gordon et al., 2001; Klijn et al., 2002). The study of several commercially available Spanish-style fermented olives revealed that they can be considered a good source of phenolic antioxidants, although their concentration depends on olive cultivar and processing method (Romero et al., 2004; Blekas et al., 2002).

Tempoyak is a traditional Malaysian fermented condiment made from the pulp of the durian fruit (*Durio zibethinus*), a native to the Malay Archipelago fruit. Durian fruit is an economically important fruit in the region of South-East Asia. The pulp is prepared from excess, poor quality or over-ripe fruits. The fermentation procedure is quite simple; the pulp is mixed with a small amount of salt and placed in a sealed container for 7 days. Only recently, the lactic acid bacteria microbiota responsible for tempoyak fermentation has been studied; *Lb. plantarum* was the predominant species whereas *Lb. fermentum*, *Lb. brevis*, *Leu. mesenteroides* and *Lb. mali* formed a secondary microbiota. Further research is required in order to elucidate the mode of predominance and the microbial succession if any (Leisner et al., 2001).

The main producers of fermented capper berries are Mediterranean countries. Pickling of capper berries is done by a quite simple technique; fruits are harvested during June and maybe July, immersed into tap water where the fermentation takes place for 5-7 days in temperatures ranging from 23-43°C and then placed into brine and consumed. Pulido et al. (2005) studied the dynamics of the microbial community responsible for the fermentation. *Enterobacteriaceae* were detected but only the first day of fermentation; the most possible reason for the population decline has been assigned to the pH decrease during the first day;

from 7.5 to 4.37 as well as the continuing decrease to a value of 3.55, at the end of the fermentation. *Lb. plantarum* was found to dominate during fermentation with *Lb. paraplantarum*, *Lb. pentosus*, *Lb. brevis*, *Lb. fermentum*, *Pediococcus pentosaceus*, *Pd. acidilactici* and *Ent. faecium* forming a secondary microbiota. The rapid pH decrease has also been identified as a possible reason for absence of *Leuconostoc* sp., often present in other fruit and vegetable fermentation.

4.2. Vegetable Fermentation

Vegetable fermentation is worldwide and mainly performed on a small scale, mainly by entrepreneurs. The current exceptions to this are sauerkraut, cucumbers and kimchi, since these are of significant commercial importance.

A basic scheme for the production of fermented vegetables consists of seven steps (Figure 2). The first step is the selection of the raw materials and depends on the desired final product, e.g. cabbage for sauerkraut production. Then, a primary processing takes place that essentially includes washing of the raw materials and might also include further treatment such as coring and trimming, cutting, peeling, blanching or cooking etc. The salting step that follows, distinguish fermented vegetables into three categories: dry-salted, brine-salted and non-salted. With dry salting, vegetables are treated with dry salt (2-2.5%) and brine is formed due to osmotic extraction of water from the already cut tissue. Brine formation is enhanced by the mechanical pressure that is applied in some cases. This brine contains fermentable carbohydrates and all the nutrients necessary for microbial growth. Brine-salting is alternatively used for vegetables that contain less moisture. As soon as vegetables are immersed into brine solution (15-20% NaCl), carbohydrates and other nutrients are extracted and fermentation begins. Some vegetables are fermented by lactic acid bacteria without prior treatment with salt. In that case, the rapid colonization of the food by lactic acid bacteria is crucial in order to inhibit growth of other genera by lowering of the pH value and restriction of the oxygen. A variation of this technique is the so-called 'pit fermentation', an ancient method for preserving excess starchy vegetables and fruits and currently applied in many places such as South Pacific, Ethiopia and Himalaya (Gashe, 1987; Tamang et al., 2005). After salting, an intermediate processing step might occur that most commonly includes rinsing of the edible material and mixing with salt, spices and flavorings or the addition of starter cultures in the case of commercially significant products. Then, the fermentation takes place and its duration depends upon many factors, with temperature being the most important. After fermentation has been completed, a secondary processing step might occur, during which the product is subjected to processes such as drying, pasteurization, etc. When the final product is ready, it is packed and consumed either as such, or after a heat treatment such as boiling or frying.

The factors that determine the outcome of vegetable fermentation are the naturally occurring microbial populations in the raw materials as well as the environmental conditions such as pH, temperature and salt concentration. The microbiota of fresh vegetables is dominated by Gram-negative aerobic bacteria and yeasts. Lactic acid bacteria only constitute a minor portion of the initial population. Most vegetables will undergo lactic acid fermentation as soon as the above mentioned environmental conditions are appropriately adjusted. During fermentation, the physicochemical parameters will inevitably change and so will the dominant microorganisms, leading most of the times to a specific and reproducible microbial succession (Harris, 1998).

Spontaneous sauerkraut fermentation, a typical dry-salted fermentation, is characterized by an initial heterofermentative stage that is followed by a homofermentative one.

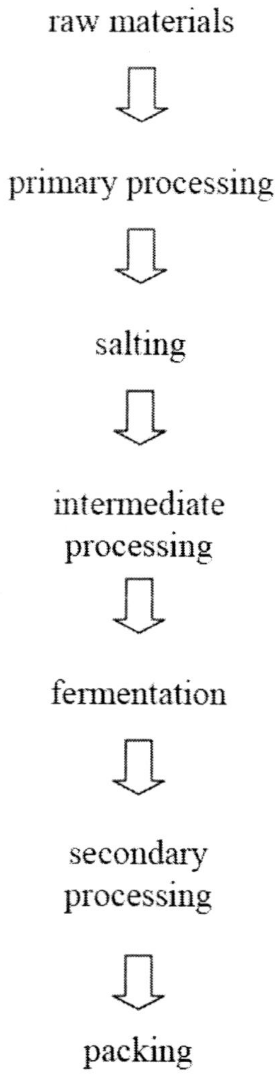

Figure 2. Basic scheme for the production of fermented vegetables.

The first stage is dominated by *Leu. mesenteroides* mainly due to its comparatively higher initial numbers and shorter generation time. During this stage, the pH drop combined with the added NaCl inhibits undesirable Gram-negative microorganisms. The carbon dioxide that is produced replaces air and creates an anaerobic atmosphere, that helps to prevent oxidation of ascorbic acid and darkening of the natural color of the cut cabbage. Moreover it facilitates growth of other lactic acid bacteria. As the fermentation proceeds, the relatively sensitive to acidic conditions *Leu. mesenteroides* is replaced sequentially by *Lb. brevis* and, at the final stage of fermentation, by *Lb. plantarum*. During this final stage, large quantities of lactic acid are formed by the remaining carbohydrates, leading to further lowering of the pH value. Sauerkraut production at 18-20°C normally leads to an acidity of 1.6-2.3% (calculated as lactic acid) and a pH value of 3.5 or less.

The biogenic amine content of sauerkraut drew the attention very early (Mayer and Pause, 1972). Since then a lot of research has taken place and several new methods have been successfully applied in its determination. Despite the quantitative wide variation that occurred, tyramine, putrescine, cadaverine, histamine as well as tryptamine, spermidine and spermine were found to be present (Taylor et al., 1978; Halasz et al., 1994; Kalac et al., 1999; Moret et al., 2005). In spontaneous sauerkraut fermentations, the duration of each stage as well as its outcome, vary greatly. The utilization of well defined starter cultures would help to counteract this problem. Despite the fact that commercial starters are available, they are not widely used (Hammes, 1991). The answer to this unwillingness could be the development of starter cultures with apparent health benefits, such as the suppression of biogenic amine production (Halasz et al., 1999; Kalac et al., 2000), the ability to prevail when NaCl content has been intensively lowered, with no negative impact on sauerkraut quality (Viander et al., 2003; Wiander and Ryhanen, 2005) and the introduction of strains producing only L(+) lactic acid. Apart from the dominant microbiota, in spontaneous sauerkraut fermentations, a secondary microbiota, such as *Ent. faecalis*, *Lb. confusus*, *Lb. curvatus*, *Lb. sakei*, *Lactococcus lactis lactis*, *Leu. fallax* and *Pd. pentosaceus* also co-exists (Harris et al., 1992; Schillinger et al., 1989; Barrangou et al., 2002), although its role remains unclear. Only recently the bacteriophage ecology of natural sauerkraut fermentation has been studied (Yoon et al., 2002; Lu et al., 2003b) and the results suggested a possible role in the microbial succession that is left to be confirmed.

Cucumber fermentation, a typical brine-salted fermentation, is considered to take place in four stages (Fleming et al., 1995a); initiation, primary fermentation, secondary fermentation and post-fermentation. Initially, the pH of the brine is approximately 5.5 and fermentable carbohydrates, mainly glucose and fructose, are present. As a result, a rapid grow of several Gram-positive and Gram-negative bacteria and yeasts occurs. This stage can last as long as 7 days during which, lactic acid bacteria and yeasts prevail over undesirable bacteria, mainly due to acidification. This stage is usually completed primarily by *Lb. plantarum* and secondarily by *Pd. pentosaceus* and *Lb. brevis*. *Leu. mesenteroides* strains are most of the times inhibited by the higher salt concentration, besides heterofermentative species are not desired during this type of fermentation (Harris, 1998). Secondary fermentation by fermentative yeasts can occur only when fermentable carbohydrates have remained. Post-fermentation depends upon the storage conditions; open tanks facilitates growth of oxidative yeasts, molds and bacteria, while anaerobic tanks does not allow this stage. Several factors, including initial load, growth rates, salt and acid tolerances account for this microbial sequence (Daeschel et al., 1987). It is often that predominance of *Lb. plantarum* might lead to concentrations of lactic acid that are too high for direct consumption of the product and might influence its texture as well (Passos et al., 1993a). To solve this, Daeschel et al. (1988) suggested a mixed fermentation by *Lb. plantarum* and fermentative yeasts in order to divert a portion of the fermentable carbohydrates into non-acidic metabolic products. Growth of *Lb. plantarum* in pure and mixed cultures was studied and modeled by Passos et al. (1993a; b). Withdrawal of NaCl from cucumber fermentation has also been attempted (Fleming et al., 1995b) and despite the blanching that proceeded fermentation by direct inoculation with *Lb. plantarum*, microbial recontamination that led to the deterioration of the quality could not be avoided.

Lu et al. (2002a; b; 2003a) provided with a new insight in the development of the cucumber fermentation microbiota by studying the effect of anions and cations on carbohydrate utilization during cucumber juice fermentation; the effect of the fruit size on the buffering capacity, sugar utilization, terminal pH and texture of the fermented end product, as well as the properties of *Lb. plantarum* bacteriophage ΦJL-1, isolated from commercial cucumber fermentation product, revealing that all these factors should be taken into consideration when developing controlled fermentation strategies.

A wide variety of brine-salted fermented vegetables exists, the majority of them coming down from Asia. Many vegetables are pickled depending on the season and availability. Among others, turnips (*Brassica rapa*) are used for the production of the traditional Italian 'brovada' with mainly *Pd. parvulus*, *Lb. plantarum* and *Lb. hilgardii* being involved to the fermentation (Maifreni et al., 2004); mustard leaves (*Brassica juncea*) are used for the production of Indonesian Sayur-Asin with *Leu. mesenteroides* and *Lb. confusus* initiating the fermentation and *Lb. plantarum* and *Pd. pentosaceus* dominating in the final stages (Puspito and Fleet, 1985); bamboo shoots (*Bambusa arundinaced*) are used for the production of Thailand's Naw-mai-dong with *Leu. mesenteroides*, *Pd. cerevisiae*, *Lb. brevis*, *Lb. buchneri* and *Lb. fermentum* being involved (Steinkraus, 1996).

Very few non-salted fermented vegetables as well as vegetables prepared by pit fermentation have been studied. Gundruk and Sinki are non-salted, prepared by pit fermentation, wilted vegetable products indigenous to the Himalayas. The former is prepared by leaves of local vegetables, mainly belonging to the *Brassica* genus while the latter by radish tap-roots. The lactic acid bacteria involved in Gundruk fermentation are *Lb. plantarum*, *Lb. casei casei*, *Lb. casei pseudoplantarum*, *Lb. fermentum*, *Leu. fallax* and *Pd. pentosaceus*. On the other hand, the lactic acid bacteria involved in sinki fermentation are *Lb. fermentum*, *Lb. brevis*, *Lb. plantarum* and *Leu. fallax* (Tamang and Sarkar, 1993; Tamang et al., 2005)

Kimchi is a general name referring to a group of fermented vegetables in Korea. It is prepared mainly by Chinese cabbage (*Brassica pekinensis*) that is accompanied by other ingredients including vegetables, seafoods, spices, fruits, seasonings and cereals (Harris, 1998). It is quite often that more specific names are used to specify the raw materials used, the processing method, season or locality. This variability consequently leads to a great diversity on the microbial population. Typically, kimchi fermentation is initiated by *Leu. mesenteroides* and terminated by *Lb. plantarum* (Choi et al., 2002a). Other lactic acid bacteria such as *Weissella confusa*, *Leu. citreum*, *Lb. sakei*, *Lb. curvatus*, *Leu. gelidum*, *Serratia marcescens*, *Lb. brevis*, *Lc. lactis* subsp. *lactis*, *Leu. gasicomitatum*, *Leu. carnosum*, *Lb. algidus* (Lee et al., 2005; Kim and Chun, 2005; Kim et al., 2002) as well as new species, namely *Ws. kimchii* (Choi et al., 2002b), *Ws. koreensis* (Lee et al., 2002), *Leu. kimchii* (Kim et al., 2000b), *Leu. inhae* (Kim et al., 2003a) have been also isolated. Due to its high importance, kimchi has been given a lot of attention. Several methods have been studied in order to extend the self life, including heat treatment, radiation, antiseptic and antimicrobial agents, herb medicine, pH adjuster and oyster shell powder (Choi et al., 2006). Gamma irradiation has also been used as a method to control enteric pathogens in early fermentation stages (Kim et al., 2004), although less than 5 kGy should be applied for proper sensorial properties, it has been documented that a dose up to 10kGy had no effect on amino acid content, physiological and antimicrobial activity (Song et al., 2004). Surprisingly, scarce information exists regarding the application of starter cultures on kimchi fermentation (Choi et al., 2003). On the other hand, Lee et al (2006) used several lactic acid bacteria isolated from kimchi fermentation as starter cultures for sausage fermentation. Literature on biogenic amine content is also limited; only Tsai et al. (2005) studied several commercially available products and despite the quantitative variation, putrescine, cadaverine, tryptamine, spermidine, histamine and tyramine were frequently detected. Successful efforts have also been done to replace as much as 50% of the NaCl with KCl (Choi et al., 1994). The applicability of a multi-channel taste sensor to the monitoring of kimchi fermentation has been studied by Kim et al (2005). An active packaging system with color indicators to monitor the ripeness has been suggested by Hong and Park (2000).

The popularity and acceptance that fermented vegetables have met, led the researches to study the fermentation of other vegetables such as onions (Roberts and Kidd, 2005), garlic (de Castro et al., 1998) carrots (Aukrust et al., 1994; Montano et al., 1997) and mixtures of carrot, cabbage, beet and onion (Gardner et al., 2001) as well as juices made by a variety of vegetables (Karovicova and Kohajdova, 2003). The outcome was promising, further research is though necessary for better assessment of the products' safety and nutritional impact.

4.3. Cereal Fermentation

Cereal grains constitute a major source of nutrients throughout the world. Depending on the climate, different cereals flourish in different geographical areas, i.e. in Europe wheat and rye are mostly used as a substrate for lactic acid fermentation, in Africa maize, sorghum, millet, tef and occasionally rice and wheat, while in Asia mainly rice. Given the large diversity of the raw materials, a great amount of cereal-based lactic acid fermented products exist. Since most of them have been a part of everyday life and given their high importance, many have been attached to special occasions, especially of religious origin, and a specific and exact ritual for their production is in many cases followed.

Many fermented cereal-based foods have been studied, mostly sourdoughs intended for breadmaking. Sourdough production can be as simple as mixing flour and water and placing in a warm place. After several enrichments with flour and water, the sourdough will be formed. There are several variations of this process depending on the desired acidity, fermentation temperature or microbial activity (Spicher and Bruemmer, 1995).

Generally, based on the technology applied for their production, sourdoughs can be classified into three types:

(1) type I sourdoughs include those produced with traditional techniques and are characterized by continuous, daily refreshments in order to keep the microorganisms in an active state. This type can be further subdivided into type Ia that include utilisation of pure starter culture from different origin, type Ib that include all spontaneously fermented sourdoughs and type Ic that include sourdoughs made in tropical regions fermented at high temperatures.

(2) type II sourdoughs are those semi-fluid preparations that have been created in order to assist the industrialised breadmaking process and serve mainly as dough acidifiers.

(3) type III sourdoughs are dried doughs in powder form, which are initiated by defined starter cultures and are used as acidifier supplements and aroma carriers during breadmaking.

The increased interest in sourdough ecosystem results from its positive effect on the quality of the final product; most of these being directly related to the sourdough microbiota. Thus, the microbiota of a wide variety of spontaneously fermented sourdoughs has been studied. In table 1 an example of the great lactic acid bacteria diversity encountered in this microecosystem is presented.

Table 1. Diversity of lactic acid bacteria in wheat and rye sourdoughs

Origin	Lactic acid bacteria	References
wheat sourdoughs		
France	***Lb. plantarum, Lb. casei, Lb. delbrueckii Lb. acidophilus, Lb. brevis, Leu. mesenteroides Pd. pentosaceus, Lb. hammesii***	Infantes and Tourneur, 1991 Valcheva et al., 2005
Germany	***Lb. delbrueckii, Lb. plantarum, Lb. casei, Lb. fermentum, Lb. buchneri, Lb. brevis***	Spicher, 1959
Greece	***Lb. sanfranciscensis, Ws. cibaria, Ent. faecium, Lb. brevis, Pd. pentosaceus, Lb. paralimentarius, Lb. zymae***	Paramithiotis, 2001 Vancanneyt et al., 2005
Italy	***Lb. acidophilus, Lb. alimentarius, Lb. brevis, Lb. sanfranciscensis, Lb. casei susp. casei, Lb. cellobiosus, Lb. belbrueckii subsp. delbrueckii, Lb. farciminis, Lb. fermentum, Lb. fructivorans, Lb. plantarum, Leu. citreum, Leu. mesenteroides subsp. dextranicum, Leu. mesenteroides subsp. mesenteroides, Lc. lactis subsp. lactis, Pd. parvulus, Pd pentosaceus, Ws. confusa, Lb. rossii, Lb. pentosus***	Gobbetti et al., 1994 Ottogalli et al., 1996 Corsetti et al., 2001; 2005 Randazzo et al., 2005
Spain	***Lb. brevis, Lb. plantarum***	Barber et al., 1983
Rye sourdoughs		
Denmark	***Lb. reuteri, Lb. panis, Lb. amylovorus***	Rosenquist and Hansen, 2000
Finland	***Lb. acidophilus, Lb. plantarum, Lb. casei***	Salovaara and Katunpaa, 1984
Germany	***Lb. amylovorus, Lb. pontis, Lb. frumenti, Lb. reuteri***	Mueller et al., 2001
Russia	***Lb. plantarum, Lb. brevis, Lb. fermentum***	Kazanskaya et al., 1983

Lactic acid bacteria such as *Lb. sanfranciscensis*, *Lb brevis* and *Lb. plantarum* as well as yeasts like *S. cerevisiae* are more often isolated by wheat or rye sourdoughs in significant numbers that suggest their fundamental role. However, other lactic acid bacteria are as well isolated at lower numbers and therefore have been characterised as secondary microbiota. Little information currently exists regarding their contribution to the establishment of this ecosystem. The key sourdough lactic acid bacterium, with very few exceptions, is considered to be *Lb. sanfrancescensis*. Absence of preference for glucose or maltose, maltose utilisation via maltose phosphorylase and utilisation of fructose as an electron acceptor are its special features that can largely justify its competitiveness and persistence in this microenvironment. In the case of the San Francisco sourdough, this lactic acid bacterium co-exists with *S. exiguus*, a maltose non-fermenting yeast; thus no antagonism for the main carbon source occurs. In the case of sourdoughs where a co-existence with *S. cerevisiae*, a maltose fermenting yeast, is evident (such as the Greek sourdoughs), antagonism for the main carbon source takes place. In that case, the property that clearly differentiates this bacterium from the secondary microbiota and has a decisive role for its effective competitiveness seems to be its ability to utilize fructose as an electron acceptor (Paramithiotis et al., 2006a). The latter affects largely the end metabolic products and consequently the quality of the resulting bread. In the presence of a carbon source other that fructose (e.g. maltose), the latter is utilised as an electron acceptor and stoichiometrically converted to mannitol with a concomitant increase in both ATP and acetic acid production through the acetate kinase reaction (Gobbetti et al., 1995; Martinez-Anaya et al., 1994; Stolz et al., 1995). Mannitol itself can influence breads flavour though its sweet taste (Korakli and Vogel, 2003). The increase of acetic acid production, apart from the taste, can also influence the microbial stability of the ecosystem.

The benefits of sourdough utilization in breadmaking are manifold. Apart from the organoleptic quality, it also exerts benefits regarding the textural properties, nutritional value and shelf life of the resulting bread.

As far as sensorial properties are concerned, the main function of sourdough is the generation of precursor molecules, i.e. sugars, organic acids, amino acids and alcohols. Lactic and acetic acid have very little direct but substantial indirect effect on flavor; it has been reported that acetic acid level of 100-200 ppm increase the intensity of other aroma compounds in wheat bread. Moreover, if the fermentation quotient, i.e. the molar ratio of lactic to acetic acid is 2.0 to 2.7, then the pleasant odour of rye bread is perceived (Roecken and Voysey, 1995).

Sourdough application will result in a partial acidification of the bread dough that will affect structure-forming components like gluten, starch and arabinoxylans and the activity of endogenous or bacterial enzymes, affecting thus the rheological properties of the dough and, inevitably, the quality of the final product (Arendt et al., 2007). The baking properties of rye and wheat flours are rather different. In wheat flour, the gluten network plays a major role in the structure, whereas in rye doughs proteins possess a minor role and the pentosans are responsible for the network development. Moreover, acidification results in the partial inactivation of the endogenous amylases; a quite important task since starch in rye is gelatinised at 55-70°C which coincides with the optimum for amylase activity. This would lead to excessive amylolysis that in the case of poor quality rye flour it would result in the complete hydrolysis of the crumb. The acidification also exerts positive effects on the structure of starch granules, leading to an increased water binding capacity (Hammes and Ganzle, 1998). Although sourdough is an essential ingredient for ensuring baking properties of doughs containing more than 20% of rye flour, its addition to wheat doughs remains optional and, in practice, is used only to obtain enhanced organoleptic properties.

The potential of sourdough in improving the nutritional properties of cereals has gained interest during the last decade. Its effect on degradation of cereal phytates and the consequent increase in mineral bioavailability as well as its potential use against to gluten-sensitive enteropathy has already been described. Sourdough utilisation could also be the answer to the problems raising, when significant amounts of cereal bran is added to the formula, in order to exert all the recognised effects of dietary fiber (Salmenkallio-Marttila et al., 2001).

Scarce literature data exists regarding the fate of certain phytochemicals during sourdough breadmaking. It has been reported that sourdough fermentation affected the folate, tocopherol, tocotrienol and thiamin content either through acidity formation or by a still unknown mechanism (Kariluoto et al., 2004; Liukkonen et al., 2003; Wennemark and Jaegerstad, 1992; Ternes and Freund, 1988; Boskov-Hansen et al., 2002). Sourdough fermentation has also been found to have a protective role for β-glucan in bread (Katina et al.,

2005) and influence the postprandial glycaemic responses (Liljeberg et al., 1995; 1996; Ostman et al., 2002).

The self life of bread is largely affected by sourdough incorporation in the recipe, regarding both staling and quality deterioration due to microbial growth. A decrease of the staling rate during storage has been reported and assigned to the proteolytic and amylolytic capacity of a given strain (Corsetti et al., 1998a). Furthermore, the delay of moulding has been initially assigned to the mixture of lactic and acetic acid produced by the lactic acid bacteria. Further studies, though, indicated that this effect was more like a strain depended property; more accurately, the antimould effect of *Lb. sanfranciscensis* strain CB1 has been assigned mainly to the produced caproic acid (Corsetti et al., 1998b), whereas the antimould effect of *Lb. plantarum* strain 21B has been assigned to phenyllactic as well as 4-hydroxyphenyllactic acid (Lavermicocca et al., 2000). Furthermore, several other antifungal metabolites such as cyclic peptides, proteinaceous compounds and 3-hydroxylated fatty acids have been recently isolated from lactic acid bacteria (Schnuerer and Mangusson, 2005). Partial prevention of rope, caused mainly by strains of *B. subtilis* and *B. licheniformis*, is also another function of sourdough. The occurrence of *Bacillus* sp. spores in the raw materials used in baking is very common. Utilization of sourdough in breadmanking is one of the additive-free methods to prevent rope formation and this has been assigned to the enhancement of spores' thermal inactivation during baking, mainly due to the presence of acetic acid as well as inhibition of their growth due to the low pH (Roecken, 1996).

Apart from sourdough's potential for healthier products that has already been discussed, a lot of research has also taken place concerning the exopolysaccharides produced by sourdough lactic acid bacteria and their potential effect on the quality of the final product. Sourdough lactic acid bacteria have been shown to produce both glucans (reuteran, dextran or mutan) and fructans (levan or inulin) (van Hijum et al., 2006; Tieking and Gaenzle, 2005). Exopolysaccharide producing lactic acid bacteria have been incorporated in the industrial practice, since they improve dough rheology as well as the textural quality of bread (Decock and Capelle, 2005). Moreover, the use of fructan-forming strains enables the production of prebiotic fructooligosaccharides and the heterooligosaccharides erlose and arabsucrose from sucrose during sourdough fermentation (Tieking et al., 2005a; b).

The production of bacteriocins and their potential contribution to the formation of this unique ecosystem has also been subjected to extensive study. Several strains have been shown to produce antimicrobial compounds of proteinaceous nature, among them *Lb. bavaricus* MI401 with bavaricin A (Larsen et al., 1993), *Lb. plantarum* ST 31 with plantaricin ST31 (Todorov et al., 1999), *Lb. sanfranciscensis* C57 with a bacteriocin-like inhibitory substance (BLIS C57) (Corsetti et al., 1996), *Lb. amylovorus* with amylovorin L (De Vuyst et al., 1996) and *Lb. reuteri* LTH 2584 with reutericyclin (Gaenzle et al., 2000). Some strains of *Lb. reuteri* isolated from sourdoughs produce antagonistic compounds not fitting the general bacteriocin definition. Reuterin (2-hydroxy-propanal) is a product of glycerol metabolism of *Lb. reuteri* inhibitory against a wide range of Gram-positive as well as Gram-negative organisms (Gaenzle, 2004). The structure, antimicrobial activity, kinetics of production and possible applications of these compounds has also been studied in depth.

A huge variety of less studied indigenous fermented cereal-based foods and beverages also exists. The significance of these products for the local populations has been recognised. In Africa, lactic acid fermented foods generally constitute a part of the daily main dishes of the average individual. For example, ogi, a maize-based lactic acid fermented gruel is the major indigenous traditional weaning food in West Africa, whereas gari, another important cassava-based lactic acid fermented product, serves as one of the three daily main dishes of most families in a large part of the continent (Oyewole, 1997). On the other hand, the diet in East-Asia is characterized by a need for protein and flavor enrichment, as it consists mainly of rice and small amounts of dairy products. Fermentation in general and especially lactic acid fermentation covers that need supplying the population with a variety of savory dishes. Products such as the various idli-type breads, made by rice and black gum, are important staple foods for the Indian and Sri Lankan people and they are consumed three or four times a week at breakfast and supper (Lee, 1997). Finally, Latin America has also a long and very important tradition in household lactic acid fermentation, but this diversity refers mainly to beverages, many of which were used as nutritive complements or for religious ceremonies since ancient times (Quevedo, 1997). Perhaps the most studied of these is pozol, a corn-based refreshing non-alcoholic beverage, consumed in South-eastern Mexico since pre-hispanic times as an important part of the daily diet of different ethnic groups (Wacher et al., 2000).

5. Lactic Acid Fermentation of Animal-Derived Material

5.1. Meat Fermentation

The production of fermented sausages includes mixing of minced meat and fatty tissue with curing agents, carbohydrates and spices, stuffing into casings, and ripening, which can be subdivided into fermentation and ageing. Variations in the type and amount of raw materials, fermentation as well as drying conditions give rise to a wide range of products. Several factors account for the selection of the meat species used; among them the eating habits and price seems to play the most important role. Thus, a mixture of pork and beef is usually used in Germany and Greece, whereas in Hungary and Italy sausages are manufactured only from pork. The fatty tissue most frequently used, especially in high-quality dry sausages, is firm pork back fat. Other fatty tissues might lead to organoleptically less acceptable products. Carbohydrates are added to the mixture since the amount of glucose and glucose-6-phosphate generated during glycogenolysis does not allow a sufficient pH value reduction. As mentioned by Talon et al. (2004), southern technology produces sausages with higher pH (5.2-5.8) values than northern sausages (4.8-4.9). Addition of carbohydrates may vary according to the eating habits and the desired characteristics of the final product. The curing agents that are used are sodium chloride, sodium nitrite and potassium nitrate. The first, which is normally added at 2.4-3.0%, apart from the apparent effect on the taste, lowers the water activity and interacts with the myofibrilar protein fraction. Sodium nitrite has a multiple function. It assists the development of curing color, inhibits the autoxidative processes leading to rancidity and contributes to the establishment of the desired Gram-positive microbiota. Sodium ascorbate is frequently added along with nitrite to enhance the development of curing colour and flavor. Potassium nitrate is still used for the manufacture of dry sausages with long ripening times at an amount less than 300 $mgkg^{-1}$. Spices have been used to enhance the taste. It has been, though, documented that manganese containing spices (e.g. red pepper) stimulate growth of lactic acid bacteria (Zaika and Kissinger, 1984) whereas others, such as garlic, contain antioxidants that may contribute to the shelf life of dry sausages. Generally, the ripening of spontaneously fermented sausages is a three-stage procedure, like the one used for the production of 'salciccia sotto sugna' (Gardini et al., 2001a; Parente et al., 2001) where the initial drying stage at 24°C, 80% relative humidity (RH) for 24 hours is followed by subsequent fermentation stages at 18-22°C, 75-89% RH for 5 days and 15-20°C, 80-85% RH for 2 months. Quite often, the second stage of fermentation is not performed at fixed conditions, e.g. the fermented sausages prepared in Greece (Figure 3) (Papamanoli et al., 2003; Drosinos et al., 2005; 2007; Coppola et al., 1998; Greco et al., 2005; Comi et al., 2005) during which, the fermentation temperature gradually decrease from 18-24°C to 11-19°C with an alteration to relative humidity from 60-94% to 60-90% within 5-7 days, according to the occasion. The same step can also be omitted, e.g. 'chorizo' (Garcia-Varona et al., 2000), 'sarchichon' (Lizaso et al., 1999) or the Naples-type salami (Mauriello et al., 2004) manufacture. Fermentation can also be carried out with only one step, such as 'sobrasada' (Rosselo et al., 1995) or 'Salame di Senise' (Baruzzi et al., 2006) production. All these variations lead to an extended diversity as far as the lactic acid microbiota is concerned (Table 2). Classification of fermented sausages can be based upon several factors, such as meat species, moisture, protein, water activity, etc. The microbiota of fresh meat mainly consists of Gram-negative, oxidase-positive rods, particularly psychrotrophic pseudomonads. Psychrotrophic Enterobacteriaceae are also present, while Gram-positive organisms including lactic acid bacteria occur only in small numbers. During processing and as the water activity is reduced to 0.96-0.97 and the oxygen consumed, a shift in the microbiota composition towards the lactic acid bacteria, staphylococci as well as Micrococcaceae takes place (Luecke, 1998). At the end of ripening, lactic acid microbiota reaches a population of 10^7-10^8 while Micrococaceae 10^4-10^5 cfug^{-1}, depending of course upon the raw materials and the ripening conditions. Their function during ripening is continuously under investigation. Lactic acid bacteria, apart from the formation of lactic acid and the concomitant inhibition of pathogenic and spoilage bacteria seem to play a role in the proteolytic as well as the lipolytic activities (Sanz et al., 1999a; b; Fadda et al., 1998; 1999; El Soda et al., 1986; Montel et al., 1998, Drosinos et al., 2007). As far as catalase-positive cocci are concerned, their main role is the reduction of nitrate. However, their contribution to proteolysis and lipolysis has also been studied (Mauriello et al., 2002; 2004; Kenneally et al., 1998; Drosinos et al., 2007). As a consequence, most commercially available starter cultures consist of a mixture of these. The lactic acid bacteria species available in this form include among others *Lb. sakei*, a lactic acid bacterium with a high degree of adaptation to the respective microenvironment (Champomier-Verges et al., 2001), *Lb. pentosus*, *Lb. curvatus*, both of them being quite frequently isolated from spontaneously fermented sausages, most of the times though forming a secondary microbiota. On the other hand, the catalase-positive cocci available include among others *St. xylosus* and *St. carnosus* both of them have been found to form the staphylococcal microbiota in a variety of fermented sausages. Some products require the presence of yeasts and moulds, providing with characteristic organoleptic properties. Thus, strains of *Debaryomyces hansenii*, *Candida famata* as well as *Penicillium nalviovense* and *Pen. chrysogenum* are commercially supplied as starter cultures.

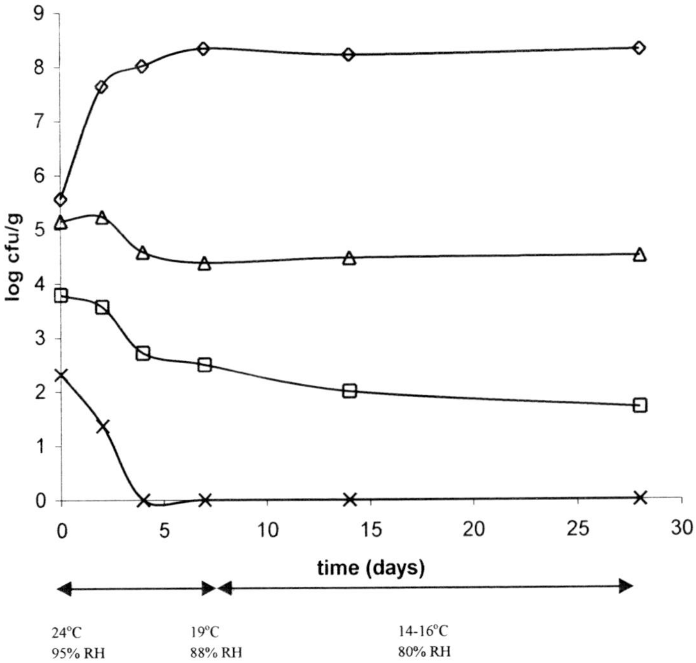

Figure 3. Evolution of lactic acid bacteria (◊), *Micrococcaceae* (□), enterococci (Δ) and enterobacteria (×) during manufacture of dry fermented sausages in Greece.

Apart from the ongoing research on the development of new starter cultures with enhanced properties, research has also been focused on issues emerging from the industrialization of sausage fermentation. Acceleration of the ripening process would result in the reduction of the storage time and consequently would increase the profit margin and the competitiveness of the end product. Several researchers have incorporated lipases and proteases in the mixture, and studied the effect on the ripening time and the sensorial quality and a general conclusion that can be drawn is that the addition of enzymes requires as well a modification of the processing parameters (Fernandez et al., 2000)

Research has been focused on issues regarding the safety of the final product, as well. The microbial stability of dry fermented sausages is an excellent example of what is referred to as 'hurdle concept'. The hurdles inhibiting or delaying growth of unwanted bacteria are low water activity, nitrite, low redox potential and acidity. The water activity is initially lowered by the addition of NaCl. As fermentation proceeds, the pH is lowered affecting negatively the water holding capacity of the meat, leading thus to a further decrease of the water activity. Usually, the drying of sausages in ripening chamber results in a final water activity lower than 0.90. Nitrite in the form of undissociated nitrous acid is able to pass the ion barrier of bacterial cell wall and disturb the function of bacterial enzymes and therefore bacterial growth. The redox potential is lowered after the first day of fermentation due to the microbial utilization of the oxygen present in the sausage matrix. This is making nitrite more effective and restricts the growth of aerobic spoilage bacteria (pseudomonads) derived from the raw meat. Lactic acid produced after a few days of fermentation, leads in the lowering of the pH value to values around 5.0. The lower external pH disturbs the homeostasis of different pathogens (e.g. *Salmonella* spp., *Clostridium* spp.) as well as spoilage bacteria, such as pseudomonads and enterococci. The undissociated state of the acid, which is uncharged and able to penetrate the cell membrane, and thus far more effective, accounts only to the 10% of the total lactic acid, due to the fact that at the pH values typical for dry sausage is only around 4.8 (Tyopponen et al., 2003). As far as the fate of the pathogens during sausage fermentation is concerned, growth of *Bacillus* and *Clostridium* spores, mainly derived from spices, are normally avoided due to the low pH and water activity. On the other hand, *St. aureus* is quite tolerant to these hurdles but luckily is a poor competitor to starter lactic acid bacteria and staphylococci, especially at fermentation temperatures of 20-25°C. However, several studies dealing with survival of *L. monocytogenes* and *E. coli* O157:H7 have revealed an increased concern. The utilisation of bacteriocinogenic starter cultures seems to be the most preferred approach against these microorganisms.

Table 2. Diversity of lactic acid bacteria in spontaneously fermented sausages

Origin	Lactic acid bacteria	References
Indonesia	***Lb. plantarum, Pd. acidilactici, Lb. farciminis***	Antara et al., 2002
Italy	***Ent. pseudoavium, Lc. lactis subsp. lactis, Lb. brevis, Lb. curvatus, Lb. paraplantarum, Lb. plantarum, Lb. sakei, Leu. citreum, Leu. mesenteroides, Lb. paracasei subsp. paracasei, Lb. graminis, Lb. alimentarius, Lb. coryneformis subsp. torquens, Pd. pentosaceus, Leu. carnosum, Leu. gelidum, Lactobacillus sp.***	Comi et al., 2005 Coppola et al., 1998 Greco et al., 2005 Parente et al., 2001 Andrighetto et al., 2001 Rantsiou et al., 2004
Greece	***Lb. plantarum, Lb. paracasei subsp. paracasei, Lb. paracasei subsp. tolerans, Lb. sakei, Lb. casei, Lb. curvatus, Lb. rhamnosus, Lb. brevis, Lb. pentosus, Lb. salivarius, Leu. mesenteroides subsp. mesenteroides, Lc. lactsc subsp. lactis, Ent. faecium, Lactobacillus sp.***	Papamanoli et al., 2003 Samelis et al., 1998 Drosinos et al., 2005; 2007
Spain	***Lb. sakei, Lb. curvatus, Lb. alimentarius, Ent. faecium, Ent. faecalis, Lactobacillus sp, Lb. plantarum, Pediococcus sp., Lb. bavaricus***	Santos et al., 1998 Garcia Fontan et al., 2007 Hugas et al., 1993
Hungary	***Lb. curvatus, Lb. plantarum Lb. sakei, Leu. citreum, Leu. mesenteroides, Ws. viridescens, Lactobacillus sp.***	Rantsiou et al., 2005
Argentina	***Lb. sakei, Lb. plantarun, Lb. pentosus, Lb. curvatus, Pd. acidilactici, Ent. faecalis, Ent. flavescens, Ent. faecium, Ent. mundtii, Ent. durans***	Fontana et al., 2005a; b

Thus, several bacteriocinogenic lactic acid bacteria have been utilized as starter cultures for the production of dry fermented sausages, bacteriocin production has been optimized and mathematical models have been created in order to predict its production under various conditions (Talon et al., 2002; 2006; Drosinos et al., 2006).

Scarce information exists regarding the survival of viruses and parasites during ripening. The infectivity of enterovirus, echovirus and poliovirus has been shown to be retained during fermentation. The source of these viruses could be human faeces entering the production line through contaminated personnel. On the other hand, foot-and-mouth disease virus is acid sensitive and thus unlikely to retain its infectivity. Several parasites can infect the animals, whose meat is used as a raw material. Among them, *Toxoplasma*, whose cysts are rapidly inactivated during fermented sausage manufacture, and *Trichinella spiralis*, whose larvae seem to be inactivated by the combination of low water activity and pH (Luecke, 1998).

The biogenic amine content of fermented sausages as well as the factors affecting their production has been studied in some detail. It is characterized by great variability that occurs as a result of a complex equilibrium between the abiotic sausage environment and the enzymatic activities of the microbial population (Suzzi and Gardini, 2003; Bover-Cid et al., 2001a). Several authors have studied the biogenic amine content of various types of fermented sausages. It seems that tyramine, putrescine and cadaverine is most often present at concentration exceeding 100 mg kg^{-1}(Suzzi and Gardini, 2003). The production of biogenic amines has been attributed to several microorganisms; among them *Enterobacteriaceae*, lactobacilli, staphylococci and enterococci. *Enterobacteriaceae* are mainly responsible for the production of biogenic amines during storage of raw materials and uncontrolled fermentation. Several lactobacilli strains belonging to the species *Lb. buchneri*, *Lb. alimentarius*, *Lb. plantarum*, *Lb. curvatus*, *Lb. farciminis*, *Lb. bavaricus*, *Lb. homohiochii*, *Lb. reuteri* and *Lb. sakei* have been found amine positive either in vitro or in situ with tyramine being quantitatively the most important biogenic amine produced (Masson et al., 1996; Montel et al., 1999; Pereira et al., 2001). As far as biogenic amine production by staphylococci is concerned, it seems that decarboxylase activity is quite common amongst *Staphylococcus* strains (Straub et al., 1995; Silla Santos, 1998; Montel et al., 1999; Matruscelli et al., 2000; Drosinos et al., 2007). Finally, many enterococci were found able to produce tyramine (Bover-Cid et al., 2001a; Masson et al., 1996; Gardini et al., 2001b). Several factors have been found to affect biogenic amine production. The effect of the pH value (Masson et al., 1999), sodium chloride content (Chander et al., 1989), sausage size (Bover-Cid et al., 1999b), temperature (Komprda et al., 2001), additives (Bover-Cid et al., 2001b; Komprda et al., 2004) and, of course, starter cultures (Bover-Cid et al., 1999a; Gardini et al., 2002; Ayhan et al., 1999; Gonzalez-Fernandez et al., 2003) have been extensively studied.

Several attempts have been made in order to re-formulate the fermented sausage recipe, mainly by substituting NaCl and fat. This task seems to be difficult, since both ingredients possess a specific role in the manufacture of dry fermented sausages. Substitution of sodium chloride has been mainly performed by potassium chloride, potassium lactate, glycine, manganese chloride, calcium chloride as well as calcium ascorbate (Gou et al., 1996; Ibanez et al., 1995; 1996; 1997; Gimeno et al., 1998; 1999; 2001) pointing out the differences observed in the organoleptic quality of the end product without studying the effect on the safety from a microbiological point of view. The reduction of fat and the addition of compounds such as inulin and dietary fible have been studied by Mendoza et al (2001) and Garcia et al (2002), respectively, with some promising results. Olive oil (Bloukas et al., 1997; Muguerza et al., 2001; 2002) as well as soy oil (Muguerza et al., 2004) has also been used as possible replacing compounds, with again some very interesting outcomes.

Since the idea of using probiotic starter cultures in sausage fermentation has developed, several lactic acid bacteria have been screened for their capacity to survive their passage through the GIT and for their possible in-site actions. Erkkila and Petaja (2000) reported that *Lb. curvatus* strain RM10 and *Pd. acidilactici* strain P2, isolated from freeze-dried commercial meat starter cultures exhibited the strongest capacity for surviving acidic conditions and 0.30% bile salts. Erkkila et al. (2001) demonstrated the suitability of three probiotic *Lb. rhamnosus* strains (GG, E-97800 and LC-705) to produce dry sausage. Pennachia et al. (2004) suggested a very effective screening procedure and consequently isolated 20 potentially probiotic *Lactobacillus* strains, 11 of which, most of them belonging to *Lb. plantarum* group, exhibited good adhesion capability to Caco-2 cell layers (Pennachia et al., 2006). The capacity of two *Lb. plantarum* strains to in vitro survive the passage through the human GIT either as freeze-dried culture or embedded in a sausage matrix fermented with this strain has been demonstrated by Klingberg and Budde (2006). Muthukumarasamy and Holley (2006) suggested that microencapsulation of probiotic strains like *Lb. reuteri* may be an option for formulation of fermented meat products carrying viable health promoting bacteria. The microencapsulation of a reuterin producing *Lb. reuteri* strain demonstrated, on the other hand, that it reduced their inhibitory action against *E. coli* O157:H7 (Muthukumarasamy and Holley, 2007). The need for in vitro study concerning the fate of staphylococci present in fermented sausages was pointed out by Paramithiotis et al. (2006b), since most of the strains studied exhibited an aptitude not only to withstand but to proliferate, as well, in the presence of bile salts even at an acidic environment as well as their ability to adhere to stainless-steel plates.

5.2. Milk Fermentation

Lactic acid fermented milk products are definitely the most studied and standardized of all lactic acid fermented products. The milk from several animal species is used for the production of cheeses and fermented milks, with cow's milk being by far the most important. Cheesemaking is an ancient practice with a wide range of early written references, such as the ones from Homer, Pliny, Athinaios and Aristotle, exhibiting its popularity as well as its importance in Greek and Roman times. An extensive list of cheese varieties exists nowadays and still there are many that have not yet been described in detail. The classification of cheeses is mainly based on moisture content and milk species. Furthermore, these groups were subdivided according to their surface microflora, blue veins, 'eye' formation, scalding temperatures and ripening microorganisms. The development of the first commercial starter cultures at the end of the 19th century was a landmark for cheesemaking. The advantages that were then offered, led to their broad appliance and today very few cheeses are produced without its addition, mainly due to safety reasons.

The preparation of most cheeses includes similar basic steps; namely milk coagulation, acidification, drainage, salting and ripening. The duration, temperature, humidity as well as specific conditions and operations of each step depends on the desired cheese type and may vary significantly from plant to plant. Coagulation is mostly achieved by rennet, which includes the enzymes chymosin and pepsin and it is derived from the fourth stomach of bovine animal or by microbial coagulants, an acid protease extracted from fermentations of selected fungal strains. Once the coagulation has been achieved, cutting of the curd takes place. Acidification proceeds as a result of the starter culture addition. The curds are then drained, salted and moulded. Ripening takes place at the desired temperature for specific period of time in ripening chambers.

The microorganisms used in cheesemaking may be divided into starter and secondary microbiota. The starter microbiota consists of lactic acid bacteria either mesophiles or thermophiles. The former, mainly *Lactococcus* sp. and *Leuconostoc* sp., grow optimally at 30-33°C and utilized usually when fermentation temperatures of 20-40°C are applied. The latter, the most important of which are *Str. salivarius* subsp. *thermophilus*, *Lb. delbrueckii* subsp. *bulgaricus*, *Lb. helveticus* and *Lb. delbrueckii* subsp. *lactis*, grow optimally at 40-45°C and used mainly at fermentation temperatures of 30-50°C.

The desired functions of these microorganisms are acidification, texture enhancement and flavor development. Acidification refers to the consumption of lactose and production of lactic acid. As a consequence, the pH is lowered. Acidification also enhances the removal of whey from the curd, reducing thus the moisture level. These functions, combined with salt addition, either as dry or through a brining step, contribute to the safety of the product, inhibiting growth of pathogenic and spoilage microorganisms. Its type of cheese has its unique textural properties and the whole manufacturing process is oriented towards the control of factors that may have an effect. Thus, factors such as milk constituents, coagulating enzymes, microbiota and manufacturing process are constantly kept under control in order to obtain the desired texture of the final product. Flavor is another specific characteristic of each cheese type. Control of milk composition, heat treatments, microbiota growth, salt addition, fermentation and storage conditions are necessary to obtain the desired flavor. Above from the lactic acid bacteria added as a starter culture, the development of a secondary microbiota is also necessary for the production of some cheeses. Its growth will inevitably lead to alteration of the product's texture and flavor, and thus its control is of primary importance. An example of important secondary microbiota are moulds such as *Penicillium camemberti* or *Pe. roqueforti*, yeasts such as *Kluyveromyces lactis* and *Candida kefir* and some bacteria such as *Corynebacteria*, *Kocuria* sp., *Lactobacillus* sp. *Pediococcus* sp. and Propionibacteria.

Fermented milks have also been a part of the human diet since ancient times. Milk fermentation with mesophilic starters (e.g. sour milk) are typical for north-east Europe while those prepared with thermophilic starters (e.g. yoghurt) are traditionally produced in central and south Asia and south-east Europe (Oberman and Libudzisz, 1998). The microorganisms involved in the production of fermented milks include lactic acid bacteria such as *Lactobacillus*, *Lactococcus*, *Streptococcus*, *Leuconostoc*, *Pediococcus* and *Bifidobacterium*, yeasts such as *Torulaspora*, *Candida*, *Saccharomyces* and *Klyuveromyces* and moulds such as *Geotrichum candidum*. A classification of traditional fermented milks has been proposed by Marshall (1984), on the basis of the dominating microbiota. This includes four types of fermented milks. The type I is typical of Scandinavian, central and eastern European countries. Mesophilic lactococci as well as leuconostocs form their dominating microbiota. They are characterized by their consistency due to the lactic acid coagulation of milk proteins and the flavor and aroma produced by citric acid and lactose fermentation. Specific attributes are also common, such as the need of slime forming strains of *Lc. lactis* ssp. *lactis* and *Lc. lactis* ssp. *cremoris* for the production of 'langfil' or 'viili'. In type II fermented milks, dominating microorganism is strains belonging to *Lactobacillus* sp. These products are typical for eastern Europe, especially Bulgaria. In this category also belongs the Japanese fermented milk under the trade name 'Yakult', which is made using the *Lb. casei* strain Shirota. Type III fermented milks include all those products, the responsible microbiota of which belong to the thermophilic group of lactic acid bacteria. These products are typical of central and eastern Europe. *Str. salivarius* subsp. *thermophilus* and *Lb. delbrueckii* subsp. *bulgaricus* are typical representatives of these microorganisms and yoghurt is the most popular amongst fermented milks produced with thermophilic lactic acid bacteria. Finally, type IV, includes these fermented milks the dominant microbiota of which comprises of both lactic acid bacteria and yeasts. These products are also called acid-alcoholic fermented milks since the main products of the microbial metabolism are lactic acid and ethanol. They are widely produced in Asia and the Middle East.

Fermented dairy products are beneficial as they represent good sources of vitamins and minerals and generally contain small amounts of lipids. Fresh fermented dairy products may supply a number of lactic acid bacteria that may provide additional health benefits. This is the 'probiotic' concept that is largely applied in dairy products. Today, a wide variety of dairy fermented products including probiotic lactobacilli exists. Most of them are supplied in highly concentrated forms containing more than $1x10^{10}$ $cfug^{-1}$ and are usually freeze-dried; spray dried or microencapsulated and incorporated mainly in fermented milks (Nighswonger et al., 1996; Sanders et al., 1996; Heller, 2001; Oliveira et al., 2001) and cheeses (Gomes and Malcata, 1998; Stanton et al., 1998; Gardiner et al., 2002). As our knowledge regarding the gastrointestinal track microbiota and the possible in-site functions of probiotic bacteria improves, the criteria are continuously updated (Saito, 2004; Saarela et al., 2000; Klaenhammer and Kullen, 1999). Currently, a huge amount of literature exists regarding the potential probiotic properties of a range of lactic acid bacteria such as *Lb. brevis* (Ronka et al., 2003), *Lb. plantarum* (Cebeci and Gurakan, 2003), *Lb. casei* (Bertazzoni Minelli et al., 2004) as well as yeasts such as *S. boulardii* (Lourens-Hattingh and Viljoen, 2001a) as well as their benefits regarding prevention or treatment of different types of diarrhoea (Guandalini et al., 2000; Marteau et al., 2001; 2002; Hilton et al., 1997), inflammatory and irritable bowel diseases (Marteau et al., 2002; Abbott, 2004; Niedzielin et al., 2001) and immunologic effects (Meydani and Ha, 2000). Concerning yogurt, apart from being an excellent source of protein, calcium, phosphorus, riboflavin (vitamin B2), thiamin (vitamin B1) and vitamin B12, and a valuable source of folate, niacin, magnesium and zinc (McKinley, 2005), other issues have also been raised. Research has shown that lactose in yoghurt is tolerated better by individuals with lactose intolerance (Gallagher et al., 1974; McDonagh et al., 1987; Dewit et al., 1988; Rosado et al., 1992). Several explanations have been given in order to fit the experimental data. It seems that the presence of living bacteria (Gilliland and Kim, 1984; Savaiano et al., 1984; Lerebours et al., 1989) or β-galactosidase activity detected in the duodenum and terminal ileum after yoghurt consumption (McDonagh et al., 1987; Marteau et al., 1990; Martini et al., 1991) are not the only factors contributing to the improved lactose tolerance by humans, and additional factors might also contribute, such as the different viscosity of yoghurt compared to milk may result in slower gastric emptying and thus, a longer transit through the gastrointestinal tract, which, in turn, may improve the absorption and reduce the lactose load in the colon (Vesa et al., 1996). Furthermore, numerous observational studies in adults; children and adolescents indicate a significant inverse relationship between calcium, or dairy consumption, and both body weight and body fat. (Davies et al., 2000; Lin et al., 2000; Periera et al., 2002; Jacqmain et al., 2003; Teegarden, 2003; Carruth and Skinner, 2001; Skinner et al., 2003).

The tendency for reducing the fat intake by humans had also reached the dairy industry. Reduction of fat content has definitely a negative effect on the organoleptic properties of cheese. Negative economic consequences such as lower cheese yield and slower ripening are also a drawback. Therefore, to improve these products, a number of alternatives have been proposed, focusing mainly on texture and flavour. These alternatives generally propose either a modification of conventional cheesemaking processes or the introduction of new technologies (Bryant et al., 1995; Katsiari and Voutsinas, 1994; Rodrigues et al., 1996; Merrill et al., 1994).

Microbial cells are sources of valuable enzymes, proteins and other bioproducts that might exert benefits to the consumer. Several attempts have been made and many cell rupture techniques having a potential for possible use in the dairy industry have been described (Geciova et al., 2002). The use of bacteriocins to induce the lysis of lactic acid bacteria is a novel approach to accelerating cheese ripening and controlling flavour development (Pelaez and Requena, 2005).

Research has also been focused in issues regarding to safety of the final product. A lot of attention has been given to the antagonistic effect of antimicrobial compounds produced by the natural microbiota present in milk and milk products as well as their possible application in the dairy industry. The lantibiotics nisin and lacticin 481 produced by wild lactococcal strains are the most abundant bacteriocins among wild lactic acid bacteria isolates (Rodrıguez et al., 2000; Wouters et al., 2002). Several other bacteriocin producing strains with a potential use in various dairy products have been studied. Among them, several *Lactobacillus* sp., (Ennahar et al., 1998; Topisirovic et al., 2006), *Enterococcous* sp. (Foulquie Moreno et al., 2003; Giraffa and Carminati, 1997), *Streptococcus* sp. (Anastasiou et al., 2007; Ivanova et al., 1998) and *Lactococcus* sp. strains (Arques et al., 2005; Rodriguez et al., 2005; Buyong et al., 1998) revealing the potentiality of using specific strains to the safety of the final product, at least from a microbiological point of view.

Cheese is, after fish, the most commonly implicated food item with biogenic amines poisoning (Stratton et al., 1991); however this refers mainly to ripened and/or long-term stored hard and semi-hard cheese (Innocente and D'Agostin, 2002) since during ripening, degradation of casein occurs leading to the accumulation of free amino acids that can be converted into biogenic amines by bacterial decarboxylases. The latter can occur either by starter microorganisms (Fernandez-Garcıa et al., 2000), or other spontaneous microbiota (Roig-Sagues et al., 2002). The biogenic amine content of various cheeses produced from raw unpasteurized milk (Fernandez-Garcıa et al., 2000; Innocente and D'Agostin, 2002; Novella-Rodrıgues et al., 2003; Pinho et al., 2001), region-specific cheeses produced from sheep or goat milk (Novella-Rodrıgues et al., 2003; Ordonez et al., 1997; Pinho et al., 2001; Valsamaki et al., 2000; Martuscelli et al., 2005) as well as cheeses from cows' milk like semi-soft cheese (Schneller et al., 1997), Swiss-type hard cheese (Petridis and Steinhart, 1996a; b) and Dutch-type hard cheese (Komprda et al., 2007) have been studied. Since biogenic amine production is a strain-specific character (Halasz et al., 1994; Innocente and D'Agostin, 2002; Valsamaki et al., 2000) a solution to this problem would be simply to select the proper lactic acid bacteria as starter cultures and to eliminate as much as possible of the native milk microbiota.

A lot of research concerning the mycotoxin, especially aflatoxin M_1, content in milk and dairy products has taken place. It has been concluded that consumption by dairy cattle of aflatoxin B_1, through contaminated feed, leads to the excretion of aflatoxin M_1 into milk (Applebaum et al., 1982). The occurrence in dairy products such as cheese (Barbieri et al., 1994; Barrios et al., 1996; Brackett and Marth, 1982a; b) and yoghurt (Martins and Martins, 2004; Kim et al., 2000a) as well as the effect of the processing conditions (Govaris et al., 2001; 2002; Blanco et al., 1993; Mendonca and Venancio, 2005; Applebaum and Marth, 1982) has also been studied with a variety of methods, leading some times to contradictory results (Mendoca and Venancio, 2005). The results obtained indicated that, at least in some cases when the content was above the legal limits, this could become a serious problem for public health and the need for strengthening the research towards effective detoxification methods.

Milk contains many components that provide critical nutritive elements, immunological protection, and biologically active substances to both neonates and adults (Severin and Wenshui, 2005). Physiologically active peptides derived from milk proteins, have been found in fermented dairy products, such as yoghurt, sour milk and cheese, have drawn special attention during the last years. These peptides although inactive within the sequence of the parent protein molecule, they are liberated by the hydrolysis of parent protein molecule, either in the GIT or by proteolytic starter cultures and have been shown *in vivo* to exert a wide range of nutritional, functional and biological activities affecting, e.g., the digestive, cardiovascular, immune and nervous systems (Korhonen and Pihlanto, 2006). It is thus very likely that in the near future research will orient to establish these effects as well as to study their fate during production of fermented dairy products.

6. Conclusion

Knowledge concerning lactic acid fermented foods has been significally expanded during the last decades. Intensive research has revealed a wide diversity of lactic acid bacteria present throughout the fermentation of several substrates. Due to their often appearance in similar products in significant populations, some of them have been characterised as native or adapted to the specific microenvironment; while others have been given the title of secondary microbiota. Research has rightfully focused on the properties that enable the former to prevail as well as their impact on the quality of the final product. As far as the secondary microbiota is concerned, it mainly consists of microorganisms capable to withstand the adverse conditions of the respective microenvironment and, in most of the cases, may have a limited direct or indirect effect on the quality of the final product. Several attempts have been made to incorporate lactic acid bacteria strains with desired properties in the production procedure; with many of them currently used in an industrial scale.

Nowadays, it is a common belief that the key principles determining the structure and function of the majority of lactic acid fermented foods are more or less understood. Thus, research on that field has weakened and focused on aspects regarding the safety and nutritional properties in order to meet the consumer demand for safe and healthy foods. Significant effort has been given in order to understand the factors that affect growth and prevalence of pathogenic microorganisms in various food commodities. The data that has been accumulated is fited within mathematical models, so that prediction of growth is enabled. As this area is enthusiastically growing, significant results are expected within the next decade. On the other hand, the demand for healthy foods has opened a new area for further research. The tendency for healthier formulations has reached every branch of lactic acid fermented foods. Attempts to reduce the utilisation of potentially harmfull, for the consumer, compounds are continuously made. Furthermore, there is a need to study the fate of compounds naturally occurring in the raw materials, such as phytochemicals, to which a series of positive effects on human health and well-being has been attributed. Due to the substantial importance, this field seems capable of overlapping other research activities in the future.

REFERENCES

Abbott, A. Gut reaction. *Nature*, 2004, 427, 284-286.

Abou-Arab, AAK. Degradation of organochlorine pesticides by meat starter in liquid media and fermented sausage. *Food Chem. Toxicol.*, 2002, 40, 33-41.

Abou-Arab, AAK. Effect of processing and storage of dairy products on lindane residues and metabolites. *J. Food Chem.*, 1999, 64, 467-473.

Abou-Arab, AAK. Effect of Ras cheese manufacturing on the stability of DDT and its metabolites. *J. Food Chem.*, 1997, 59, 115-119.

Adams, MR; Nicolaides, L. Review of the sensitivity of different pathogens of fermentation. *Food Control*, 1997, 8, 227-239.

Adhikari, K; Mustapha, A; Grun, IU; Fernando, L. Viability of microencapsulated Bifidobacteria in set yogurt during refrigerated storage *J. Dairy Sci.*, 2000, 83, 1946-1951.

Afman, L; Mueller, M. Nutrigenomics: from molecular nutrition to prevention of disease. *J. Am. Diet. Assoc.*, 2006, 106, 569-576.

Ames, BN. Micronutrient deficiencies. A major cause of DNA damage. *Ann. NY Acad. Sci.*, 1999, 889, 87-106.

Anastasiou, R; Georgalaki, M; Manolopoulou, E; Kandarakis, I; De Vuyst, L; Tsakalidou, E. The performance of *Streptococcus macedonicus* ACA-DC 198 as starter culture in Kasseri cheese production. *Int. Dairy J.*, 2007, 17, 208-217.

Andrighetto, C; Zampese, L; Lombardi, A. RAPD-PCR characterization of lactobacilli isolated from artisanal meat plants and traditional fermented, sausages of Veneto region (Italy). *Lett. Appl. Microbiol.*, 2001, 33, 26-30.

Angelov, A; Gotcheva, V; Kuncheva, R; Hristozova, T. Development of a new oat-based probiotic drink. *Int. J. Food Microbiol.*, 2006, 112, 75-80.

Antara, NS; Sujaya, IN; Yokota, A; Asano, K; Aryanta, WR; Tomita, F. Identification and succession of lactic acid bacteria during fermentation of urutan, a Balinese indigenous fermented sausage. *World J. Microbiol. Biotechnol.*, 2002, 18, 255-262.

Applebaum, RS; Brackett, RE; Wiseman, DV; Marth, EH; Aflatoxin: Toxicity to dairy cattle and occurence in milk and milk products - a review. *J. Food Prot.*, 1982, 45, 752-777.

Applebaum, RS; Marth, EH. Fate of aflatoxin M_1 in cottage cheese. *J. Food Prot.*, 1982, 45, 903-904.

Arendt, EK; Ryan, LAM; Dal Bello, F. Impact of sourdough on the texture of bread. *Food Microbiol.*, 2007, 24, 165-174.

Arino, A; Herrera, A; Conchello, MP; Lazaro, R; Perez, C. Hexachloro-cyclohexane residues in meat products after processing. *J. Food Comp. Anal.*, 1993, 6, 55-61.

Arino, A; Herrera, A; Conchello, MP; Perez, C. Hexachlorobenzene residues in Spanish meat products after cooking, curing and long-term ripening. *J. Food Prot.*, 1992, 55, 920-923.

Arino, A; Lazaro, R; Conchello, P; Bayarri, S; Herrera, A. The effect of processing on incurred residues on DDE in meat products. *Food Addit. Contam.*, 1995, 12, 559-566.

Arques, JL; Rodriguez, E; Gaya, P; Medina, M; Guamis, B; Nunez, M. Inactivation of *Staphylococcus aureus* in raw milk cheese by combinations of high-pressure treatments and bacteriocin producing lactic acid bacteria. *J. Appl. Microbiol.*, 2005, 98, 254-260.

Aukrust, TW; Blom, H; Sandtorv, F; Slinde, E. Interaction between starter culture and raw material in lactic acid fermentation of sliced carrot. *Lebensm. Wiss. u. Technol.*, 1994, 27, 337-341.

Auroma, OI; Deiana, M; Jenner, A; Halliwell, B; Kaur, H; Banni, S; Corongiu, F; Dessi, MA; Aeschbach, R. Effect of hydroxytyrosol found in extra virgin olive oil on oxidative DNA damage and on low-density lipoprotein oxidation. *J. Agric. Food Chem.*, 1998, 46, 5181-5187.

Ayhan, K; Kolsarici, N; Ozkan, GA. The effects of a starter culture on the formation of biogenic amines in Turkish soudjoucks. *Meat Sci.*, 1999, 53, 183-188.

Barber, S; Baguena, R; Martinez-Anaya, MA; Torner, MJ. Microflora de la masa madre panaria. I. Identificacion y propiedades funcionales de microorganismos de masas madre industriales, elaboradas con harina de trigo. *Rev. Agr. Tecnol. Alim.*, 1983, 23, 552-562.

Barbieri, G; Bergamini, C; Ori, E; Reska, P. Aflatoxin M_1 in Parmeasan cheese. HPLC determination. *J. Food Sci.*, 1994, 59, 1313-1331.

Barrangou, R; Yoon, S-S; Breidt Jr, F; Fleming, HP; Klaenhammer, TR. Identification and characterization of *Leuconostoc fallax* strains isolated from an industrial sauerkraut fermentation. *Appl. Environm. Microbiol.*, 2002, 68, 2877-2884.

Barrios, MJ; Gualda, MJ; Cabanas, JM; Medina, LM; Jordano, R. Occurrence of aflatoxin M_1 in cheeses from the South of Spain. *J. Food Prot.*, 1996, 59, 898-900.

Baruzzi, F; Matarante, A; Caputo, L; Morea, M. Molecular and physiological characterization of natural microbial communities isolated from a traditional Southern Italian processed sausage. *Meat Sci.*, 2006, 72, 261-269.

Bata, A; Lasztity, R. Detoxification of mycotoxin-contaminated food and feed by microorganisms. *Trends Food Sci. Technol.*, 1999, 10, 223-228.

Bayarri, S; Conchello, P; Arino, AA; Lazaro, R; Herrera, A. Evaluation of an analytical method for an in-vitro study of degradation of organochlorine compounds by meat starter microorganisms. *Pest. Sci.*, 1997, 50, 120-126.

Bayarri, S; Herrera, A; Conchello, MP; Arino, AA; Lazaro, R; Yague, C. Influence of meat processing and meat starter microorganisms on the degradation of organochlorine contaminants. *J. Agric. Food Chem.*, 1998, 46, 3187-3193.

Bernardeau, M; Guguen, M; Vernoux, JP. Beneficial lactobacilli in food and feed: long-term use, biodiversity and proposals for specific and realistic safety assessments. *FEMS Microbiol. Rev.*, 2006, 30, 487-513.

Bertazzoni Minelli, E; Benini, A; Marzotto, M; Sbarbati, A; Ruzzenente, O; Ferrario, R; Hendriks, H; Dellaglio, F. Assessment of novel probiotic *Lactobacillus casei* strains for the production of functional dairy foods. *Int. Dairy J.*, 2004, 14, 723-736.

Betoret, N; Puente, L; Dıaz, MJ; Pagan, MJ; Garcıa, MJ; Gras, ML; Martınez-Monzo, J; Fito, P. Development of probiotic-enriched dried fruits by vacuum impregnation. *J. Food Eng.*, 2003, 56, 273-277.

Birt, DF; Hendrich, S; Wang, WQ. Dietary agents in cancer prevention: flavonoids and isoflavonoids. *Pharmacol. Ther.*, 2001, 90, 157-177.

Blanco, JL; Carrion, BA; Liria, N; Diaz, S; Garcia, ME; Dominguez, L; Suarez, G. Behavior of aflatoxins during manufacture and storage of yoghurt. *Milchwissenchaft*, 1993, 48, 385-387.

Blekas, G; Vassilakis, C; Harizanis, C; Tsimidou, M; Boskou, DG. Biophenols in table olives. *J. Agric. Food Chem.*, 2002, 50, 3688-3699.

Bloukas, JG; Paneras, ED; Fournitzis, GC. Effect of replacing pork backfat with olive oil on processing and quality characteristics of fermented sausages. *Meat Sci.*, 1997, 45, 133-144.

Boskov-Hansen, H; Andersen, MF; Nielsen, LM; Bäck-Knudsen, K-E; Meyer, AS; Christensen, LP; Hansen, A. Changes in dietary fibre, phenolic acids and activity of endogenous enzymes during rye bread making. *Eur. Food Res. Technol.*, 2002, 214, 33-42.

Bosque, PJ; Ryou, C; Telling, G; Peretz, D; Legname, G; DeArmond, SJ; Prusiner, SB. Prions in skeletal muscle. *Proc. Natl. Acad. Sci. USA*, 2002, 99, 3812-3817.

Bover-Cid, S; Hugas, M; Izquierdo-Pulido, M; Vidal-Carou, MC. Amino acid-decarboxylase activity of bacteria isolated from fermented pork sausages. *Int. J. Food Microbiol.*, 2001a, 66, 185-189.

Bover-Cid, S; Izquierdo-Pulido, M; Vidal-Carou, MC. Effect of proteolytic starter cultures of *Staphylococcus* spp. on biogenic amine formation during the ripening of dry fermented sausages. *Int. J. Food Microbiol.*, 1999a, 46, 95-104.

Bover-Cid, S; Miguelez-Arrizado, MJ; Vidal-Carou, MC. Biogenic amine accumulation in ripened sausages affected by the addition of sodium sulphite. *Meat Sci.*, 2001b, 59, 391-396.

Bover-Cid, S; Schoppen, S; Izquierdo-Pulido, M; Vidal-Carou, MC. Relationship between biogenic amine contents and the size of dry fermented sausages. *Meat Sci.*, 1999b, 51, 305-311.

Brackett, RE; Marth, EH. Fate of aflatoxin M_1 in Cheddar cheese and in process spread. *J. Food Prot.*, 1982a, 45, 549-552.

Brackett, RE; Marth, EH. Fate of aflatoxin M_1 in Parmesan and Mozzarella cheese. *J. Food Prot.*, 1982b, 45, 597-600.

Brattstrom, L. Vitamins as homocysteine-lowering agents. *J. Nutr.*, 1996, 126, 1276S-1280S.

Bryant, A; Ustunol, Z; Stefe, J. Texture of Cheddar cheese as influenced by fat reduction. *J. Food Sci.*, 1995, 60, 1216-1219.

Buyong, N; Kok, J; Luchansky, JB. Use of a genetically enhanced, pediocin-producing starter culture, *Lactococcus lactis* subsp. *lactis* MM217, to control *Listeria monocytogenes* in Cheddar cheese. *Appl. Environm. Microbiol.*, 1998, 64, 4842-4845.

Caggia, C; Randazzo, CL; Di Salvo, M; Romeo, F; Giudici, P. Occurrence of *Listeria monocytogenes* in green table olives. *J. Food Prot.*, 2004, 67, 2189-2194.

Canas, BJ; Havery, DC; Robinson, LR; Sullivan, MP; Joe, FL; Diachenko, GW. Ethyl carbamate levels in selected fermented foods and beverages. *J. AOAC*, 1989, 72, 873-876.

Cannon, JP; Lee, TA; Bolanos, JT; Danziger, LH. Pathogenic relevance of *Lactobacillus*: a retrospective review of over 200 cases. *Eur. J. Clin. Microbiol Infect. Dis.*, 2005, 24, 289-294.

Carruth, BR; Skinner, JD. The role of dietary calcium and other nutrients in moderating body fat in preschool children. *Int. J. Obesity*, 2001, 25, 559-566.

Cebeci, A; Gurakan, C. Properties of potential probiotic *Lactobacillus plantarum* strains. *Food Microbiol.*, 20, 2003, 511-518.

Champomier-Verges, M-C; Chaillou, S; Cornet, M; Zagorec, M. *Lactobacillus sakei*: recent developments and future prospects. *Res. Microbiol.*, 2001, 152, 839-848.

Chander, H; Batish, VH; Babu, S; Singh, RS. Factors affecting amine production by a selected strain of *Lactobacillus bulgaricus*. *J. Food Sci.*, 1989, 54, 940-942.

Charalampopoulos, D; Pandiella, SS; Webb, C. Growth studies of potentially probiotic lactic acid bacteria in cereal-based substrates. *J. Appl. Microbiol.*, 2002, 92, 851-859.

Chen, C-C; Walker, A. Probiotics and prebiotics: Role in clinical disease states. *Adv. Pediatrics*, 2005, 52, 77-113.

Choi, A; Yeon, J; Kim, M; Lee, J-H. Reevaluation of the change of *Leuconostoc* species and *Lactobacillus plantarum* by PCR during Kimchi fermentation. *J. Microbiol. Biotechnol.*, 2002a, 12, 166-171.

Choi, H-J; Cheigh, C-I; Kim, S-B; Lee, J-C; Lee, D-W; Choi, S-W; Park, J-M; Pyun, Y-R. *Weissella kimchii* sp. nov., a novel lactic acid bacterium from kimchi. *Int. J. System. Evol. Microbiol.*, 2002b, 52, 507-511.

Choi, I-K; Jung, S-H; Kim, B-J; Park, S-Y; Kim, J; Han, H-U. Novel *Leuconostoc citreum* starter culture system for the fermentation of kimchi, a fermented cabbage product. *Anton. van Leeuwen.*, 2003, 84, 247-253.

Choi, S-Y; Beuchat, LR; Perkins, LM; Nakayama, T. Fermentation and sensory characteristics of kimchi containing potassium chloride as a partial replacement for sodium chloride. *Int. J. Food Microbiol.*, 1994, 21, 335-340.

Choi, YM; Whang, JH; Kim, JM; Suh, HJ. The effect of oyster shell powder on the extension of the shelf-life of Kimchi. *Food Control*, 2006, 17, 695-699.

Chorianopoulos, NG; Boziaris, IS; Stamatiou, A; Nychas, G-JE. Microbial association and acidity development of unheated and pasteurized green-table olives fermented using glucose or sucrose supplements at various levels. *Food Microbiol.*, 2005, 22, 117-124.

Comi, G; Urso, R; Iacumin, L; Rantsiou, K; Cattaneo, P; Cantoni, C; Cocolin, L. Characterisation of naturally fermented sausages produced in the North East of Italy. *Meat Sci.*, 69, 2005, 381-392.

Commane, D; Hughes, R; Shortt, C; Rowland I. The potential mechanisms involved in the anti-carcinogenic action of probiotics. *Mutation Res.*, 2005, 591, 276-289.

Coppola, R; Giagnacovo, B; Iorizzo, M; Grazia L. Characterization of lactobacilli involved in the ripening of soppressata molisana, a typical southern Italy fermented sausage. *Food Microbiol.*, 1998, 15, 347-353.

Corsetti, A; Gobbetti, M; Balestrieri, F; Paoletti, F; Russi, L; Rossi, J. Sourdough lactic acid bacteria effects on bread firmness and staling. *J. Food Sci.*, 1998a, 63, 347-351.

Corsetti, A; Gobbetti, M; Rossi, J; Damiani, P. Antimould activity of sourdough lactic acid bacteria, identification of a mixture of organic acids produced by *Lactobacillus sanfrancisco* CB1. *Appl. Microbiol. Biotechnol.*, 1998b, 50, 253-256.

Corsetti, A; Gobbetti, M; Smacchi, E. Antimicrobial activity of sourdough lactic acid bacteria: isolation of a bacteriocin like inhibitory substance from *Lactobacillus sanfrancisco* C57. *Food Microbiol.*, 1996, 13, 447-456.

Corsetti, A; Lavermicocca, P; Morea, M; Baruzzi, F; Tosti, N; Gobbetti, M. Phenotypic and molecular identification and clustering of lactic acid bacteria and yeasts from wheat (species *Triticum durum* and *Triticum aestivum*) sourdoughs of southern Italy. *Int. J. Food Microbiol.*, 2001, 64, 95-104.

Corsetti, A; Settanni, L; van Sinderen, D; Felis, GE; Dellaglio, F; Gobbetti, M. *Lactobacillus rossii* sp. nov., isolated from wheat sourdough. *Int. J. System Evolut. Microbiol.*, 2005, 55, 35-40.

Crittenden, RG; Martinez, NR; Playne, MJ. Synthesis and utilisation of folate by yoghurt starter cultures and probiotic bacteria. *Int. J. Food Microbiol.*, 2003, 80, 217-222.

Cummings, JH; Macfarlane, GR; Englyst, HN. Prebiotic digestion and fermentation. *Am. J. Clin. Nutr.*, 2001, 73, 415S-420S.

Daeschel, MA; Andersson, RE; Fleming, HP. Microbial ecology of fermenting plant materials. *FEMS Microbiol. Rev.*, 1987, 46, 357-367.

Daeschel, MA; Fleming, HP; McFeeters, RF. Mixed culture fermentation of cucumber juice with *Lactobacillus plantarum* and yeast. *J. Food Sci.*, 1988, 53, 862-864.

Davies, KM; Heaney, RP; Recker, RR; Lappe, JM; Barger-Lux, MJ; Rafferty, K; Hinders, S. Calcium intake and body weight. *J. Clin. Endocrinol. Metab.*, 2000, 85, 4635-4638.

De Castro, A; Montano, A; Casado, F-J; Sanchez, A-H; Rejano, L. Utilization of *Enterococcus casseliflavus* and *Lactobacillus pentosus* as starter cultures for Spanish-style green olive fermentation. *Food Microbiol.*, 2002, 19, 637-644.

De Castro, A; Montano, A; Sanchez, AH; Rejano, L. Lactic acid fermentation and storage of blanched garlic. *Int. J. Food Microbiol.*, 1998, 39, 205-211.

De Mejia, EG; Prisecaru, VI. Lectins as bioactive plant proteins: a potential in cancer treatment. *Crit. Rev. Food Sci. Nutr.*, 2005, 45, 425-445.

De Vries, MC; Vaughan, EE; Kleerebezem, M; de Vos, WM. *Lactobacillus plantarum*-survival, functional and potential probiotic properties in the human intestinal tract. *Int. Dairy J.*, 2006, 16, 1018-1028.

De Vuyst, L; Callewaert, R; Pot, B. Characterization of the antagonistic activity of *Lactobacillus amylovorus* DCE 471 and large-scale isolation of its bacteriocin amylovorin L471. *System. Appl. Microbiol.*, 1996, 19, 9-20.

Decock, P; Capelle, S. Bread technology and sourdough technology. *Trends Food Sci. Technol.*, 2005, 16, 113-120.

Dewit, O; Pochart, P; Desjeux, JF. Breath hydrogen concentration and plasma glucose, insulin and free fatty acid levels after lactose, milk, fresh or heated yoghurt ingestion by healthy young adults with or without lactose malabsorption. *Nutrition*, 1988, 4, 131-135.

Di Cagno, R; De Angelis, M; Auricchio, S; Greco, L; Clarke, C; De Vincenzi, M; Giovannini, C; D'Archivio, M; Landolfo, F; Parrilli, G; Minervini, F; Arendt, E; Gobbetti, M. Sourdough bread made from wheat and non-toxic flours and started with selected lactobacilli is tolerated in celiac sprue patients. *Appl. Environm. Microbiol.*, 2004, 70, 1088-1096.

Di Cagno, R; De Angelis, M; Lavermicocca, P; De Vincenzi, M; Giovannini, C; Faccia, M; Gobbetti, M. Proteolysis by sourdough lactic acid bacteria: effects on wheat flour protein fractions and gliadin peptides involved in human cereal intolerance. *Appl. Environ. Microbiol.*, 2002, 68, 623-633.

Dollear, FG; Mann, GE; Codifer, LP; Gardner, JK; Koltun, SP; Vix, HLE. Elimination of aflatoxins from peanut meal. *J. Am. Oil Chem. Soc.*, 1968, 45, 862-865.

Dormont, D; Prions, BSE and food. *Int. J. Food Microbiol.*, 2002, 78, 181-189.

Drosinos, EH; Mataragas, M; Veskovic-Moracanin, S; Gasparik-Reichardt, J; Hadziosmanovic, M; Alagic, D. Quantifying nonthermal inactivation of *Listeria monocytogenes* in European fermented sausages using bacteriocinogenic lactic acid bacteria or their bacteriocins: a case study for risk assessment. *J. Food Prot.*, 2006, 69, 2648-2663.

Drosinos, EH; Mataragas, M; Xiraphi, N; Moschonas, G; Gaitis, F; Metaxopoulos, J. Characterization of the microbial flora from a traditional Greek fermented sausage. *Meat Sci.*, 2005, 69, 307-317.

Drosinos, EH; Paramithiotis, S; Kolovos, G; Tsikouras, I; Metaxopoulos, I. Phenotypic and technological diversity of lactic acid bacteria and staphylococci isolated from traditionally fermented sausages in Southern Greece. *Food Microbiol.*, 2007, 24, 260-270.

Duran Quintana, MC; Garcya Garcya, P; Garrido Fernandez, A. Establishment of conditions for green table olive fermentation at low temperature. *Int. J. Food Microbiol.*, 1999, 51, 133-143.

Eka, OU. Effect of fermentation on the nutrient status of locust beans. *Food Chem.*, 1980, 5, 303-308.

El Soda, M; Korayem, M; Ezzat, N. The esterolytic and lipolytic activities of lactobacilli. III: Detection and characterisation of the lipase system. *Milchwissenschaft*, 1986, 41, 353-355.

El-Nezami, H; Haskard, C; Salminen, E; Mykkanen, H; Ahokas, J; Salminen, S. Lactic acid bacteria and bifidobacteria can reduce dietary exposure to aflatoxins. *Brit. J. Nutr.*, 2002a, 88, S119-S121.

El-Nezami, H; Kankaanpaa, P; Salminen, S; Ahokas, J. Ability of dairy strains of lactic acid bacteria to bind a common food carcinogen, aflatoxin B_1. *Food Chem. Toxicol.,* 1998, 36, 321-326.

El-Nezami, H; Polychronaki, N; Salminen, S; Mykkanen, H. Binding rather than metabolism may explain the interaction of two food-grade *Lactobacillus* strains with zearalenone and its derivative α-zearalenol. *Appl. Environm. Microbiol.,* 2002b, 68, 3545-3549.

Ennahar, S; Assobhei, O; Hasselmann, C. Inhibition of *Listeria monocytogenes* in a smear-surface soft cheese by *Lactobacillus plantarum* WHE92. *J. Food Prot.,* 1998, 61, 186-191.

Erkkila, S; Petaja, E. Screening of commercial meat starter cultures at low pH and in the presence of bile salts for potential probiotic use. *Meat Sci.*, 2000, 55, 297-300.

Erkkila, S; Suihko, M-L; Eerola, S; Petaja, E; Mattila-Sandholm, T. Dry sausage fermented by *Lactobacillus rhamnosus* strains. *Int. J. Food Microbiol.*, 2001, 64, 205-210.

Evans, DA; Hirsch, JB; Dushenkov, S. Phenolics, inflammation and nutrigenomics. *J. Sci. Food Agric.*, 2006, 86, 2503-2509.

Fadda, S; Sanz, Y; Vignolo, G; Aristoy, M-C; Oliver, G; Toldra, F. Characterization of muscle sarcoplasmic and myofibrillar protein hydrolysis caused by *Lactobacillus plantarum*. *Appl. Environ. Microbiol.*, 1999, 65, 3540-3546.

Fadda, S; Vignolo, G; Holgado, APR; Oliver, G. Proteolytic activity of *Lactobacillus* strains isolated from dry-fermented sausages on muscle sarcoplasmic proteins. *Meat Sci.*, 1998, 49, 1-18.

Faller, J; Huhnerfuss, H; Konig, WA; Krebber, R; Ludwig, P. Do marine bacteria degrade hexachlorocyclohexane stereoselectively? *Environm. Sci. Technol.*, 1991, 25, 676-678.

Farag, RS; Rashed, MM; Abo-Hagger, AA. Aflatoxin destruction by microwave heating. *Int. J. Food Sci. Nutr.*, 1996, 47, 197-208.

Fernandez, AG; Garcia, PG; Balbuena, MB. Olive fermentations. In: Reed G, Nagodawithana TW editors. *Biotechnology second, completely revised edition, vol. 9, Enzymes, biomass, food and feed.* Weinheim: VCH; 1995; 593-628.

Fernandez, M; Ordonez, JA; Bruna, JM; Herranz, B; de la Hoz, L. Accelerated ripening of dry fermented sausages. *Trends Food Sci. Technol.*, 2000, 11, 201-209.

Fernandez-Garcia, E; Tomillo, J; Nunez, M. Formation of biogenic amines in raw milk Hispanico cheese manufactured with proteinases and different levels of starter culture. *J. Food Prot.*, 2000, 63, 1551-1555.

Fleming H.P., Kyung K.H., Breidt F. 1995a. Vegetable fermentations. In: Reed G, Nagodawithana TW editors. *Biotechnology second, completely revised edition, vol. 9, Enzymes, biomass, food and feed.* Weinheim: VCH; 1995a; 629-662.

Fleming, HP; McDonald, LC; McFeeters, RF; Thompson, RL; Humphries, EG. Fermentation of cucumbers without sodium chloride. *J. Food Sci.*, 1995b, 60, 312-315.

Fontana, C; Cocconcelli, PS; Vignolo, G. Monitoring the bacterial population dynamics during fermentation of artisanal Argentinean sausages. *Int. J. Food Microbiol.*, 2005b, 103, 131-142.

Fontana, C; Vignolo, G; Cocconcelli, PS. PCR–DGGE analysis for the identification of microbial populations from Argentinean dry fermented sausages. *J. Microbiol. Meth.*, 2005a, 63, 254-263.

Forssen, KM; Jagerstad, MI; Wigertz, K; Witthoeft, CM. Folates and dairy products: a critical update. *J. Am. Coll Nutr.*, 2000, 19, 100S-110S.

Foulquie Moreno, MR; Rea, M; Cogan, T; De Vuyst L. Applicability of a bacteriocin-producing *Enterococcus faecium* as a co-culture in Cheddar cheese manufacture. *Int. J. Food Microbiol.*, 2003, 81, 73-84.

Gaenzle, MG. Reutericyclin: biological activity, mode of action, and potential applications. *Appl. Microbiol. Biotechnol.*, 2004, 64, 326-332.

Gaenzle, MG; Hoeltzel, A; Walter, J; Jung, G; Hammes, WP. Characterization of reutericyclin produced by *Lactobacillus reuteri* LTH2584. *Appl. Environm. Microbiol.*, 2000, 66, 4325-4333.

Gallagher, CR; Mollesson, AL; Caldwell, JH. Lactose intolerance and fermented dairy products. *J. Am. Diet Assoc.*, 1974, 65, 418-419.

Garcia Fontan, MC; Lorenzo, JM; Parada, A; Franco, I; Carballo, J. Microbiological characteristics of androlla, a Spanish traditional.pork sausage. *Food Microbiol.*, 2007, 24, 52-58.

Garcia, ML; Domınguez, R; Galvez, MD; Casas, C; Selgas, MD. Utilization of cereal and fruit fibres in low fat dry fermented sausages. *Meat Sci.*, 2002, 60, 227-236.

Garcia, PG; Barranco, CR; Quintana, MCD; Fernandez, AG. Biogenic amine formation and zapatera spoilage of fermented green olives: effect of storage temperature and debittering process. *J. Food Prot.*, 2004, 67, 117-123.

Garcia, PG; Brenes-Balbuena, M; Hornero-Mendez, D; Garcia-Borrego, A; Fernandez, AG. Content of biogenic amines in table olives. *J. Food Prot.*, 2000, 63, 111-116.

Garcia-Varona, M; Santos, EM; Jaime, I; Rovira, J. Characterisation of *Micrococcaceae* isolated from different varieties of chorizo. *Int. J. Food Microbiol.*, 2000, 54, 189-195.

Gardiner, GE; Bouchier, P; O'Sullivan, E; Kelly, J; Collins, JK; Fitzgerald, G; Ross, RP; Stanton, C. A spray-dried culture for probiotic Cheddar cheese manufacture. *Int. Dairy J.*, 2002, 12, 749-756.

Gardini, F; Martuscelli, M; Caruso, MC; Galgano, F; Crudele, MA; Favati, F; Guerzoni, ME; Suzzi, G. Effects of pH, temperature and NaCl concentration on the growth kinetics, proteolytic activity and biogenic amine production of *Enterococcus faecalis*. *Int. J. Food Microbiol.*, 2001b, 64, 105-117.

Gardini, F; Martuscelli, M; Crudele, MA; Paparella, A;, Suzzi, G. Use of *Staphylococcus xylosus* as a starter culture in dried sausages: effect on the biogenic amine content. *Meat Sci.*, 2002, 61, 275-283.

Gardini, F; Suzzi, G; Lombardi, A; Galgano, F; Crudele, MA; Andrighetto, C; Schirone, M; Tofalo, R. A survey of yeasts in traditional sausages of southern Italy. *FEMS Yeast Res.*, 2001a, 1, 161-167.

Gardner, NJ; Savard, T; Obermeier, P; Caldwell, G; Champagne, CP. Selection and characterization of mixed starter cultures for lactic acid fermentation of carrot, cabbage, beet and onion vegetable mixtures. *Int. J. Food Microbiol.*, 2001, 64, 261-275.

Gashe, BA. Kocho fermentation. *J. Appl. Bacteriol.*, 1987, 62, 473-477.

Geciova, J; Bury, D; Jelen, P. Methods for disruption of microbial cells for potential use in the dairy industry, a review. *Int. Dairy J.*, 2002, 12, 541-553.

George, L; Mills, JL; Johansson, ALV; Nordmark, A; Olander, B; Granath, F; Cnattingius, S. Plasma folate levels and risk of spontaneous abortion. *J. Am. Med. Assoc.*, 2002, 288, 1867-1873.

Gilliland, SE; Kim, H. Effect of viable starter culture bacteria in yoghurt on lactose utilization in humans. *J. Dairy Sci.*, 1984, 67, 1-6.

Gilliland, SE; Nelson, CR; Maxwell, C. Assimilation of cholesterol by *Lactobacillus acidophilus*. *Appl. Environm. Microbiol.*, 1985, 49, 377-385.

Gimeno, O; Astiasaran, I; Bello, J. A mixture of potassium, magnesium and calcium chlorides as a partial replacement of sodium chloride in dry fermented sausages. *J. Agric. Food Chem.*, 1998, 46, 4372-4375.

Gimeno, O; Astiasaran, I; Bello, J. Calcium ascorbate as a potential partial substitute for NaCl in dry fermented sausages: effect on colour, texture and hygienic quality at different concentrations. *Meat Sci.*, 2001, 57, 23-29.

Gimeno, O; Astiasaran, I; Bello, J. Influence of partial replacement of NaCl with KCl and $CaCl_2$ on texture and colour of dry fermented sausages. *J. Agric. Food Chem.*, 1999, 47, 873-877.

Giovannini, C; Straface, E; Modesti, D; Coni, E; Cantafora, A; de Vincenci, M; Malorni, W; Masella, R. Tyrosol, the major olive oil biophenol, protects against oxidized-LDL-induced injury in caco-2-cells. *J. Nutr.*, 1999, 129, 1269-1277.

Giraffa, G; Carminati, D. Control of *Listeria monocytogenes* in the rind of Taleggio, a surface-smear cheese, by a bacteriocin from *Enterococcus faecalis* 7C5. *Sci. Aliments*, 1997, 17, 383-391.

Gobbetti, M; Corsetti, A; Rossi, J. Maltose-fructose co-fermentation by *Lactobacillus brevis* subsp. *lindneri* CB1 fructose-negative strain. *Appl. Microbiol. Biotechnol.*, 1995, 42, 939-944.

Gobbetti, M; Corsetti, A; Rossi, J; La Rosa, F; De Vincenzi, S. Identification and clustering of lactic acid bacteria and yeasts from wheat sourdoughs of central Italy. *Int. J. Food Sci.*, 1994, 6, 85-94.

Gomes, AM; Malcata, FX. Development of probiotic cheese manufactured from goat milk: response surface analysis via technological manipulation. *J. Dairy Sci.*, 1998, 81, 1492-1507.

Gonzalez-Fernandez, C; Santos, EM; Jaime, I; Rovira, J. Influence of starter cultures and sugar concentrations on biogenic amine contents in chorizo dry sausage. *Food Microbiol.*, 2003, 20, 275-284.

Gordon, MH; Paiva-Martins, F; Almeida, M. Antioxidant activity of hydroxytyrosol acetate compared with that of other olive oil polyphenols. *J. Agric. Food Chem.*, 2001, 49, 2480-2485.

Gou, P; Guerrero, L; Gelabert, J; Arnau, J. Potassium chloride, potassium lactate and glycine as sodium chloride substitutes in fermented sausages and in dry-cured pork loin. *Meat Sci.*, 1996, 42, 37-48.

Govaris, A; Roussi, V; Koidis, PA; Botsoglou, NA. Distribution and stability of aflatoxin M_1 during processing, ripening and storage of telemes cheese. *Food Addit. Contam.*, 2001, 18, 437-443.

Govaris, A; Roussi, V; Koidis, PA; Botsoglou, NA. Distribution and stability of aflatoxin M_1 during production and storage of yoghurt. *Food Addit. Contam.*, 2002, 19, 1043-1050.

Greco, M; Mazzette, R; De Santis, EPL; Corona, A; Cosseddu, AM. Evolution and identification of lactic acid bacteria isolated during the ripening of Sardinian sausages. *Meat Sci.*, 2005, 69, 733-739.

Gregory, JF. Case study: Folate bioavailability. *J. Nutr.*, 2001, 131, 1376S-1382S.

Guandalini, S; Pensabene, L; Zikri, MA; Dias, JA; Casali, LG; Hoekstra, H; Kolacek, S; Massar, K; Micetic-Turk, D; Papadopoulou, A; de Sousa, JS; Sandhu, B; Szajewska, H; Weizman, Z. *Lactobacillus* GG administered in oral rehydration solution to children with acute diarrhea: a multicenter European trial. *J. Pediatric Gastroenterol. Nutr.*, 2000, 30, 54-60.

Hagler, WM; Hutchins, JE; Hamilton, PB. Destruction of aflatoxin in corn with sodium bisulfite. *J. Food Prot.*, 1982, 45, 1287-1291.

Haider, K. Degradation and metabolization of lindane and other hexachlorocyclohexane isomers by anaerobic and aerobic soil microorganisms. *Z. Naturforschung Sect.*, 1979, 34, 1066-1069.

Haider, K; Jagnow, R; Kohnen, R; Lim, SM. Degradation of chlorinated benzenes, phenols and cyclohexane derivatives by benzene and phenol utilizing soil bacteria under aerobic conditions. *Arch. Microbiol.*, 1974, 96, 183-200.

Halasz, A; Barath, A; Holzapfel, WH. The influence of starter culture selection on sauerkraut fermentation. *Z. Lebensm. Unters. Forsch. A*, 1999, 208, 434-438.

Halasz, A; Barath, A; Simon-Sarkadi, L; Holzapfel, W. Biogenic amines and their production by microorganisms in food. *Trends Food Sci. Technol.*, 1994, 5, 42-49.

Halliwell, B. Dietary polyphenols: Good, bad, or indifferent for your health? *Cardiovasc. Res.*, 2007, 73, 341-347.

Hammes, WP. Fermentation of non-dairy foods. *Food Biotechnol.*, 1991, 5, 293-303.

Hammes, WP; Ganzle, MG. Sourdough breads and related products. In: Woods BJB, editor. *Microbiology of Fermented Foods.* London: Blackie Academic and Professional, 1998; 199-216.

Harris, LJ. The microbiology of vegetable fermentations. In: Wood, BJB editor. *Microbiology of Fermented Foods.* London: Blackie Academic and Professional; 1998; 45-72.

Harris, LJ; Fleming, HP; Klaenhammer, TR. Characterisation of two nisin-producing *Lactococcus lactis* subsp. *lactis* strains isolated from a commercial sauerkraut fermentation. *Appl. Environm. Microbiol.*, 1992, 58, 1484-1489.

Harty, DW; Oakey, HJ; Patrikakis, M; Hume, EB; Knox, KW. Pathogenic potential of lactobacilli. *Int. J. Food Microbiol.*, 1994, 24, 179-189.

Heller, KJ. Probiotic bacteria in fermented foods: product characteristics and starter organisms. *Am. J. Clin. Nutr.*, 2001, 73, 374S-379S.

Hilton, E; Kolakowski, P; Singer, C; Smith, M. Efficacy of *Lactobacillus* GG as a diarrheal preventive in travellers. *J. Travel. Med.*, 1997, 4, 41-43.

Hong, S-I; Park, W-S. Use of color indicators as an active packaging system for evaluating kimchi fermentation. *J. Food Engin.*, 2000, 46, 67-72.

Hoppner, K; Lampi, B. Total folate, pantothenic acid and biotin content of yogurt products. *Can. Inst. Food Sci .Technol.*, 1990, 23, 223-225.

Hugas, M; Garriga, M; Aymerich, T; Monfort, JM. Biochemical characterization of lactobacilli isolated from dry fermented sausages. *Int. J. Food Microbiol.*, 1993, 18, 107-113.

Ibanez, C; Quintanilla, L; Astiasaran, I; Bello, J. Dry fermented sausages elaborated with *Lactobacillus plantarum - Staphylococcus carnosus*. Part II. Effect of partial replacement of NaCl with KCl on the proteolytic and insolubilization processes. *Meat Sci.*, 1997, 46, 277-284.

Ibanez, C; Quintanilla, L; Cid, C; Astiasaran, I; Bello, J. Part I. Effect of partial replacement of NaCl with KCl on the stability and the nitrosation process. *Meat Sci.*, 1996, 44, 227-234.

Ibanez, C; Quintanilla, L; Irigoyen, A; Garcıa-Jalon, I; Cid, C; Astiasaran, I; Bello, J. Partial replacement of sodium chloride with potassium chloride in dry fermented sausages: influence on carbohydrate fermentation and the nitrosation process. *Meat Sci.*, 1995, 40, 45-53.

Infantes, M; Tourneur, C. Etude de la flore lactique de levains naturels de panification provenant de differentes regions francaises. *Sci. Alim.*, 1991, 11, 527-545.

Innocente, N; D'Agostin, P. Formation of biogenic amines in a typical semihard Italian cheese. *J. Food Prot.*, 2002, 65, 1498-1501.

Isolauri, E; Salminen, S; Ouwehand, AC. Probiotics. *Best Pract. Res. Clin. Gastroenterol.*, 2004, 18, 299-313.

Ivanova, I; Miteva, V; Stefanova, Ts; Pantev, A; Budakov, I; Danova, S; Moncheva, P; Nikolova, I; Dousset, X; Boyaval, P. Characterization of a bacteriocin produced by *Streptococcus thermophilus* 81. *Int. J. Food Microbiol.*, 1998, 42, 147-158.

Jacqmain, M; Doucet, E; Despres, J-P; Bouchard, C; Tremblay, A. Calcium intake, body composition, and lipoprotein-lipid concentrations in adults. *Am. J. Clin. Nutr.*, 2003, 77, 1448-1452.

Jaegerstad, M; Jastrebova, J; Svensson, U. Folates in fermented vegetables-a pilot study. *Lebensm.-Wiss. u.-Technol.*, 2004, 37, 603-611.

Kailasapathy, K. Microencapsulation of probiotic bacteria: technology and potential applications. *Curr. Issues Intestinal Microbiol.*, 2002, 3, 39-48.

Kailasapathy, K. Survival of free and encapsulated probiotic bacteria and their effect on the sensory properties of yoghurt. *Lebensm.-Wiss. u.-Technol.*, 2006, 39, 1221-1227.

Kalac, P; Spicka, J; Krizek, M; Pelikanova, T. The effects of lactic acid bacteria inoculants on biogenic amines formation in sauerkraut. *Food Chem.*, 2000, 70, 355-359.

Kalac, P; Spicka, J; Krizek, M; Steidlova, S; Pelikanova, T. Concentrations of seven biogenic amines in sauerkraut. *Food Chem.*, 1999, 67, 275-280.

Kariluoto, S; Aittamaa, M; Korhola, M; Salovaara, H; Vahteristo, L; Piironen, V. Effects of yeasts and bacteria on the levels of folates in rye sourdoughs. *Int. J. Food Microbiol.*, 2006, 106, 137-143.

Kariluoto, S; Vahteristo, L; Salovaara, H; Katina, K; Liukkonen, K-H; Piironen, V. Effect of baking method and fermentation on folate content of rye and wheat breads. *Cereal Chem.*, 2004, 81, 134-139.

Karlovsky, P. Biological detoxification of fungal toxins and its use in plant breeding, feed and food production. *Nat. Toxins*, 1999, 7, 1-23.

Karovicovα, J; Kohajdovα, Z. Lactic acid fermented vegetable juices. *Hort Sci.*, 2003, 30, 152-158.

Katina, K; Arendt, E; Liukkonen, K-H; Autio, K; Flander, L; Poutanen K. Potential of sourdough for healthier cereal products. *Trends Food Sci. Technol.*, 2005, 16, 104-112.

Katsiari, MC; Voutsinas, LP. Manufacture of low-fat Feta cheese. *Food Chem.*, 1994, 49, 53-60.

Kazanskaya, LN; Afanasyeva, OV; Patt, VA. Microflora of rye sours and some specific features of its accumulation in bread baking plants of the USSR. In: Holas J, Kratochvil F editors. *Developments in food science. Progress in cereal chemistry and technology.* London: Elsevier; 1983; 759-763.

Kedia, G; Wang, R; Patel, H; Pandiella, SS. Use of mixed cultures for the fermentation of cereal-based substrates with potential probiotic properties. *Process Biochem.*, 2007, 42, 65-70.

Kenneally, PM; Leuschner, RG; Arendt, EK. Evaluation of the lipolytic activity of starter cultures for meat fermentation purposes. *J. Appl. Microbiol.*, 1998, 84, 839-846.

Kim, B; Lee, J; Jang, J; Kim, J; Han, H. *Leuconostoc inhae* sp. nov., a lactic acid bacterium isolated from kimchi. *Int. J. System Evol. Microbiol.*, 2003a, 53, 1123-1126.

Kim, D-H; Song, H-P; Yook, H-S; Ryu, Y-G; Byun, M-W. Isolation of enteric pathogens in the fermentation process of Kimchi (Korean fermented vegetables) and its radicidation by gamma irradiation. *Food Control*, 2004, 15, 441-445.

Kim, EK; Shon, DH; Ryu, D; Park, JW; Hwang, HJ; Kim, YB. Occurrence of aflatoxin M_1 in Korean dairy products determined by ELISA and HPLC. *Food Addit. Contam.*, 2000a, 17, 59-64.

Kim, HJ; Camilleri, M; McKinzie, S; Lempke, MB; Burton, DD; Thomforde, GM; Zinsmeister, AR. A randomized controlled trial of a probiotic, VSL#3, on gut transit and symptoms in diarrhoeapredominant irritable bowel syndrome. *Aliment. Pharmac. Ther.*, 2003b, 17, 895-904.

Kim, HS; Gilliland, S. *Lactobacillus acidophilus* as a dietary adjunct for milk to aid lactose digestion in humans. *J. Dairy Sci.*, 1983, 66, 959-966.

Kim, J; Chun, J; Han, H-U. *Leuconostoc kimchii* sp. nov., a new species from kimchi. *Int. J. System Evol. Microbiol.*, 2000b, 50, 1915-1919.

Kim, M; Chun, J. Bacterial community structure in kimchi, a Korean fermented vegetable food, as revealed by 16S rRNA gene analysis. *Int. J. Food Microbiol.*, 2005, 103, 91-96.

Kim, N; Park, K-R; Park, I-S; Cho, Y-J; Bae, YM. Application of a taste evaluation system to the monitoring of Kimchi fermentation. *Biosens. Bioelectron.*, 2005, 20, 2283-2291.

Kim, T-W; Lee, J-Y; Jung, S-H; Kim, Y-M; Jo, J-S; Chung, D-K; Lee, H-J; Kim, H-Y. Identification and distribution of predominant lactic acid bacteria in Kimchi, a Korean traditional fermented food. *J. Microbiol. Biotechnol.*, 2002, 12, 635-642.

Kim, Y-I. Folate and carcinogenesis: Evidence, mechanism and implications. *J. Nutr. Biochem.*, 1999, 10, 66-88.

Klaenhammer, TR; Kullen, MJ. Selection and design of probiotics. *Int. J. Food Microbiol.*, 1999, 50, 45-57.

Klijn, MJJ; Schouw, YT; Wilson, PWF; Grobbee, DE; Jacques, PF. Dietary intake of phytoestrogens is associated with a favorable metabolic cardiovascular risk profile postmenopausl U.S. women: the Framingham study. *J. Nutr.*, 2002, 132, 276-282.

Klingberg, TD; Budde, BB. The survival and persistence in the human gastrointestinal tract of five potential probiotic lactobacilli consumed as freeze-dried cultures or as probiotic sausage. *Int. J. Food Microbiol.*, 2006, 109, 157-159.

Komprda, T; Neznalova, J; Standara, S; Bover-Cid, S. Effect of starter culture and storage temperature on the content of biogenic amines in dry fermented sausage polican. *Meat Sci.*, 2001, 59, 267-276.

Komprda, T; Smela, D; Novicka, K; Kalhotka, L; Sustova, K; Pechova, P.Content and distribution of biogenic amines in Dutch-type hard cheese. *Food Chem.*, 2007, 102, 129-137.

Komprda, T; Smela, D; Pechova, P; Kalhotka, L; Stencl, J; Klejdus, B. Effect of starter culture, spice mix and storage time and temperature on biogenic amine content of dry fermented sausages. *Meat Sci.*, 2004, 67, 607-616.

Koo, HL; DuPond, HL. Current and future developments in travelers' diarrhea therapy. *Expert Rev. Anti-Infective Therapy*, 2006, 4, 417-427.

Korakli, M; Gaenzle, MG; Vogel, RF. Metabolism by bifidobacteria and lactic acid bacteria of polysaccharides from wheat and rye, and exopolysaccharides produced by *Lactobacillus sanfranciscensis*. *J. Appl. Microbiol.*, 2002, 92, 958-965.

Korakli, M; Vogel, RF. Purification and characterisation of mannitol dehydrogenase from *Lactobacillus sanfranciscensis*. *FEMS Microbiol. Lett.*, 2003, 220, 281-286.

Korhonen, H; Pihlanto, A. Bioactive peptides: Production and functionality, *Int. Dairy J.*, 2006, 16, 945-960.

Kozubek, A; Tyman, JHP. Resorcinolic lipids, the natural non-isoprenoid phenolic amphiphiles and their biological activity. *Chem. Rev.*, 1999, 99, 1-25.

Land, MH; Rouster-Stevens, K; Woods, CR; Cannon, ML; Cnota, J; Shetty, AK. *Lactobacillus* sepsis associated with probiotic therapy. *Pediatrics*, 2005, 115, 178-181.

Larsen, AG; Vogensen, FK; Josephsen, J. Antimicrobial activity of lactic acid bacteria isolated from sour doughs: purification and characterization of bavaricin A, a bacteriocin produced by *Lactobacillus bavaricus* MI401. *J. Appl. Bacteriol.*, 1993, 75, 113-122.

Lavermicocca, P; Valerio, F; Evidente, A; Lazzaroni, S; Corsetti, A; Gobbetti, M. Purification and characterization of novel antifungal compounds from the sourdough *Lactobacillus plantarum* strain 21B. *Appl. Eviron. Microbiol.*, 2000, 66, 4084-4090.

Lee, C-H. Lactic acid fermented foods and their benefits in Asia. *Food Control*, 1997, 8, 259-269.

Lee, J-S; Heo, G-Y; Lee, JW; Oh, Y-J; Park, JA; Park, Y-H; Pyun, Y-R; Ahn, JS. Analysis of kimchi microflora using denaturing gradient gel electrophoresis. *Int. J. Food Microbiol.*, 2005, 102, 143-150.

Lee, J-S; Lee, KC; Ahn, J-S; Mheen, T-I; Pyun, Y-R; Park, Y-H. *Weissella koreensis* sp. nov., isolated from kimchi. *Int. J. System Evolut. Microbiol.*, 2002, 52, 1257-1261.

Lee, J-Y; Kim, C-J; Kunz, B. Identification of lactic acid bacteria isolated from kimchi and studies on their suitability for application as starter culture in the production of fermented sausages. *Meat Sci.*, 2006, 72, 437-445.

Lee, LS. Aflatoxin. *J. Am. Oil Chem. Soc.*, 1989, 66, 1398-1413.

Leisner, JJ; Vancanneyt, M; Rusul, G; Pot, B; Lefebvre, K; Fresi, A; Tee, LK. Identification of lactic acid bacteria constituting the predominating microflora in an acid-fermented condiment (tempoyak) popular in Malaysia. *Int. J. Food Microbiol.*, 2001, 63, 149-157.

Lerebours, E; Ndam, CN; Lavoine, A; Hellot, MF; Antoine, JM; Colin, R. Yoghurt and fermented-then-pasteurized milk: effects of short-term and long-term ingestion on lactose absorption and mucosal lactase activity in lactase-deficient subjects. *Am. J. Clin. Nutr.*, 1989, 49, 823-827.

Liljeberg, H; Bjorck, I. Delayed gastric emptying rate as a potential mechanism for lowered glycemia after eating sourdough bread: Studies in humans and rats using test products with added organic acids or an organic salt. *Am. J. Clin. Nutr.*, 1996, 64, 883-893.

Liljeberg, H; Loenner, C; Bjorck, I. Sourdough fermentation or addition of organic acids or corresponding salts to bread improves nutritional properties of starch in healthy humans. *J. Nutr.*, 1995, 125, 1503-1511.

Lin, MY; Young, CM. Folate levels in cultures of lactic acid bacteria. *Int. Dairy J.*, 2000, 10, 409-413.

Lin, YC; Lyle, RM; McCabe, LD; McCabe, GP; Weaver, CM; Teegarden, D. Dairy calcium is related to changes in body composition during a two-year exercise intervention in young women. *J. Am. Coll. Nutr.*, 2000, 19, 754-760.

Liukkonen, K-H; Katina, K; Wilhelmson, A; Myllymaki, O; Lampi, A-M; Kariluoto, S; Piironen, V; Heinonen, S-M; Nurmi, T; Adlercreutz, H; Peltoketo, A; Pihlava, J-M; Hietaniemi, V; Poutanen, K. Process-induced changes on bioactive compounds in whole grain rye. *Proc. Nutr. Soc.*, 2003, 62, 117-122.

Lizaso, G; Chasco, M; Beriain, J. Microbiological and biochemical changes during ripening of salchichon, a Spanish dry cured sausage. *Food Microbiol.*, 1999, 16, 219-228.

Lourens-Hattingh, A; Viljoen, BC. Growth and survival of a probiotic yeast in dairy products. *Food Res. Int.*, 2001a, 34, 791-796.

Lourens-Hattingh, A; Viljoen, BC. Yogurt as probiotic carrier food. *Int. Dairy J.*, 2001b, 11, 1–17.

Lu, Z; Breidt, F; Fleming, HP; Altermann, E; Klaenhammer, TR. Isolation and characterization of a *Lactobacillus plantarum* bacteriophage, ΦJL-1, from a cucumber fermentation. *Int. J. Food Microbiol.*, 2003a, 84, 225-235.

Lu, Z; Breidt, F; Plengvidhya, V; Fleming, HP. Bacteriophage ecology in commercial sauerkraut fermentations. *Appl. Environm Microbiol.*, 2003b, 69, 3192-3202.

Lu, Z; Fleming, HP; McFeeters, RF. Effects of fruit size on fresh cucumber composition and the chemical and physical consequences of fermentation. *J. Food Sci.*, 2002b, 67, 2934-2939.

Lu, Z; Fleming, HP; McFeeters, RF; Yoon, SA. Effects of anions and cations on sugar utilization in cucumber juice fermentation. *J. Food Sci.*, 2002a, 67, 1155-1161.

Luecke, F-K. Fermented sausages. In: Wood BJB, editor. *Microbiology of fermented foods*. Second edition. London, UK: Blackie Academic and Professional; 1998; 441-483.

Maifreni, M; Marino, M; Conte, L. Lactic acid fermentation of *Brassica rapa*: chemical and microbial evaluation of a typical Italian product (brovada). *Eur. Food Res. Technol.*, 2004, 218, 469-473.

Marshall VM. Flavour development in fermented milks. In: Davies FL, Law BA editors. *Advances in the Microbiology and Biochemistry of Cheese and Fermented Milk*. London: Elsevier Applied Science; 1984; 153-186.

Marteau, P; Pochart, P; Flourie, B; Pellier, P; Santos, L; Desjeux, F; Rambaud, J. Effect of chronic ingestion of a fermented dairy product containing *Lactobacillus acidophilus* and *Bifidobacterium bifidum* on metabolic activities of the colonic flora in humans. *Am. J. Clin. Nutr.*, 1990, 52, 685-688.

Marteau, P; Seksik, P; Jian, R. Probiotics and intestinal health effects: a clinical perspective. *Br. J. Nutr.*, 2002, 88, S51-S57.

Marteau, PR; de Vrese, M; Cellier, CJ; Schrezenmeir, J. Protection from gastrointestinal diseases with the use of probiotics. *Am. J. Clin. Nutr.*, 2001, 73, 430S-436S.

Martensson, O; Oste, R; Holst, O. The effect of yoghurt culture on the survival of probiotic bacteria in oat-based, non-dairy products. *Food Res. Int.*, 2002, 35, 775-784.

Marth, EH; Doyle, MP. Update on molds: Degradation aflatoxin. *Food Technol.*, 1979, 33, 81-87.

Martin, R; Jimenez, E., Olivares, M; Marin, ML; Fernandez, L; Xaus, J; Rodriguez, JM. *Lactobacillus salivarius* CECT 5713, a potential probiotic strain isolated from infant feces and breast milk of a mother-child pair. *Int. J. Food Microbiol.*, 2006, 112, 35-43.

Martinez-Anaya, MA; Llin, ML; Macias, MP; Collar, C. Regulation of acetic acid production by homo- and heterofermentative lactobacilli in whole wheat sourdoughs. *Z. Lebensm. Unters. Forsch.*, 1994, 199, 186-190.

Martini, MC; Kukielka, D; Savaino, DA. Lactose digestion from yoghurt: influence of meal and additional lactose. *Am. J. Clin. Nutr.*, 1991, 53, 1253-1258.

Martins, ML; Martins, HM. Aflatoxin M_1 in yoghurts in Portugal. *Int. J. Food Microbiol.*, 2004, 91, 315-317.

Martuscelli, M; Crudele, MA; Gardini, F; Suzzi, G. Biogenic amine formation and oxidation by *Staphylococcus xylosus* strains from artisanal fermented sausages. *Lett. Appl. Microbiol.*, 2000, 31, 228-232.

Martuscelli, M; Gardini, F; Torriani, S; Mastrocola, D; Serio, A; Chaves-Lopez, C; Schirone, M; Suzzi, G. Production of biogenic amines during the ripening of Pecorino Abruzzese cheese. *Int. Dairy J.*, 2005, 15, 571-578.

Marzotto, M; Maffeis, C; Paternoster, T; Ferrario, R; Rizzotti, L; Pellegrino, M; Dellaglio, F; Torriani, S. *Lactobacillus paracasei* A survives gastrointestinal passage and affects the fecal microbiota of healthy infants. *Res. Microbiol.*, 2006, 157, 857-866.

Masson, F; Johansson, G; Montel, MC. Tyramine production by a strain of *Carnobacterium divergens* inoculated in meat-fat mixture. *Meat Sci.*, 1999, 52, 65-69.

Masson, F; Talon, R; Montel, MC. Histamine and tyramine production by bacteria from meat products. *Int. J. Food Microbiol.*, 1996, 32, 199-207.

Mauriello, G; Casaburi, A; Blaiotta, G; Villani, F. Isolation and technological properties of coagulase negative staphylococci from fermented sausages of Southern Italy. *Meat Sci.*, 2004, 67, 149-158.

Mauriello, G; Casaburi, A; Villani, F. Proteolytic activity of *Staphylococcus xylosus* strains on pork myofibrillar and sarcoplasmic proteins and use of selected strains in the production of Naples type salami. *J. Appl. Microbiol.*, 2002, 92, 482-490.

Mayer, K; Pause, G. Biogene amine in sauerkraut. *Lebensm.-Wiss. u.-Technol.*, 1972, 5, 108-109.

McDonagh, FE; Hitchins, AD; Wong, NP; Well, P; Bodwell, CE. Modification of sweet acidophilus milk to improve the utilisation by lactose intolerant persons. *Am. J. Clin. Nutr.*, 1987, 45, 570-574.

McKinley, MC. The nutrition and health benefits of yoghurt. *Int. J. Dairy Technol.*, 2005, 58, 1-12.

Mendonca, C; Venancio, A. Fate of aflatoxin M_1 in cheese whey processing. *J. Sci. Food. Agric.*, 2005, 85, 2067-2070.

Mendoza, E; Garcıa, ML; Casas, C; Selgas, MD. Inulin as fat substitute in low fat, dry fermented sausages. *Meat Sci.*, 2001, 57, 387-393.

Merrill, RK; Oberg, CJ; McMahon, DJ. A method for producing reduced fat Mozzarella cheese. *J. Dairy Sci.*, 1994, 77, 1783-1789.

Meydani, SN; Ha, W-K. Immunologic effects of yoghurt. *Am. J. Clin. Nutr.*, 2000, 71, 861-872.

Middleton, E; Kandaswami, C; Theoharides, TC. The effects of plant flavonoids on mammalian cells: implications for inflammation, heart disease, and cancer. *Pharmacol. Rev.*, 2000, 52, 673-751.

Mirna, A; Coretti, K. Influence of processing on the degradation of pesticides in meat products. *Meat Sci.*, 1979, 3, 97-108.

Mithen, RF; Dekker, M; Verkerk, R; Rabot, S; Johnson, IT. The nutritional significance, biosynthesis and bioavailability of glucosinolates in human foods. *J. Sci. Food Agric.*, 2000, 80, 967-984.

Montano, A; Sanchez, AH; De Castro, A. Controlled fermentation of Spanish-type green olives. *J. Food Sci.*, 1993, 58, 842-852.

Montano, A; Sαnchez, AH; Rejano, L; De Castro, A. Processing and storage of lye - treated carrots fermented by a mixed starter culture. *Int. J. Food Microbiol.*, 1997, 35, 83-90.

Montel, MC; Masson, F; Talon, R. Bacterial role in flavour development. *Meat Sci.*, 1998, 49, S111-S123.

Montel, MC; Masson, F; Talon, R. Comparison of biogenic amine content in traditional and industrial French dry sausages. *Sci. Aliments*, 1999, 19, 247-254.

Moret, S; Smela, D; Populin, T; Conte, LS. A survey on free biogenic amine content of fresh and preserved vegetables. *Food Chem.*, 2005, 89, 355-361.

Motarjemi, Y. Impact of small scale fermentation technology on food safety in developing countries. *Int. J. Food Microbiol.*, 2002, 75, 213-229.

Mueller, MRA; Wolfrum, G; Stolz, P; Ehrmann, MA; Vogel, RF. Monitoring the growth of *Lactobacillus* species during a rye flour fermentation. *Food Microbiol.*, 2001, 18, 217-227.

Muguerza, E; Fista, G; Ansorena, D; Astiasaran, I; Bloukas, JG. Effect of fat level and partial replacement of pork backfat with olive oil on processing and quality characteristics of fermented sausages. *Meat Sci.*, 2002, 6, 397-404.

Muguerza, E; Gimeno, O; Ansorena, D; Astiasaran, I. New formulations for healthier dry fermented sausages: a review. *Trends Food Sci. Technol.*, 2004, 15, 452-457.

Muguerza, E; Gimeno, O; Ansorena, D; Bloukas, JG; Astiasaran, I. Effect of replacing pork backfat with preemulsified olive oil on lipid fraction and sensory quality of Chorizo de Pamplona-A traditional Spanish fermented sausage. *Meat Sci.*, 2001, 59, 251-258.

Muthukumarasamy, P; Holley RA. Microbiological and sensory quality of dry fermented sausages containing alginate-microencapsulated *Lactobacillus reuteri*. *Int. J. Food Microbiol.*, 2006, 111, 164-169.

Muthukumarasamy, P; Holley, RA. Survival of *Escherichia coli* O157:H7 in dry fermented sausages containing micro-encapsulated probiotic lactic acid bacteria. *Food Microbiol.*, 2007, 24, 82-88.

Niedzielin, K; Kordecki, H; Birkenfeld, BA. Controlled, double-blind, randomised study on the efficacy of *Lactobacillus plantarum* 299V in patients with irritable bowel syndrome. *Eur. J. Gastroenterol. Hepatol.*, 2001, 13, 1143-1147.

Nighswonger, BD; Brashears, MM; Gilliland, SE. Viability of *Lactobacillus acidophilus* and *Lactobacillus casei* in fermented milk products during refrigerated storage. *J. Dairy Sci.*, 1996, 79, 212-219.

Nout, MJR. Fermented foods and food safety. *Food Res. Int.*, 1994, 27, 291-298.

Nout, MJR; Motarjemi, Y. Assessment of fermentation as a household technology for improving food safety: a joint FAO/VVHO workshop. *Food Control*, 1997, 8, 221-226.

Novella-Rodrıgues, S; Veciana-Nogues, MT; Izquerdo-Pulido, M; Vidal-Carou, MC. Distribution of biogenic amines and polyamines in cheese. *J. Food Sci.*, 2003, 68, 750-755.

Oberman, H; Libudzisz, Z. Fermented milks. In: Woods BJB, editor. *Microbiology of Fermented Foods.* London: Blackie Academic and Professional; 1998; 308-350.

Oliveira, MN; Sodini, I; Remeuf, F; Corrieu, G. Effect of milk supplementation and culture composition on acidification, textural properties and microbiological stability of fermented milks containing probiotic bacteria. *Int. Dairy J.*, 2001, 11, 935-942.

Ordonez, JA; Ibanez, FC; Torre, P; Barcina, Y. Formation of biogenic amines in Idiazabal ewe's-milk cheese: effect of ripening, pasteurization, and starter. *J. Food Prot.*, 1997, 60, 1371-1375.

Ostman, E; Nilsson, M; Liljeberg-Elmstahl, H; Molin, G; Bjorck, I. On the effect of lactic acid on blood glucose and insulin responses to cereal products: mechanistic studies in healthy subjects and in vitro. *J. Cereal Sci.*, 2002, 36, 339-346.

Ottogalli, G; Galli, A; Foschino, R. Italian bakery products obtained with sour dough: characterization of the typical microflora. *Adv. Food Sci.*, 1996, 18, 131-144.

Oyewole OB. Lactic fermented foods in Africa and their benefits. *Food Control*, 1997, 8, 289-297.

Palacios, MC; Sanz, Y; Haros, M; Rosell, CM. Application of *Bifidobacterium* strains to the breadmaking process. *Process Biochem.*, 2006, 41, 2434-2440.

Papamanoli, E; Tzanetakis N; Litopoulou-Tzanetaki, E; Kotzekidou, P. Characterization of lactic acid bacteria isolated from a Greek dry-fermented sausage in respect of their technological and probiotic properties. *Meat Sci.*, 2003, 65, 859-867.

Paramithiotis, S. *Study on the symbiosis of wild yeast and lactic acid bacteria strains in sourdough.* Greece, Agricultural University of Athens, PhD thesis; 2001.

Paramithiotis, S; Gioulatos, S; Tsakalidou, E; Kalantzopoulos, G. Interactions between *Saccharomyces cerevisiae* and lactic acid bacteria in sourdough. *Process Biochem.*, 2006a, 41, 2429-2433.

Paramithiotis, S; Melissari, I; Drosinos, EH. *In vitro* assessment of properties associated with the survival through the gastro-intestinal tract of staphylococci isolated from traditional sausage fermentation. *Food Microbiol.*, 2006b, 23, 663-671.

Parente, E; Grieco, S; Crudele, MA. Phenotypic diversity of lactic acid bacteria isolated from fermented sausages produced in Basilicata (Southern Italy). *J. Appl. Microbiol.*, 2001, 90, 943-952.

Park, DL; Lee, LS; Price, RL; Pohland, AE. Review of the decontamination of aflatoxins by ammoniation: Current status and regulation. *J. AOAC*, 1988, 71, 685-703.

Passos, FV; Fleming, HP; Ollis, DF; Hassan, HM; Felder, RM. Modeling the specific growth rate of *Lactobacillus plantarum* in cucumber extract. *Appl. Microbiol. Biotechnol.*, 1993a, 40, 143-150.

Passos, FV; Ollis, DF; Fleming, HP; Hassan, HM; Felder, RM. Modeling the cucumber fermentation: growth of *Lactobacillus plantarum*. *J. Ind. Microbiol.*, 1993b, 12, 341-345.

Pedersen, NS; Smith, E. Prion diseases: Epidemiology in man, *APMIS*, 2002, 110, 14-22.

Pelaez, C; Requena, T. Exploiting the potential of bacteria in the cheese ecosystem. *Int. Dairy J.*, 2005, 15, 831-844.

Pennacchia, C; Ercolini, D; Blaiotta, G; Pepe, O; Mauriello, G; Villani, F. Selection of *Lactobacillus* strains from fermented.sausages for their potential use as probiotics. *Meat Sci.*, 2004, 67, 309-317.

Pennacchia, C; Vaughan, EE; Villani, F. Potential probiotic *Lactobacillus* strains from fermented sausages: Further investigations on their probiotic properties *Meat Sci.*, 2006, 73, 90-101.

Pereira, CI; Barreto Crespo, MT; San Romao, MV. Evidence for proteolytic activity and biogenic amines production in *Lactobacillus curvatus* and *Lactobacillus homohiochii*. *Int. J. Food Microbiol.*, 2001, 68, 211-216.

Peric, M; Raseta, J; Visacki, MS; Spiric, A. Degradation of organochlorine pesticides as influenced by micrococci isolated from fermented sausages. *Technol. Mesa*, 1980, 21, 132-133.

Periera, MA; Jacobs, DR; Van Horn, L; Slattery, ML; Kartashov, AI; Ludwig, DS. Dairy consumption, obesity, and the insulin resistance syndrome in young adults: the CARDIA Study. *J. Am. Med. Assoc.*, 2002, 287, 2081-2089.

Petridis, KD; Steinhart, H. Biogene Amine in der Hartkaeseproduktion: I. Einfluss verschiedener Parameter auf den Amingehalt im Endprodukt am Beispiel von Emmentaler Kaese. *Deutsche Lebensm.-Rundschau*, 1996a, 92, 114-120.

Petridis, KD; Steinhart, H. Biogene Amine in der Hartkaeseproduktion: II. Stufenkontroll-Studie einer standardisierten Emmentalerkaese-Produktion. *Deutsche Lebensm.-Rundschau*, 1996b, 92, 142-146.

Pinho, O; Ferreira, IMPLVO; Mendes, E; Oliveira, BM; Ferreira, M. Effect of temperature on evolution of free amino acid and biogenic amine contents during storage of Azetao cheese. *Food Chem.*, 2001, 75, 287-291.

Pulido, RP; Ben Omar, N; Abriouel, H; Lopez, RL; Martinez Canamero, M; Galvez, A. Microbiological study of lactic acid fermentation of caper berries by molecular and culture-dependent methods. *Appl. Environm. Microbiol.*, 2005, 71, 7872-7879.

Puspito, H; Fleet, GH. Microbiology of sugan asin fermentation. *Appl. Microbiol. Biotechnol.*, 1985, 22, 442-445.

Quensen-III, JF; Boyd, SA; Tiedje, JT. Dechlorination of four commercial polychlorinated biphenyl mixture (Aroclors) by anaerobic microorganisms from sediments. *Appl. Environm. Microbiol.*, 1990, 56, 2360-2369.

Quevedo, F. Food safety and fermented foods in selected Latin American countries. *Food Control*, 1997, 8, 299-302.

Radberg, K; Biernat, M; Linderoth, A; Zabielski, R; Pierzynowski, SG; Westrom, BR. Enteral exposure to crude red kidney bean lectin induces maturation of the gut in suckling pigs. *J. Anim. Sci.*, 2001, 79, 2669-2678.

Randazzo, CL; Heilig, H; Restuccia, C; Giudici, P; Caggia, C. Bacterial population in traditional sourdough evaluated by molecular methods. *J. Appl. Microbiol.*, 2005, 99, 251-258.

Randazzo, CL; Restuccia, C; Romano, AD; Caggia, C. *Lactobacillus casei*, dominant species in naturally fermented Sicilian green olives. *Int. J. Food Microbiol.*, 2004, 90, 9-14.

Rantsiou, K; Comi, G; Cocolin, L. The *rpo*B gene as a target for PCR-DGGE analysis to follow lactic acid bacteria population dynamics during food fermentations. *Food Microbiol.*, 2004, 21, 481-487.

Rantsiou, K; Drosinos, EH; Gialitaki, M; Urso, R; Krommer, J; Gasparik-Reichardt, J; Toth, S; Metaxopoulos, I; Comi, G; Cocolin, L. Molecular characterization of *Lactobacillus* species isolated from naturally fermented sausages produced in Greece, Hungary and Italy. *Food Microbiol.*, 2005, 22, 19-28.

Rao, DR; Reddy, AV; Pulusani, SR; Cornwell, PE. Biosynthesis and utilization of folic acid and vitamin B12 by lactic acid cultures in skim milk. *J. Dairy Sci.*, 1984, 67, 1169-1174.

Reddy, NR; Pierson, MD. Reduction in antinutritional and toxic components in plant foods by fermentation. *Food Res. Int.*, 1994, 27, 281-290.

Reuter, G; Klein, G; Goldberg, M. Identification of probiotic cultures in food samples. *Food Res. Int.*, 2002, 35, 117-124.

Roberts, JS; Kidd, DR. Lactic acid fermentation of onions. *Lebensm.-Wiss. u.-Technol.*, 2005, 38, 185-190.

Rodrıguez, E; Calzada, J; Arques, JL; Rodrıguez, JM; Nunez, M; Medin, M. Antimicrobial activity of pediocin-producing *Lactococcus lactis* on *Listeria monocytogenes*, *Staphylococcus aureus* and *Escherichia coli* O157:H7 in cheese. *Int. Dairy J.*, 2005, 15, 51-57.

Rodrıguez, E; Gonzalez, B; Gaya, P; Nunez, M; Medina, M. Diversity of bacteriocins produced by lactic acid bacteria isolated from raw milk. *Int. Dairy J.*, 2000, 10, 7-15.

Rodriguez, J; Requena, T; Goudedranche, H; Maubois, JL; Juarez, M. Accelerated ripening of reduced-fat semi-hard cheese from a mixture of cow's, goat's and ewe's ultrafiltered milk by using a Lac- Prt- strain of *Lactococci*. *Lait*, 1996, 76, 515-524.

Roecken, W. Applied aspects of sourdough fermentation. *Adv. Food Sci. (CMTL)*, 1996, 5/6, 212-216.

Roecken, W; Voysey, PA. Sourdough fermentation in breadmaking. *J. Appl. Bacteriol. Sym. Sup.*, 1995, 79, 38S-48S.

Roig-Sagues, AX; Molina, AP; Hernandes-Herrero, MM. Histamine and tyramine-forming microorganisms in Spanish traditional cheeses. *Eur. Food Res. Technol.*, 2002, 215, 96-100.

Romero, C; Brenes, M; Yousfi, K; Garcia, P; Garcia, A; Garrido, A. Effect of cultivar and processing method on the contents of polyphenols in table olives. *J. Agric. Food Chem.*, 2004, 52, 479-484.

Ronka, E; Malinen, E; Saarela, M; Rinta-Koski, M; Aarnikunnas, J; Palv, A. Probiotic and milk technological properties of *Lactobacillus brevis*. *Int. J. Food Microbiol.*, 2003, 83, 63-74.

Rosado, JL; Solomons, NW; Allen, LH. Lactose digestion from unmodified, low-fat and lactose-hydrolysed yoghurt in adult lactose maldigesters. *Eur. J. Clin. Nutr.*, 1992, 46, 61-67.

Rosenquist, H; Hansen, A. The microbial stability of two bakery sourdoughs made from conventionally and organically grown rye. *Food Microbiol.*, 2000, 17, 241-250.

Ross, AB; Shepherd, MJ; Schuepphaus, M; Sinclair, V; Alfaro, B; Kamal-Eldin, A; Aman, P. Alkylresorcinols in cereals and cereal products. *J. Agric. Food Chem.*, 2003, 51, 4111-4118.

Rossello, C; Barbas, JI; Berna, A; Lopez, N. Microbial and chemical changes in 'sobrasada' during ripening. *Meat Sci.*, 1995, 40, 379-385.

Rubia-Soria, A; Abriouel, H; Lucas, R; Ben Omar, N; Martinez-Canamero, M; Galvez, A. Production of antimicrobial substances by bacteria isolated from fermented table olives. *World J. Microbiol. Biotechnol.*, 2006, 22, 765-768.

Saarela, M; Mogensen, G; Fonden, R; Matto, J; Mattila-Sandholm, T. Probiotic bacteria: safety, functional and technological properties. *J. Biotechnol.*, 2000, 84 197-215.

Sair, AI; D'Souza, DH; Jaykus, LA. Human enteric viruses as causes of foodborne disease. *Compr. Rev. Food Sci. Food Safety*, 2002, 1, 73-89.

Saito, T. Selection of useful probiotic lactic acid bacteria from the *Lactobacillus acidophilus* group and their applications to functional foods. *Anim. Sci. J.*, 2004, 75, 1-13.

Salmenkallio-Marttila, M; Katina, K; Autio, K. Effect of bran fermentation on quality and microstructure of high-fibre wheat bread. *Cereal Chem.*, 2001, 78, 429-435.

Salminen, MK; Tynkkynen, S; Rautelin, H; Saxelin, M; Vaara, M; Ruutu, P; Sarna, S; Valtonen, V; Jarvinen, A. *Lactobacillus* bacteremia during a rapid increase in probiotic use of *Lactobacillus rhamnosus* GG in Finland. *Clin. Infect. Dis.*, 2002, 35, 1155-1160.

Salovaara, H; Katunpaa, H. An approach to the classification of Lactobacilli isolated from Finnish sour rye dough ferments. *Acta Alim. Pol.*, 1984, 10, 231-239.

Samelis, J; Metaxopoulos, J; Vlassi, M; Pappa, A. Stability and safety of traditional Greek salami - a microbiological ecology study. *Int. J. Food Microbiol.*, 1998, 44, 69-82.

Sanders, ME; Walker, DC; Walker, KM; Aoyama, K; Klaenhammer, TR. Performance of commercial cultures in fluid milk applications. *J. Dairy Sci.*, 1996, 79, 943-955.

Santos, EM; Gonzalez-Fernandez, C; Jaime, I; Rovira, J. Comparative study of lactic acid bacteria house flora isolated in different varieties of 'chorizo'. *Int. J. Food Microbiol.*, 1998, 39, 123-128.

Sanz, Y; Fadda, S; Vignolo, G; Aristoy, M-C; Oliver, G; Toldra, F. Hydrolytic action of *Lactobacillus casei* CRL 705 on pork muscle sarcoplasmic and myofibrillar proteins. *J. Agric. Food Chem.*, 1999a, 47, 3441-3448.

Sanz, Y; Fadda, S; Vignolo, G; Aristoy, M-C; Oliver, G; Toldra, F. Hydrolysis of muscle myofibrillar proteins by *Lactobacillus curvatus* and *Lactobacillus sake*. *Int. J. Food Microbiol.*, 1999b, 53, 115-125.

Savaiano, DA; El Anouar, AA; Smith, DE; Levitt, MD. Lactose malabsorption from yoghurt, pasteurized yoghurt, sweet acidophilus milk, and cultured milk in lactase-deficient individuals. *Am. J. Clin. Nutr.*, 1984, 40, 1219-1223.

Schillinger, U; Holzapfel, W; Kandler, O. Nucleic acid hybridization studies on *Leuconostoc* and heterofermentative lactobacilli and description of *Leuconostoc amelibiosum* sp. nov. *System Appl. Microbiol.*, 1989, 12, 48-55.

Schneller, R; Good, P; Jenny, M. Influence of pasteurised milk, raw milk and different ripening cultures on biogenic amine concentrations in semi-soft cheeses during ripening. *Z. Lebensm. Unters. Forsch. A*, 1997, 204, 265-272.

Schnuerer, J; Magnusson, J. Antifungal lactic acid bacteria as biopreservatives. *Trends Food Sci. Technol.*, 2005, 16, 70-78.

Severin, S; Wenshui, X. Milk biologically active components as nutraceuticals. *Crit. Rev. Food Sci. Nutr.*, 2005, 45, 645-656.

Shetty, PH; Jespersen, L. *Saccharomyces cerevisiae* and lactic acid bacteria as potential mycotoxin decontaminating agents. *Trends Food Sci. Technol.*, 2006, 17, 48-55.

Silano, M; De Vincenzi, M. Bioactive antinutritional peptides derived from cereal prolamins: a review. *Nahrung*, 1999, 43, 175-184.

Silla Santos, MH. Amino acid decarboxylase capability of microorganisms isolated in Spanish fermented meat products. *Int. J. Food Microbiol.*, 1998, 39, 227-230

Skinner, JD; Bounds, W; Carruth, BR; Ziegler, P. Longitudinal calcium intake is negatively related to children's body fat indexes. *J. Am. Diet Assoc.*, 2003, 103, 1626-1631.

Sommer, R; Rommel, M; Lewetzow, R. Die Ueberlebensdauer von Toxoplasmencysten in Fleisch und Fleischerzeugnissen. *Fleischwirtschaft*, 1965, 45, 454-457.

Song, H-P; Kim, D-H; Yook, H-S; Kim, M-R; Kim, K-S; Byun, M-W. Nutritional, physiological, physicochemical and sensory stability of gamma irradiated Kimchi (Korean fermented vegetables). *Rad. Phys. Chem.*, 2004, 69, 85-90.

Song, L; Thornalley, PJ. Effect of storage, processing and cooking on glucosinolate content of *Brassica* vegetables. *Food Chem. Toxicol.*, 2007, 45 ,216-224.

Spicher, G, Bruemmer, J-M. Baked goods. In: Reed G, Nagodawithana TW editors. Biotechnology second, completely revised edition, vol. 9, *Enzymes, biomass, food and feed.* Weinheim: VCH; 1995; 243-319.

Spicher, G. Die Mikroflora des Sauerteiges. I. Mitteilung: Untersuchungen uber die Art der in Sauerteigen anzutreffenden stabchenformigen Milchsaurebakterien (Genus *Lactobacillus* Beijerinck). *Zeitblatt fur Bakteriologie* II Abt, 1959, 113, 80-106.

Spiric, A; Visacki, V; Raseta, J; Peric, M; Rajkovic, N. Kinetics of decomposition of DDT in meat products under the action of selected micrococci. *Technol. Mesa*, 1981, 22, 344.

Spyropoulou, KE; Chorianopoulos, NG; Skandamis, PN; Nychas, G-JE. Survival of *Escherichia coli* O157:H7 during the fermentation of Spanish-style green table olives conservolea variety supplemented with different carbon sources. *Int. J. Food Microbiol.*, 2001, 66, 3-11.

Stanton, C; Lynch, PB; Collins, JK; Fitzgerald, G; Ross, RP. Probiotic cheese. *Int. Dairy J.*, 1998, 8, 491-496.

Steinkraus, KH. Classification of fermented foods: worldwide review of household fermentation techniques. *Food Control*, 1997, 8, 311-317.

Steinkraus, KH. Fermentations in world food processing. *Compr. Rev. Food Sci. Food Safety*, 2002, 1, 23-32.

Steinkraus, KH. *Handbook of indigenous fermented foods*. Second edition, revised and expanded. New York, USA: Marcel Dekker Inc; 1996.

Stolz, P; Boecker, G; Hammes, WP; Vogel, RF. Utilization of electron acceptors by lactobacilli isolated from sourdough I. *Lactobacillus sanfrancisco*. *Z. Lebensm. Unters. Forsch.*, 1995, 201, 91-96.

Stratton, JE; Hutkins, RV; Taylor, SL. Biogenic amines in cheese and other fermented foods, a review. *J. Food Prot.*, 1991, 54, 460-470.

Straub, BW; Kicherer, M; Schilcher, SM; Hammes, WP. The formation of biogenic amines by fermentation organisms. *Z. Lebensm. Unters. Forsch.*, 1995, 201, 79-82.

Styriak, I; Concova, E. Microbial binding and biodegradation of mycotoxins. *Vet. Human Toxicol.*, 2002, 44, 358-361.

Suzzi, G; Gardini, F. Biogenic amines in dry fermented sausages: a review. *Int. J. Food Microbiol.*, 2003, 88, 41-54.

Talon, R; Leroy, S. Latest developments in meat bacterial starters. In: Nollet LML, Toldra F editors. *Advanced technologies for meat processing*. New York: Taylor and Francis; 2006; 401-418.

Talon, R; Leroy-Setrin, S; Fadda, S. Bacterial starters involved in the quality of fermented meat products. In: Toldra F, editor. *Research advances in the quality of meat and meat products*. Trivandrum India: Research Signpost; 2002; 175-191.

Talon, R; Leroy-Setrin, S; Fadda, S. Dry fermented sausages. In: Hui YH, Meunier-Goddik L, Hansen AS, Josephsen J, Nip W-K, Stanfield PS, Toldra F editors. *Handbook of food and beverage fermentation technology*. New York: Marcel Dekker Inc; 2004; 397-416.

Tamang, JP; Sarkar, PK. Sinki: a traditional lactic acid fermented radish tap root product. *J. Gen. Appl. Microbiol.*, 1993, 39, 395-408.

Tamang, JP; Tamang, B; Schillinger, U; Franz, CMAP; Gores, M; Holzapfel, WH. Identification of predominant lactic acid bacteria isolated from traditionally fermented vegetable products of the Eastern Himalayas. *Int. J. Food Microbiol.*, 2005, 105, 347-356.

Taylor, SL; Leatherwood, M; Lieber, ER. Histamine in sauerkraut. *J. Food Sci.*, 1978, 43, 1030-1032.

Teegarden, D. Calcium intake and reduction in weight or fat mass. *J. Nutr.*, 2003, 133, 249S-251S.

Ternes, W; Freund, W. Effects of different doughmaking techniques on thiamin content of bread. *Getr. Mehl Brot*, 1988, 42, 293-297.

Thomzig, A; Kratzel, C; Lenz, G; Krueger, D; Beekes, M. Widespread PrpSc accumulation in muscles of hamsters orally infected with scrapie. *EMBO Rep.*, 2003, 4, 530-533.

Tieking, M; Ehrmann, MA; Vogel, RF; Gaenzle, MG. Molecular and functional characterization of a levansucrase from *Lactobacillus sanfranciscensis*. *Appl. Microbiol. Biotechnol.*, 2005a, 66, 663-665.

Tieking, M; Gaenzle, MG. Exopolysaccharides from cereal associated lactobacilli. *Trends Food Sci. Technol.*, 2005, 16, 79-84.

Tieking, M; Kuhnl, W; Gaenzle, MG. Evidence for formation of hetero-oligosaccharides by *Lactobacillus sanfranciscensis* during growth in wheat sourdough. *J. Agric. Food Chem.*, 2005b, 53, 2456-2461.

Todorov, S; Onno, B; Sorokine, O; Chobert, JM; Ivanova, I; Dousset, X. Detection and characterization of a novel antibacterial substance produced by *Lactobacillus plantarum* ST 31 isolated from sourdough. *Int. J. Food Microbiol.*, 1999, 48, 167-177.

Topisirovic, L; Kojic, M; Fira, D; Golic, N; Strahinic, I; Lozo, J. Potential of lactic acid bacteria isolated from specific natural niches in food production and preservation. *Int. J. Food Microbiol.*, 112, 2006, 230-235.

Tricon, S; Willersw, S; Smitw, HA; Burneyz, PG; Devereux, G; Frew, AJ; Halkenz, S; Hostz, A; Nelsonk, M; Shaheenz, S; Warner, JO; Calder, PC. Nutrition and allergic disease. *Clin. Exp. Allergy Rev.*, 2006, 6, 117-188.

Trujillo, E; Davis, C; Milner, J. Nutrigenomics, proteomics, metabolomics, and the practice of dietetics. *J. Am. Diet. Assoc.*, 2006, 106, 403-413.

Tsai, Y-H; Kung, H-F;. Lin, Q-L; Hwang, J-H; Cheng, S-H; Wie, C-I; Hwang, D-F. Occurrence of histamine and histamine-forming bacteria in kimchi products in Taiwan. *Food Chem.*, 2005, 90, 635-641.

Tyopponen, S; Petaja, E; Mattila-Sandholm, T. Bioprotectives and probiotics for dry sausages. *Int. J. Food Microbiol.*, 2003, 83, 233-244.

Untermann, F. Microbial hazards in food. *Food Control,* 1998, 9, 119-126.

Ushiyama, A; Tanaka, K; Aiba, Y; Shiba, T; Takagi, A; Mine, T; Koga, Y. *Lactobacillus gasseri* OLL2716 as a probiotic in clarithromycin-resistant *Helicobacter pylori* infection. *J. Gastroenterol. Hepatol.*, 2003, 18, 986-991.

Valcheva, R; Korakli, M; Onno, B; Prevost, H; Ivanova, I; Ehrmann MA; Dousset, X; Gaenzle, MG; Vogel, RF. *Lactobacillus hammesii* sp. nov., isolated from French sourdough. *Int. J. Syst. Evol. Microbiol.*, 2005, 55, 763-767.

Valsamaki, K; Michaelidou, A; Polychroniadou, A. Biogenic amine production in Feta cheese. *Food Chem.*, 2000, 71, 259-266.

van Hijum, SAFT; Kralj, S; Ozimek, LK; Dijkhuizen, L; van Geel-Schutten, IGH. Structure-function relationships of glucansucrase and fructansucrase enzymes from lactic acid bacteria. *Microbiol. Mol. Biol. Rev.*, 2006, 70, 157-176.

Vancanneyt, M; Neysens, P; De Wachter, M; Engelbeen, K; Snauwaert, C; Cleenwerck, I; Van der Meulen, R; Hoste, B; Tsakalidou, E; De Vuyst, L; Swings, J. *Lactobacillus acidifarinae* sp. nov. and *Lactobacillus zymae* sp. nov., from wheat sourdoughs. *Int. J. System. Evol. Microbiol.*, 2005, 55, 615-620.

Vega Leal-Sanchez, M; Ruiz-Barba, JL; Sanchez, AH; Rejano, L; Jimenez-Diaz, R; Garrido, A. Fermentation profile and optimization of green olive fermentation using *Lactobacillus plantarum* LPCO10 as a starter culture. *Food Microbiol.*, 2003, 20, 421-430.

Vesa, TH; Marteau, P; Zidi, S; Pochart, P; Rambaud, JC. Digestion and tolerance of lactose from yoghurt and different semi-solid fermented dairy products containing *Lactobacillus acidophilus* and bifidobacteria in lactose maldigesters. Is bacterial lactase important? *Eur. J. Clin. Nutr.*, 1996, 50, 730-733.

Viander, B; Maki, M; Palva, A. Impact of low salt concentration, salt quality on natural large-scale sauerkraut fermentation. *Food Microbiol.*, 2003, 20, 391-395.

Visioli, F; Galli, C; Plasmati, E; Viappiani, S; Hernandez, A; Colombo, C; Sala, A. Olive phenol hydroxytyrosol prevents passive smoking-induced oxidative stress. *Circulation,* 2000, 102, 2169-2171.

Wacher, C; Canas, A; Barzana, E; Lappe, P; Ulloa, M; Owens, JD. Microbiology of Indian and Mestizo pozol fermentations. *Food Microbiol.*, 2000, 17, 251-256.

Waites, WM; Arbuthnott, JP. Foodborne illness: an overview. *Lancet*, 1990, 336, 722-725.

Wennemark, B; Jaegerstad, M. Breadmaking and storage of various wheat fractions affect vitamin E. *J. Food Sci.*, 1992, 57, 1205-1209.

Wiander, B; Ryhanen, E-L. Laboratory and large-scale fermentation of white cabbage into sauerkraut and sauerkraut juice by using starters in combination with mineral salt with a low NaCl content. *Eur. Food Res. Technol.*, 2005, 220, 191-195.

Wouters, JTM; Ayad, EHE; Hugenholtz, J; Smit, G. Microbes from raw milk for fermented dairy products. *Int. Dairy J.*, 2002, 12, 91-109.

Yao, LH; Jiang, YM; Shi, J; Tomas-Barberan, FA; Datta, N; Singanusong, R; Chen, SS. Flavonoids in food and their health benefits. *Plant. Food Hum. Nutr.*, 2004, 59, 113-122.

Yoon, KY; Woodams, EE; Hang, YD. Fermentation of beet juice by beneficial lactic acid bacteria. *Lebensm.-Wiss. u.-Technol.*, 2005, 38, 73-75.

Yoon, KY; Woodams, EE; Hang, YD. Production of probiotic cabbage juice by lactic acid bacteria. *Biores. Technol.*, 2006, 97, 1427-1430.

Yoon, SS; Barrangou-Poueys, R; Breidt, F; Klaenhammer, TR; Fleming, HP. Isolation and characterization of bacteriophages from fermenting sauerkraut. *Appl. Environm. Microbiol.*, 2002, 68, 973-976.

Young, JC; Subryan, LM; Potts, D; McLaren, ME; Gobran, FH. Reduction in levels of deoxynivalenol in contaminated wheat by chemical and physical treatment. *J. Agric. Food Chem.*, 1986, 34, 461-465.

Yousef, AE; Marth, EH. Use of ultraviolet energy to degrade aflatoxin M_1 in raw or heated milk with and without added peroxide. *J. Dairy Sci.*, 1989, 69, 2243-2247.

Zaika, LL; Kissinger, JC. Fermentation enhanced by spices: identification of active component. *J. Food Sci.*, 1984, 49, 5-9.

Zarate, G; Nader-Macias, ME. Influence of probiotic vaginal lactobacilli on in vitro adhesion of urogenital pathogens to vaginal epithelial cells. *Lett. Appl. Microbiol.*, 2006, 43, 174-180.

Zhang, H; Li, J. Detoxification of moniliformin. *Acta Microbiol. Sin.*, 1994, 34, 119-123.

In: Food Microbiology Research Trends
Editor: Marta V. Palino, pp. 93-129
ISBN 978-1-60021-723-4

Chapter 3

CONCEPTUAL AND METHODOLOGICAL FOUNDATIONS FOR DEVELOPING MICROBIAL RISK ASSESSMENT MODELS

F. Pérez-Rodríguez, R. M. García-Gimeno and G. Zurera
Department of Food Science and Technology. University of Cordoba,
Campus de Rabanales. 14014 Córdoba, Spain

1. MICROBIAL RISK ASSESSMENT

According to the definition provided by the FAO (Food Agriculture Organization), Risk Assessment (RA) is a process consisting of four stages: Hazard Identification (HI), Hazard Characterisation (HC), Exposure Assessment (EA) and Risk Characterisation (RC); the result of each stage is the starting point for the next stage (FAO/WHO, 1995). Lamerding and Fazil (2000) defined RA for microbial hazards (MRA) as the process that provides an estimation of the probability and impact of an adverse health effect attributable to potentially contaminated food.

RA techniques have been developed over the years by disciplines such as Civil Engineering, Economics and Investment, and Toxicology. This experience has been used as a foundation to establish an approach to food safety. RA was first applied to the field of food microbiology just a few years ago, since it needed to be developed to a greater extent for this purpose. The dynamic and complex nature of biological systems (natural variability, growth, death, survival, etc) that are subjected to MRA adds an extra layer of difficulty not found in other types of RA (e.g., chemical and environmental hazards) where the object studied is an inert system, which does not undergo modifications (Lammerding and Fazil, 2000). Hence, models must be developed in order to simulate or predict the behaviour of microorganisms as accurately as possible.

Several organisations and governments have drawn up guidelines for MRA. The FAO/WHO consulted experts and combined all the information gathered in a document that sets forth the basic guidelines for Risk Assessment (CAC, 1999). Similarly the ILSI (International Life Science Institute) held meetings in order to draft a general framework for

MRA (ILSI, 2000). However, the most widely used guidelines are those proposed by the FAO/WHO, partly due to the fact that they were developed on the basis of well-known and established toxicological and environmental risk assessment guidelines.

When carrying out MRA, analysis should be conducted prior to the formal application of the four stages as defined by the FAO (CAC, 1999). This prior stage is referred to by several authors (Lindqvist et al., 2002; European Commission, 2003) as a risk profile. A risk profile can be defined as a quick initial qualitative answer to the questions and problems raised (risk managers), or as a first step in the formal development of a RA. It is mainly based on interaction between the risk assessor and risk manager, with the aim of clearly establishing the objective and object of the MRA (Morgan et al., 1990). Furthermore, this pre-stage can also include additional aspects related with risk characterisation and levels of exposure to the hazard.

An MRA does not forcibly have to include a complete analysis of the entire food chain if it is not considered to be relevant to the risk. In other words, the objective and object will be sufficient to respond specifically to the questions and problems raised. Hence, if the hazard and risk are limited to a specific point in the food chain, for example, during the pasteurisation process, the four formal stages of the MRA will be restricted a priori to this stage. In certain cases, an RA can be provided without having to carry out all 4 stages included in a formal MRA (CAC, 1999). For example, in certain cases, it is only necessary to conduct the EA stage, which determines levels of exposure to the microorganism.

The sphere of application of the results is another factor that must be taken into account prior to conducting a full formal MRA. This is important when it comes to determining the way in which the risk is to be expressed (cases of illness/year, medical costs, probability of falling ill/portion, etc), since this will make it easier for the risk managers to interpret. This aspect will be examined in greater detail in the section on Risk Characterisation.

1.1. Hazard Identification

The CAC (Codex Alimentarius Commission) (1999) defines hazard as "a biological, chemical or physical agent in, or condition of, food with the potential to cause an adverse health effect".

Hazard Identification (HI) is the stage that studies the existence of a relationship between a hazard (chemical or biological), food or food group and an adverse health effect, based on information gathered from different scientific fields (Epidemiology, Medicine, Food Technology and Microbiology, studies on consumption patterns, etc.). In many cases, this stage is not treated with the importance it deserves and the necessary resources are not provided to carry out proper and effective hazard identification. An exhaustive HI must be based on a qualitative risk assessment (i.e. risk profile) carried out by a panel of experts. This assessment may or may not be based on the RA guidelines drawn up by the FAO/WHO (1995), although it is true that this process has certain features in common with the HACCP method, given its methodology and qualitative character (ICMSF, 2002). At this level, interaction between the risk assessor and risk manager is crucial (Lammerding and Fazil, 2000), since the manager must decide, in light of the results provided by the panel of experts, whether the hazard identified is relevant in relation to public health. The existence of a hazard does not necessary imply a public health problem, but rather it reveals high incidence in the population and/or a serious adverse health effect. When the hazard is identified in this initial stage, the risk manager must decide whether further analysis is required, for example a quantitative MRA, or whether all the questions have already been answered by the panel of experts.

For microbiological risk, the purpose of this stage is to identify the pathogenic microorganisms that might be present in food (Jouve, 2002). However, over the last few years, the MRAs published have not, a priori, required a thorough HI, since most of the microorganisms examined to date are well-known foodborne pathogens.

Because hazard identification is easy to perform, this stage has been practically reduced to the compilation of data about the characteristics of the pathogen (virulence, physiology, transmissibility, etc), risk populations and the adverse health effects associated with its ingestion, data that will be used to develop the EA and HC stages. For example, information about risk populations and the variable responses of the pathogen is extremely useful when building dose-response models and choosing specific scenarios for the different risk groups in the HC stage.

1.2. Hazard Characterisation

In the Hazard Characterisation (HC) stage, information is compiled about the nature of the adverse health effect associated with the ingestion of a hazard. According to the CAC (1999), this stage follows on from HI: the need to gather information about the cause/effect relationship between microorganisms, carried out in the HI stage, is partially linked with the development of the HC stage. However, the chief purpose of HC is to develop a dose-response model (D-R). This type of model determines the relationship between the level of exposure (dose) to a chemical, biological or physical agent and the severity of the adverse health effects (response) (ICMSF, 2002). D-R models can indicate probability of infection or illness, depending on the type of information used to build them.

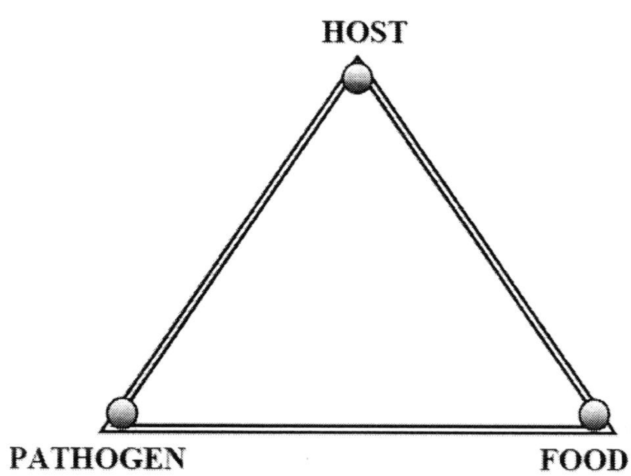

Figure 1. Epidemiological triangle.

This stage also encompasses the study of the factors that are decisive in the development of illness or an infectious process. The epidemiological triangle (Figure 1) is commonly used to represent this aspect; each vertex corresponds to a key factor in the development of the illness/infection: aspects about the pathogen, host and food matrix (WHO/FAO, 2003).

The development of D-R models is based on data obtained from experimental studies using humans, animals or *in vitro*. Models can be developed for different risk groups (children, the elderly, immuno-compromised individuals, etc). These models can also be completed using information gathered in relation to the epidemiological triangle. Experts can, at the same time, judge the reliability of the model or the need to include modifications based on the information compiled.

1.3. Exposure Assessment

The FAO/WHO (1995) defines Exposures Assessment (EA) as "The qualitative and/or quantitative evaluation of the likely intake of a biological agent via food, as well as exposure from other sources if relevant". Besides, Lammerding and Fazil (2000) describe EA as "the estimation of the probability that an individual or population will be exposed to a microbial hazard and the probable number of microorganisms ingested". According to Haas et al. (1999), however, EA is also "an attempt to determine the size and nature of the population exposed and the pathway, concentrations and distribution of microorganisms and the duration of the exposure", Based on these definitions, the following elements are required to conduct an EA:

1. Characterisation of the population exposed - size of the population studied, risk population (the elderly, children, pregnant women, immuno-compromised individuals, etc.), and consumption patterns.
2. Transmission pathway of the microorganism throughout the food chain. For this purpose, information can be compiled about the parameters of the process (sterilisation temperature, refrigeration temperature, storage time, etc.), and concentration of pathogens and indicator microorganisms in the different steps in the food chain.
3. Determination of the levels of exposure at the time of consumption, expressed as values of prevalence and concentration. They can be specific values or distributions, depending on whether a probabilistic or deterministic approach has been taken.

Consumption patterns are an essential part of an EA. For example, the size of the portion, annual/weekly consumption ratios and the circumstances under which the food is manufactured or prepared can be decisive in levels of exposure. Furthermore, considerations about regional, cultural and social differences should also be included since they might be linked to different patterns of consumption. Moreover, the existence of risk groups or subpopulations, with differences in their consumption patterns, could be significant in the determination of risk.

The application of models is crucial owing to the practical impossibility of determining values of prevalence and concentration at the time of consumption. Furthermore, in certain situations, the result of a growth process (storage) or an inactivation process (heat treatment) could lead to changes in the levels of prevalence and concentration, which cannot be detected experimentally given the limitations of analysis techniques. In these cases, the use and development of models can help to quantify the effect on pathogen levels, by estimating changes in these parameters throughout the different steps in the food chain. For viruses and parasites, models deal chiefly with processes of decontamination and/or inactivation, since this type of microorganism has no growth capacity in food itself. However, in the case of bacteria, given their capacity to multiply in food, the models also deal with growth.

There are cases in which EA is performed using only the final levels of exposure to the microorganism (prevalence and concentration), without taking into consideration, therefore, the effect of the different intermediary stages undergone by the food (e.g., heat treatment, freezing, etc). This limits its application exclusively to simple risk estimations or the establishment of a risk ranking. However, EA models that assess exposure levels "from the farm to the table" include information and models of the different processes undergone by the food. This introduces a greater degree of detail into the EA and makes it possible to study the factors involved in each of the steps or processes in terms of their effect on the final risk. This kind of EA is based on a structured scheme with models, where the different stages and processes throughout the food chain are described in a compartmentalised way (primary production, processing, point of sale, domestic environment).

The data required for this stage can be obtained from numerous sources:

- A review of the quality assurance and process data
- Sampling and *in situ* inspection data.
- Information about product design and processing
- Equipment and specifications of ingredients
- Raw material quality records.
- Specific analyses of food products on line, and finished food products.

1.4. Risk Characterisation

According to the CAC (1999), "Risk is a function of the probability of an adverse health effect and the severity of that effect, consequential to a hazard(s) in food".

Risk Characterisation (RC) basically involves using the information from the previous stages to obtain an estimation of risk. In other words, the results of the EA (concentration and prevalence) are combined (extrapolated) with the D-R model(s) obtained in the HC stage. As a result, risk is estimated, which in the first instance could be the probability of getting ill or being infected.

Risk can be considered from different perspectives (units of risk expression). The probability values of getting ill can be transformed in order to estimate the number of people who will fall ill every year or the number of cases requiring medical care, even going as far as estimating the economic cost associated. It is important to highlight that risk managers are better able to interpret the results when the choice of units in which to express risk takes account of the sphere of application. For example, when dealing with public health issues, risk is related to the population as a whole rather than individuals and should therefore be expressed as the number of cases/year instead of the individual probability of getting ill.

The economic dimension of risk is also important. Developed countries such as the US have high levels of health spending as a result of foodborne illnesses. In the year 2000, the ERC (Economic Research Service) (ERC, 2000), estimated the cost associated with 5 pathogen bacteria as $6.9 billion (Table 1). The economic aspect of risk is considered when it comes to prioritising risks; drawing up risk reduction strategies, which must maintain a balance between cost and benefit; and in the general development of food safety policies.

Table 1. Annual costs in the US derived from foodborne pathogens

Pathogen	Estimated annual foodborne illnesses*			Costs
	Cases	Hospitalizations	Deaths	
				2 x 10^12 dollars
Campylobacter spp	1,963,141	10,539	99	1.2
Salmonella	1,341,873	15,608	553	2.4
Escherichia. coli O157	62,458	1,843	52	0.7
Escherichia. coli, non O157 STEC	31,229	921	26	0.3
Listeria monocytogenes	2,493	2,298	499	2.3
Total	3,401,194	31,209	1,229	6.9

* Data obtained by the Center for Disease Control and Prevention in the United States.

2. Development of Microbial Risk Assessment Models for Food

MRA is based on the application of models, fundamental in the stages of Exposure Assessment (EA) and Hazard Characterisation (HC) (Lammerding and Paoli, 1997).

A model is an abstraction of reality, which captures the functional essence of a system, with sufficient detail so that it can be used in research and experimentation instead of the real system, thus diminishing risk and cutting down on the time and cost. Basically, a model is made up of a series of input variables and output variables and a set of relationships between these variables (Figure 2). Output would be the result of the model, which in the context of MRA is the estimation of risk and its impact.

Models can be classified as: physical, mathematical and computational. The first type refers to models that are developed in the laboratory, also called experimental models. There are clear examples of this kind of models in scientific literature: experimental approaches to MRA and more specifically to EA. For example, Kennedy et al. (2005) examined the presence of *Staphylococcus aureus* and other pathogenic microorganisms in domestic fridges and recorded the profile of several domestic fridges in order to use this information subsequently in a laboratory study of the growth of *S. aureus*. This project is an example of an EA model created in a laboratory, where the final result was the number of *S. aureus* cells that would reach the consumer following domestic refrigeration. Even though these types of models are frequent in literature, (Gill et al., 2002), the amount of resources (time, material and people) required to apply them means that they can only deal with very specific scenarios. In other words, experimental models would not be able to tackle a complete EA that models the transmission of the microorganism throughout all the steps in the food chain.

Mathematical models use physical and chemical laws to describe the behaviour of a real system in mathematical terms (Dym, 2004); they are closely linked to computational models, which are applied as a calculation tool in the resolution of mathematics.

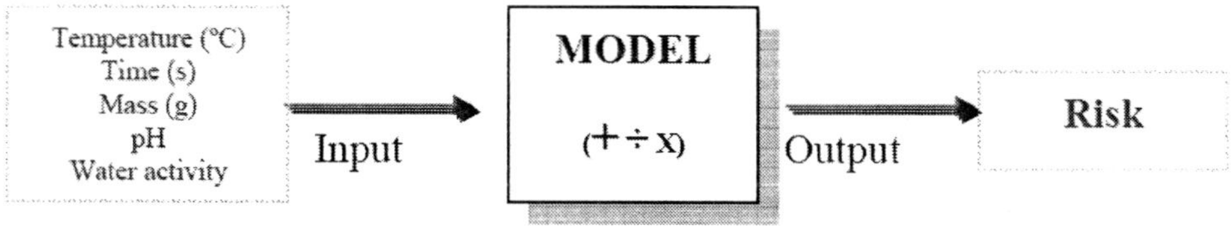

Figure 2. Schematic representation of the components of a model.

Their combined application saves a considerable amount of time and resources in comparison with experimental models. Furthermore, mathematical models make it much easier to assess the effect of possible modifications in the system. This procedure is essential in order to gain detailed information about how the object of study works. In our case, this could be compared to knowledge about how certain factors along the food chain (storage temperature, heat treatment, etc.) affect or influence the final risk (e.g., the probability of getting ill).

2.1. Quantitative Microbial Risk Assessment Models (QMRA) for Food

MRA models can be qualitative and/or quantitative. In the qualitative approach, the different stages are performed using categorical information for the variables; in this case, the output of the MRA is not a number but rather a risk classification (e.g., high risk or low risk). Quantitative MRA (QMRA) on the other hand, assigns numerical values to the variables, which makes the assessment more complete and accurate; however, in exchange, it requires more and better information.

The choice between the two types, quantitative or qualitative, is determined by the number of resources available and the level of accuracy and precision required (Lammerding and Fazil, 2000). Hence, when there is limited time or information available to perform the analysis, the qualitative approach is the most appropriate. Furthermore, qualitative MRA is used as a previous step to QMRA (e.g., risk profile), helping the assessors to focus on the truly necessary information and establish the scope and objectives of the QMRA that are of interest to risk managers.

QMRA can be deterministic (also called point-estimate) or probabilistic (FAO/WHO, 1995; CAC, 1999). A QMRA model is deterministic when the variables are defined by single values. The arithmetic mean is the most commonly used value, although other statistics could be also used (percentile, geometric mean, median, etc). The use of a single value for the model variables is not the most accurate or comprehensive way of defining the *physical system* being modelled. The explanation lies in the fact that the *physical system* contains intrinsic (natural) variability associated with the heterogeneity of the system (different microbial populations, different storage temperatures, etc.) and extrinsic variability, which is the consequence of the limitations of the observer, also referred to as experimental variability or uncertainty (measuring errors, sampling errors, etc). These two elements are eliminated when a specific value such as the mean is used. The importance of this fact is illustrated graphically in Figure 3, where the use of a specific value (mean) for concentration (log cfu/g) leads to a clear loss of information (Lammerding and Fazil, 2000): for example, the extreme values of concentration, which are so decisive in microbial risk (Zwietering and Perez-Rodriguez, 2006), are excluded from consideration when the mean value is used.

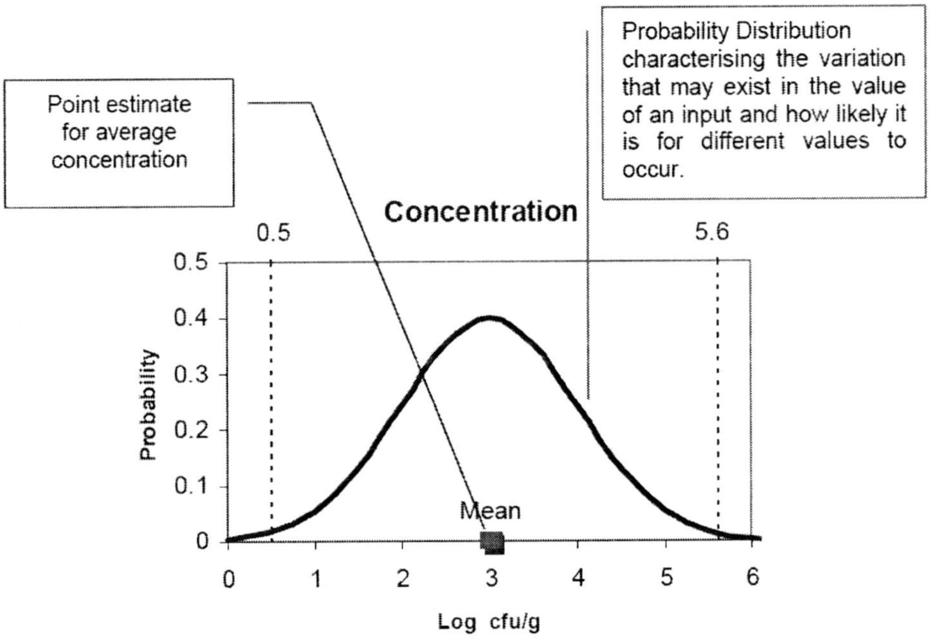

Figure 3. Comparison between the use of a specific value (mean) and a distribution to characterize a range of concentrations (adapted from Lamerding and Fazil, 2000).

The probabilistic approach, through the use of probability distributions that describe the probability associated with each value (or event), makes it possible for the QMRA (Vose, 2000a) to reflect all the possible combinations between the values of the model variables.

2.2. Use of Probability Distributions in QMRA

In order to carry out calculations using probability distributions in a probabilistic QMRA, specific statistical techniques must be used: method of moments, analytical method and numerical method.

The first method refers to a series of probability laws that enable the addition, subtraction and multiplication of the mean and variance of a normal distribution. However, this method has certain limitations that prevent widespread use: the impossibility of representing other mathematical operations (division, power, etc); and the condition that all variables must follow a normal distribution, which is not so in most cases.

Analytical methods are based on the symbolic calculation of the area of the probability functions (probability distributions). The results obtained through such calculations are extremely accurate. The disadvantage of this method is that for certain probability functions or combinations, there is no integration solution.

Finally, the most widespread methods are the numerical methods represented by the Monte Carlo simulation techniques and the Latin Hypercube sampling method. The first technique involves randomly sampling each probability distribution in the model, producing hundreds or thousands or values for each distribution, which are then used individually to operate in a traditional way. Each value or event sampled by the simulation is called iteration. The sampling is carried out in such a way that, at the end, the group of data obtained reproduces the original distribution shape. Figure 4 provides a graphic representation of the simulation process of a probabilistic model. The greater the number of iterations, the more accurate the estimation of the area; in other words, an increase in the number of iterations indicates greater probability of covering (sampling) the whole distribution area.

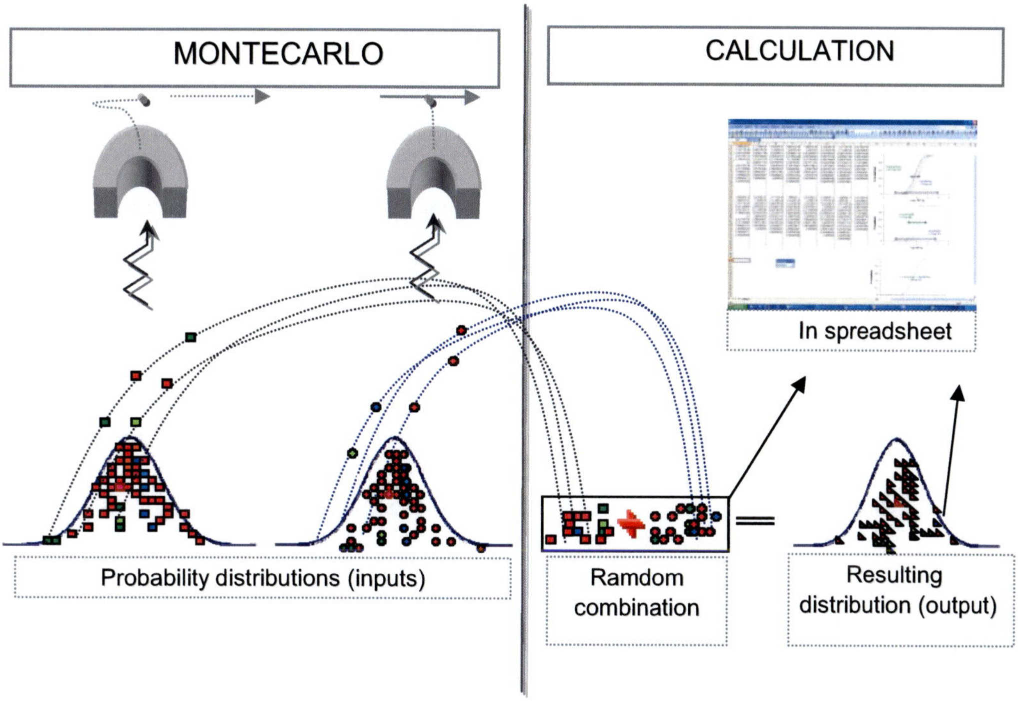

Figure 4. Graphic representation of a Monte Carlo simulation. Monte Carlo acts like a 'magnet' attracting and extracting the distribution values one by one (iteration). The distribution values with higher probabilities will be attracted more strongly and will therefore be collected more often. Then, all the values are entered on a spread sheet and combined randomly, according to the mathematical relation indicated ("+"; "-" ; "/"; etc.), in order to obtain the resulting distribution (output).

In simple terms, Monte Carlo simulation techniques are mathematical algorithms that aim to imitate the sampling of a real system. The Latin-Hypercube sampling method is similar to the Monte Carlo technique, only more efficient, given that the sampling process is artificially forced so that all the distribution areas are sampled with fewer iterations (shorter simulation time). Given its characteristics, Latin-Hypercube is particularly appropriate i) for simulations carried out on a PC, where calculation resources are more limited, and ii) when researchers are interested in sampling very low probability areas of the distributions.

2.3. Importance of Considering Variability and Uncertainty in QMRA

The probability distributions of a probabilistic model can represent uncertainty and/or variability. These two terms refer to totally different concepts. Variability represents the heterogeneity of the population, as the result of the physical system, which is irreducible through further measures or studies. Variability has been described as "random uncertainty", "stochastic variability" and "interindividual variability". Uncertainty, on the other hand, represents a lack of knowledge and can be, at times, reduced through further studies or measures. This is also termed "fundamental uncertainty", "epistemic uncertainty" and "degree of belief".

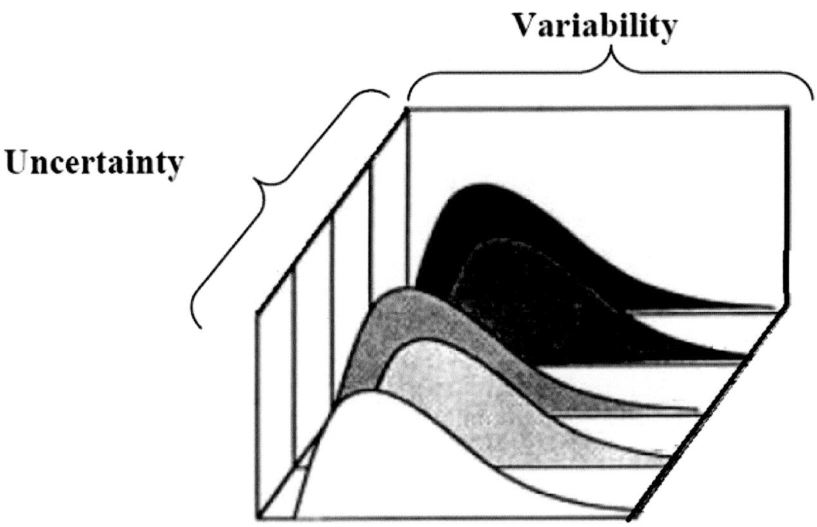

Figure 5. Representation of variability and uncertainty for a hypothetical variable.

In relation to these two components, variability and uncertainty, QMRA models can be classified as first or second order models (with one or two dimensions, respectively). First order or one-dimensional models do not establish a separation between both components in the variables, so that a single distribution describes variability and uncertainty without differentiating. Second order or two-dimensional models, on the other hand, apply more than one distribution for any one variable. For instance, in Figure 5, each distribution individually describes variability and as a whole, they describe uncertainty.

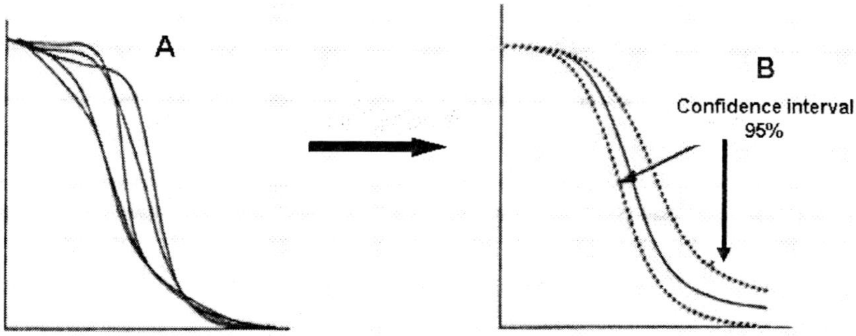

Figure 6. Representation of variability and uncertainty of the variable concentration for a pathogen at the time of consumption.

Figure 6 A represents a hypothetic variable, pathogen concentration at the time of consumption. Each cumulative curve corresponds to a possible distribution for the variable, and all together they represent the uncertainty about the true distribution of the pathogen concentration at the time of consumption. The results, in these cases, can be summarised numerically by obtaining confidence intervals (90%, 95%, etc.) from different fractions throughout the variable as shown in Figure 6 B.

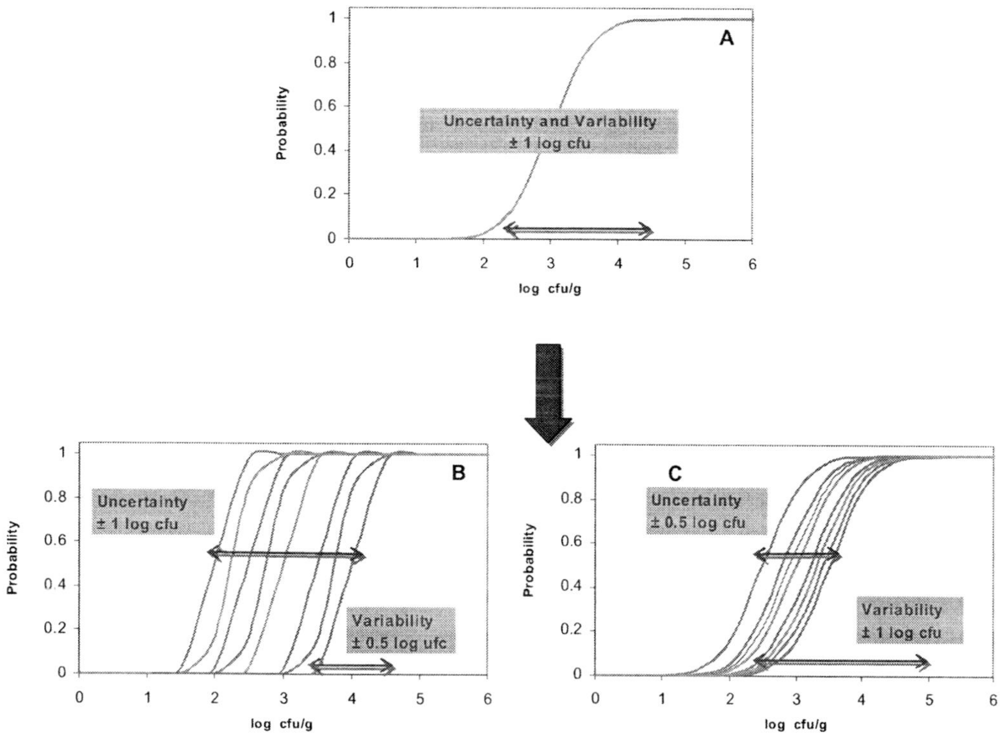

Figure 7. Graphical representation of the output of a hypothetical Quantitative Microbial Risk Assessment (QMRA): (A) of a first order model without separation of variability and uncertainty; and of a second order model, in which variability and uncertainty are separate with (B) uncertainty range >> variability range ; (C) uncertainty range << variability range.

Knowledge of the degree of uncertainty about the estimation helps to weight the options in decision-making processes. In order to illustrate this statement, here is an example in which the statistical aspects have been simplified in order to aid comprehension.

Imagine that the final estimation of a first order QMRA is the concentration of a pathogen (log cfu/g), with a mean value of 3 log cfu/g and variation of ± 1 log cfu/g.

These values indicate that the final concentration can vary between 2 and 4 log cfu/g, where the most common value is 3 log cfu/g. However, given that this is a first order model, we do not know whether this variation corresponds to the real variation that exists between samples (variability), analysis error (uncertainty), or both. Furthermore, in the latter situation, we also would not know which range corresponds with each component. Figure 7 A represents the distribution of concentrations obtained with a first order model.

For a second order model, on the other hand, the variation range ± 1 log could be characterised or broken down into a variability component and an uncertainty component as shown in Figure 7 B-C. So, given the fact that variability cannot be reduced, we could be faced with two types of extreme situation. For example, one possible case (Figure 7 B) would be ± 0.5 log variability and ± 1 log uncertainty. As a result of the high degree of uncertainty, before making a decision, further analysis and study (more information) might be recommended in order to reduce this component. However, if forced to make a decision in this situation, the information about the two components is still valuable, since the manager in this case is aware of the reliability of the data, and as a consequence, of the level of consistency of the decision to be made. Another example, the opposite of the previous case, (Figure 7 C), might be ± 1 log variability and ± 0.5 log uncertainty. The decision derived from these results would be more consistent and reliable than in the previous case, since the data are more precise and accurate (greater certainty about the result).

In both types of cases, the application of sensitivity analysis techniques can be extremely useful when it comes to supporting and assessing the decisions to be made. Sensitivity analysis helps to identify the variables (parameters) that contribute to the uncertainty of the model. For example, if variability and uncertainty are separate (Figure 7 B-C), managers who know which variables or parameters are the significant sources of uncertainty will be able to allocate resources more efficiently in order to study the issue, which would reduce the uncertainty of the results (Cullen and Frey, 1999; Zwietering and van Gerwen, 2000).

Second order models need a significantly greater amount of resources (technicians, quality and amount of data, time etc.) in comparison with first order models. Hence, the risk assessor and manager must be certain - before undertaking this type of analysis - that the problem raised cannot be resolved using another simpler approach.

2.4. Advances in the Development of QMRA

Although the outline proposed by the CAC (1999) establishes general guidelines for performing a QMRA, no procedures or methodologies have yet been validated or published about how to structure and tackle a QMRA systematically based on models. However, this fact has not prevented scientists or institutions from performing QMRAs. For instance, Lindqvist et al. (2002) conducted a quantitative assessment of *S. aureus* in raw milk cheese, based on the application of growth models and concentration and prevalence distributions for the pathogen. Furthermore, Bemrah et al. (1998) performed a quantitative assessment for

Listeria monocytogenes in raw milk cheese, based on growth and D-R models. Both were implemented on a spreadsheet and simulated using the Latin-Hypercube method.

In the institutional sphere, there are two organisations that are particularly prominent in the development of this new field: the USDA (United States Department of Agriculture); and the multinational FAO. The USDA has published several studies on QMRA, mainly through the Food Drug Administration (FDA) and the Food Safety and Inspection Service (FSIS), which carried out the majority of the research work, following a very strict methodology based on initial consultations between risk assessors and managers to determine the object and objective of QMRA and on continual review procedures to improve the QMRA study. The best example is the QMRA performed for *L. monocytogenes* in different ready to eat foods, which was especially aimed at creating a risk ranking for the different categories of food (FDA/USDA/CDC, 2003). In this case, the study focused on obtaining the estimation of risk, and not detail was considered on different processes or steps in the food chain. The risk estimation values were used as the criteria to establish a risk ranking of the different food products considered in the study. The mathematical model was developed in a spreadsheet format, applying the programming language Visual Basic. Another interesting study is the QMRA of *Escherichia coli* in minced beef (FSIS/USDA, 2001) (Table 2). In this study, models of growth, inactivation and D-R were used to yield the final risk estimation. An interesting aspect of the latter study is the use of a modular structure, which divides the modelling process into stages along the food chain (production, slaughter house and preparation). The FAO, on the other hand, has carried out a great many risk assessments: *Salmonella* spp in roasting chickens and eggs; *L. monocytogenes* in ready to eat food; *Vibrio* spp. in fish and seafood; *Campylobacter* spp. in roasting chickens; and *Enterobacter sakazakii* and other microorganisms in powdered baby milk (Table 2). In the FAO's assessments, apart from estimating risk and getting a more in-depth understanding of the factors influencing risk, there was also an underlying motive: to assess modelling methodologies in QMRA. This fact is especially evident for D-R models, which deserve major analysis (FAO/WHO, 2003) given their great impact on the estimation of risk.

Despite these advances, there are still significant gaps, which will only be filled when the techniques and methods used for QMRA are validated and standardised. It is important to keep in mind that both the FAO and the EU - through its White Book on Food Safety - have indicated the need to improve the methodology and techniques used in QMRA. The ultimate goal is to draw up guidelines that will enable QMRAs to be performed in a more routine and efficient way, as well as making the results more comparable between different assessments. Systematisation processes might be a first step in the development of more applicable and comparable models. Furthermore, the standardisation of models could also be the key to understanding the results: familiarisation with the methodology would enable risk managers, who have to interpret the results, to understand more easily the key aspects of QMRA. The studies led by Cassin et al. (1998) and Nauta et al. (2001) are examples of attempts to standardise and unify the methodologies through a systematisation of the modelling process, an aspect that will be explained later on in this section.

In the development of a QMRA model, the major modelling burden lies in the EA stage. The process of designing and developing an EA model is comparable to putting together the pieces of a puzzle. First of all, you should examine the completed image that the puzzle represents so that you know where the pieces should go. The pieces of the puzzle are compartmentalised representations of a real system, in other words, unit divisions of steps and processes along the food chain: manufacturing, transportation, retail point and domestic preparation. Each of them can be described using mathematical models. Just like when you put the pieces of a puzzle together, the models do this by establishing mathematical relationships (division, addition, power, etc), yielding the complete image of the puzzle, i.e. the mathematical representation of all the food steps and processes that are the object of the EA.

In summary, the EA model should; first of all, define the object to be examined; secondly, break it down into sub-processes in order to facilitate its mathematical description; and finally, establish the relationship between all of them in order to obtain the final estimation that is the object of the EA.

Table 2. Some published studies on Quantitative Risk Microbial Assessment (QRMA)

Orden[a]	Producto	Microorganismo(s)	Validación[b]	Referencia
1	Raw milk	*Listeria monocytogenes*	NO	Bemrah et al. 1998
1	Ground beef hamburguers	*Escherichia coli* O157:H7	NO	Cassin et al. 1998
2	Ground beef	*Escherichia coli* O157:H7	NO	FSIS/USDA, 2001
2	Chicken	*Campylobacter* spp. themophilus	NO	Rosenquist et al. 2003
1	Turkey products	*Salmonella* spp.	NO	Bemrah et al. 2003
1	Cheese made from raw milk	*Staphylococcus aureus*	YES	Lindqvist et al. 2002
1	Deli-meat products	*Listeria monocytogenes*	NO	Chen et al. 2003
2	Ready-to-eat products	*Listeria monocytogenes*	YES	FDA/USDA/CDC, 2003
1	Pork products	*Salmonella* spp.	YES	Giovannini et al. 2004
1	Whole chicken	*Salmonella* spp.	NO	Oscar, 2004
2	Deli-meat products	*Listeria monocytogenes*	YES	FSIS/USDA, 2003
2	Animal-foods	*Salmonella* spp.	NO	Hald et al. 2004
1	Soft cheeses made of raw milk	*Listeria monocytogenes*	NO	Sanna et al. 2004
1	Pasteurized liquid eggs	*Salmonella Enteritidis*	NO	Whiting and Buchanan, 1997
1	Broilers	*Campylobacter* spp.	NO	Hartnett et al. 2002
2	Ready-to-eat products	*Listeria monocytogenes*	NO	FAO/WHO, 2004
2	Raw oyster	*Vibrio vulnificus*	YES	FAO/WHO, 2005
2	Eggs and broiler chicken	*Salmonella* spp.	NO	FAO/WHO, 2002
1	Steak tartare	*Escherichia. coli* O157	YES	Nauta et al. 2001
2	Ready-to-eat meat/poultry products	*Salmonella* spp.	NO	FSIS/USDA, 2005a
2	Shell eggs and egg produts	*Salmonella enteritidis* y *Salmonella* spp.	NO	FSIS/USDA, 2005b
2	Ready-to-Eat and Partially Cooked Meat and Poultry Products	*Clostridium perfringens*	NO	FSIS/USDA, 2005c
2	Raw oyster	*Vibrio parahaemolyticus*	YES	CFSAN/FDA, 2005

[a] First order model (1) and second order model (2); [b] Validation process (YES/NO).

Cassin et al. (1998) aimed to integrate the methodological outline of MRA established by the CAC (1999) with the application of "Predictive Microbiology" and the development of scenarios, in order to provide an objective assessment of the hygienic characteristics of a food process. The Process Risk Model (PRM), as the authors term this approach, has two parts. The first describes the behaviour of the microorganism throughout the steps of processing, handling and consumption of the food, which would give rise to the EA stage. Therefore, the second part corresponds to a D-R model, which estimates the health risk associated with consuming a food.

The methodology described by Nauta (2005, and 2001) and Nauta et al. (2001), on the other hand, goes a step further both in terms of detail and systematisation. The Modular Process Risk Model (MPRM) is a development of the PRM proposed by Cassin et al. (1998), focusing solely on the EA stage. Basically, this approach proposes that any step throughout the food chain can be described mathematically through six basic processes: growth, inactivation, mixing, partitioning, removal and cross contamination (Table 3). Once again the concept of compartmentalisation appears, as mentioned previously, but in this case a greater degree of simplification is achieved. Through the application of six types of models, an accurate estimation can be made of the levels of exposure to a pathogen from a specific food process. According to Nauta (2001) and Nauta et al. (2001), the model of EA should be divided into small parts, as many as necessary in order to express the logic between the variables and thus model as accurately as possible the relationship between them, all in accordance with the purposes or objectives of the assessment. The main advantage of this method lies in the systematisation and simplification of the model creation process, which might explain why it was the subject of a technical report produced by the European Commission (2003) aimed at laying the foundations of EA methodology in a framework of QMRA.

Table 3. Qualitative effect of the six basic processes proposed by Nauta (2001) and Nauta et al. (2001) on prevalence, concentration and portion size. "=": no effect; "+": increase; "-": decrease (adapted from European Commission, 2003).

Process	Effect on the fraction of contaminated units	Effect on the total number cells over all units	Effect on unit size
Growth	=	+	=
Inactivation	-	-	=
Mixing	+	=	+
Partitioning	-	=	-
Removal	-	-	=
Cross-contamination	+	=/+	=

2.5. Types of Models in Quantitative Microbial Risk Assessment

There are many different types of models used in risk assessments. The most important are the *growth, inactivation and survival models*, which can be considered the cornerstone of the development of Quantitative Microbial Risk Assessment (QMRA); in fact, a QMRA model could just comprise growth and inactivation models. These kinds of models are the main object of study in "Predictive Microbiology". Despite emerging as a field as far back as

1920, owing to interest in producing "sterile" canned food (Bigelow et al., 1920), it has only been in the last twenty years that the most significant advances have been made, which have made models available to industries and official bodies aimed at improving process control.

Furthermore, recently, other models have emerged in this same field, called *cross contamination* models, which have not been developed to such a great extent, but have been proved to be important in the estimation of risk for certain pathogens and scenarios (Reij et al., 2004; Beumer et al., 2003). Cross contamination models deal with the transfer capacity of microorganisms between surfaces, environment (knob, air, hands, etc) and food, predicting the number of microorganisms transferred from one surface to another.

D-R models are the main argument in the development of the HC stage, and a fundamental tool for translating exposure values obtained in the EA stage into risk units in the RC stage. Even as a contrast, they are also an important source of uncertainty given the lack of knowledge about the processes involved in the mechanisms that take place during infection processes leading to illness. This section includes a classification of the mathematical functions and their theoretical basis.

There are also other types of models applied to QMRA. For instance, there are models that describe the behaviour of food handlers, which incorporate into the QMRA patterns of consumption, the level of hygiene during preparation, and the association with sociocultural factors, gender, age etc. (Christensen et al. 2005). The use of these models in the EA stage makes it possible to identify the risk factors associated with sociocultural aspects. Probability distributions are also applied in the EA stage, in order to describe how microorganisms are distributed in the food (concentration and prevalence); in many cases, a poor choice of probability distribution (e.g., the chosen distribution does not fit data) can lead to significant accuracy errors in the model's estimation. Hence, it is important to analyse the distribution data of the microorganism in the food, and then decide which distribution best fits these data (mathematically and theoretically) (Haas et al., 1999). Finally, other types of quantitative models are also applied during the EA stage, which reflect changes in concentration and prevalence in the food as the result of processing, mixing, partitioning and removal (Nauta, 2005, 2001; and Nauta et al., 2001).

2.5.1. Growth, Inactivation and Survival Models

These models use information about the characteristics of the microorganism, product and process to predict changes in the concentration (and prevalence) of the microorganism in the food (Legan et al., 2002), using mathematical equations.

Ever since it emerged as a field of study, "Predictive Microbiology" has shown a great deal of interest in differentiating between two types of model: *empirical* and *mechanistic*. The conceptual difference is clear, although less so in practice. An empirical model is developed through experimental measures without considering the underlying physical, physiological and biochemical processes from a mathematical point of view. The development of a mechanistic model, on the other hand, is based on theoretical knowledge of microbial behaviour and its translation into mathematical terms. This latter model is more difficult to design owing to the lack of knowledge about the processes involved in microbial kinetics. In spite of all this, it is still rare to find an example of a purely mechanistic model, as is the case with the model developed by Cayley et al. (1992), which links the growth of a strain of *E. coli* under conditions of osmotic stress with the intracellular accumulation of betaine and proline and the thermodynamics of osmoregulation. This dichotomy of approaches is always present in debates and although it is assumed that mechanistic models are the main objective of scientific research, in "Predictive Microbiology", the vast majority of them are empirical or quasi-mechanistic, with a few exceptions. Hence, this section will focus on empirical and quasi-mechanistic models and discuss their advantages and disadvantages when applied to QMRA.

An initial classification of Predictive Microbiology is based on the level of description and development of models, leading to *primary, secondary* and *tertiary* models. Primary models are mathematical equations that describe changes in the number of bacteria over time under specific environmental and culturing conditions. Secondary models are equations that describe how the growth parameters of primary models vary. Finally, tertiary models are applications of one or more primary and secondary models through the use of computer programmes. One of the most commonly used programmes is the Pathogen Modeling Program (PMP) developed by the USDA and available free on the Internet (www.arserrc.gov/mfs/*pathogen*.htm). This programme provides estimations for pathogenic microorganisms during inactivation and growth processes in function of different environmental variables. More recently, a database has been developed, Combase, which provides information on microbial growth and inactivation, and is available on the Internet (http://wyndmoor.arserrc.gov/combase/).

Based on the nature of the microbial response, models can be classified as: *growth* models, *inactivation* models and *survival* models.

Growth Models

Growth models deal with the range of conditions that permit growth. This type of model describes the increase in microbial concentrations over time (kinetics models); the conditions that permit growth or non-growth (growth/non-growth limit model) as well as growth probability (probabilistic model).

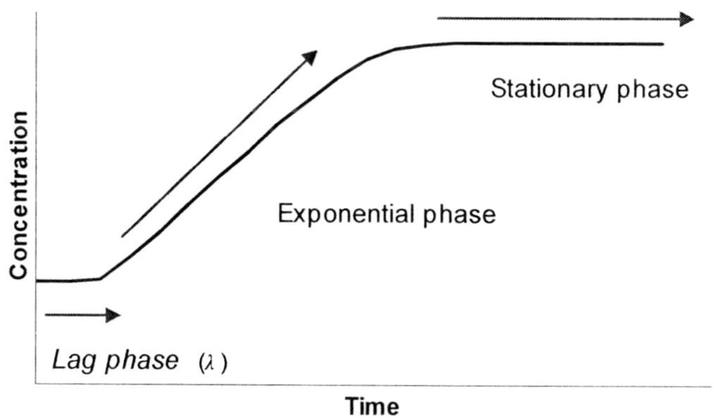

Figure 8. Microbial growth curve, encompassing four phases: Lag phase; Log or Exponential phase; and Stationary phase.

Kinetic models are based on the growth curves obtained for different factorial conditions (Figure 8). Primary kinetic models are mathematical equations that describe the change in microbial populations over time using the parameters growth rate, lag time and maximum population density (MPD). Among the different proposals, the most commonly used equations are the modified Gompertz equation, the logistic equation and the Baranyi equation (see Table 4). The first two equations are based on the Sigmoidal or S-shape of the microbial growth curve, and the third one, on the development of the differential equation that describes exponential growth phase (Figure 8).

Secondary models establish mathematical relation between the parameters derived from the primary models (growth rate and lag time) and environmental factors. There is a wide variety of models. Table 5 shows the most widely-used models in scientific literature.

Polynomial models, also called Response Surface Models (RSM), are the most widely used secondary models in Predictive Microbiology, firstly because of the good fit to experimental data and secondly because their mathematical development is straightforward. However, certain limitations make them less preferable to other types of models. RSM can only be applied in the interpolation region, since extrapolations produce erroneous and illogical predictions. Furthermore, the interpolation region is much narrower than for other types of models. So, although they are easy to implement both in spreadsheets and other computer formats, the narrow interpolation region and the impossibility of carrying out extrapolations reduce the possibility of including new ranges of factors in QMRA models. As an alternative, there are Square Root Models, such as the Ratkowsky model or the Bêlerádek model (McMeekin et al., 1993), developed on the basis of the model proposed by Ratkowsky et al. (1982):

$$\sqrt{\mu_{max}} = b \cdot (T - T_{min}) \tag{1}$$

where μ_{max} is the maximum specific growth rate expressed in cfu/ml, b is a constant and T is the temperature. The parameter T_{min} is the minimum theoretical temperature at which growth is detected.

This mathematical basis could be extended to include the effect of other environmental factors (water activity, pH, CO_2, etc). For instance, McMeekin et al. (1987) proposed a growth model for *Staphylococcus xylosus* in function of temperature and water activity, adding a multiplicative term for the new factor:

$$\sqrt{\mu_{max}} = b(T - T_{min})\sqrt{(a_w - a_{w\,min})} \tag{2}$$

The same example, but for the factor pH, was developed by Adams et al. (1991) who found the same multiplicative effect for pH.

The gamma concept was introduced by Zwietering et al. (1992). This modelling approach is based on two premises:

1. It has been observed that many factors affect bacterial growth independently. Therefore, their total effect can be represented by multiplying the individual effect of each factor.
2. The effect on growth rate can be expressed as a fraction of the maximum growth rate (when environmental factors are at the optimum level).

Under optimum conditions, the growth rate is at its maximum. However, according to the gamma concept, this value is reduced to sup-optimal levels as the consequence of the accumulative inhibitory effect of the factors. The effect of each factor is represented by a gamma function, which is adimensional with a value between 0 and 1. When there is no inhibition, the function takes the value of 1 and when the inhibitory effect is at its maximum, it takes 0. The multiplication of the gamma functions for each factor provides the total accumulated inhibitory effect. Hence, the growth rate at different levels of the factors is determined by the following equation (gamma model):

$$\mu_{\max} = \mu_{opt}\delta(T)\cdot\delta(a_w)\cdot\delta(pH) \tag{3}$$

The gamma model is highly applicable in QMRA, since the mathematical basis means that it can be completed with terms from other models to include new factors. In a QMRA model, this means that modifications can be made easily, in order to assess new factors, without having to change the basic model. One of the characteristics of an MRA is that it should be dynamic and flexible, in the sense of including new information and new approaches that enable the process to be improved continually.

Other more sophisticated models such as Artificial Neuronal Networks have shown an excellent data fit, with good prediction capacity. However, their application to QMRA models can be technically very difficult in terms of implementation, given the complexity of the functions used in this type of model. This is especially true in models developed in spreadsheets, where the amount of functions, parameters and transformations for each escalation means that it is more costly and takes longer to implement.

Although most secondary models are specific to growth rate, scientific literature also describes models for lag time, and to a lesser extent, Maximum Population Density (MPD), which is the maximum concentration achieved by a microorganism.

Lag phase models have been developed to a greater extent and their inclusion in QMRA could have a great impact on the result. In many cases, with limiting conditions in the food, whether or not a microorganism would reach or exceed a certain critical level could be determined from the length of the lag phase.

Table 4. Commonly used primary models to describe microbial growth curves

Model	Equation	Where:
Exponential	$\log(n) = \log(n_0) - \mu_{\max} t$	n= count/g at time t n_o= count/g at time t=0 t = time (h) μ_{max} = maximum growth rate (h^{-1})
Lag-exponential	$\log(n) = \log(n_0) - \mu_{\max}(t - \lambda)$	λ= lag time *(h)*
Baranyi	$n(t) = n_0 + \mu_{\max} A(t) + \ln\left(1 + \frac{e^{\mu_{\max} A(t)} - 1}{e^{n_{\max} - n_0}}\right)$	$n_{\max}$= maximum population density (count/g) $A(t)$ = is an adjustment function, considered to account for the physiological state of the cells. Its role is to define the lag phase $A(t) = t + \frac{1}{\mu_{\max}} \ln\left(\frac{e^{-\mu_{\max}(t)} + \left(e^{\lambda\mu_{\max} \cdot 1}\right)^{-1}}{1 + \left(e^{\lambda\mu_{\max} \cdot 1}\right)^{-1}}\right)$
Gompertz	$\log(n) = A + C\exp\left[-\exp\left(-B(t - M)\right)\right]$	B = maximum growth rate (h^{-1}); M= time at maximum growth rate takes place (h); A = upper asymptote of $\log_{10}$ bacterial count; C = the difference between A and the lower asymptotic $\log_{10}$ bacterial counts as t increases indefinitely.

Table 5. Commonly used primary models to describe microbial inactivation curves (adapted from Legan et al., 2002).

Model	Equation	Where
Exponential	$\log(n) = \log(n_0) - kt$	n= count/g at time t t = time k = rate constant $1/k = D$ = time need to reach a decimal reduction count/g
Logistic	$\log(n) = \log(n_0) + \frac{a}{1+\exp(b-ct)}$	n = count/g; n_0= count/g when t= zero; t = time (hours); a, b and c are all fit parameters; t = time (hours)
Weibull accumulative	$s = \exp(-bt^n)$ *or* $\log(s) = -bt^n$	s = survival fraction (n/n_0) at time t b = scale parameter n = shape parameter
Gompertz	$L(t) = A + C\exp\left[-\exp\left(-B(t-M)\right)\right]$	$L(t) = \log_{10}$ bacterial count at time t; B = the relative maximum death rate (h^{-1}); M= time at which death rate takes place (h); A = upper asymptote of $\log_{10}$ bacterial count; C = the difference between A and the lower asymptotic $\log_{10}$ bacterial counts as t increases indefinitely.

Table 6. Examples of secondary microbial inactivation models (adapted from Legan et al., 2002)

Model	Equation	where	Commentaries
Polynomial	$C_0+C_1V_1+C_2V_2+C_3V_1V_2+C_4(V_1)^2$	All *C*s are constants and all *V* are variables	Straightforward to apply by multiple linear regressions. No knowledge of process needed. No biological meaning to parameters.
Arrhenius type	$k=A\ \exp(-E_a/RT)$	*k* is death rate constant; *A* is 'collision factor'; E_a is inactivation energy of the system; *R* is the universal gas constant (8.31 J/mol/K); *T* is temperature in Kelvin	Parameters have no biological meaning. It has been adapted to include other parameters (e.g. pH) but does not predict limiting values for variables
Concept τ	$\tau = \dfrac{T_2 - T_1}{\log D_{T_1} - \log D_{T_2}}$	τ is the temperature necessary for a 10-fold change in D value; D is time for 10-fold reduction in survivors; T_1 is lower temperature; T_2 is higher temperature	It assumes a first order kinetic

Given the relationship between lag time and growth rate (Baranyi and Robert, 1994), some QMRAs have described this parameter using the simple equation:

$$\text{lag}=A+B/\mu_{max} \quad (4)$$

where A and B are regression parameters and μ_{max} is the maximum growth rate.

The downside to these models is that they are much less precise and accurate in their predictions, because lag time is not only influenced by environmental factors: pre-incubation conditions (environmental conditions prior to the experiment) also affect this kinetic parameter (Mellefont and Ross 2003; Swinnen et al., 2004; Baranyi et al., 1995; McMeekin et al., 1993). This inability to make reliable predictions, together with the high degree of variability associated with the parameter, means that in certain cases, the stochastic approach is considered the most appropriate; using a probability distribution instead of a mathematical model for lag time (Metris et al., 2005).

MPD models are considered in the literature to be of lesser importance. When applied to QMRA, the use of MPD models that do not include the effect of competitive microbial flora should be avoided, since they can overestimate the MPD by between 1 and 2 logarithms. Although, there are models that include the additional effect of competitive microbial flora, QMRA studies rarely include these types of models. In many cases, QMRA models use more straightforward alternatives:

1. Use of data about MPD obtained in food (challenge tests) and fitted to a linear model: MPD = K*T+C, where K and C are regression parameters and T is the growth temperature, considered to be the most influential factor on MPD. Other similar approach to describe MPD can be the use of a "logical" temperature-dependent MPD model (Figure 9).
2. Based on the fact that only high levels of competitive microbial flora have a significant effect on the growth of pathogens (Malakar et al., 2003), kinetic models of competitive flora have been applied in order to ascertain their concentration during the storage processes and, as a function of this, to apply a reduction factor on the growth rate of the pathogen.

Another important aspect in terms of kinetic models is the modelling of growth in dynamic conditions, i.e., when the levels of factors change as a function of time. Bernaerts et al. (2004) highlighted the need to develop these types of models, since those applied in static conditions can make mistakes in their predictions. However, even though they can be found in bibliography (McKellar, 2001; Baranyi and Roberts, 1994), they have not yet been applied to QMRA, probably due to their mathematical complexity and the lack of models available for certain microorganisms.

Growth/non-growth models define the limits of a set of conditions that permit or do not permit the growth of the microorganism. Temperature is probably the most significant factor in the determination of the line dividing the zone of growth and non-growth. This fact becomes more relevant in the case of refrigeration temperatures where microbial response is particularly variable (Figure 10). However, it might also be interesting to define the limits of growth for factors such as pH, water activity, citric acid, etc. Another variation on these models are those that estimate the time required for growth to occur, based on qualitative data about growth/non-growth, gathered at certain time intervals. Given the presence of censored data, survival analysis regression techniques are required.

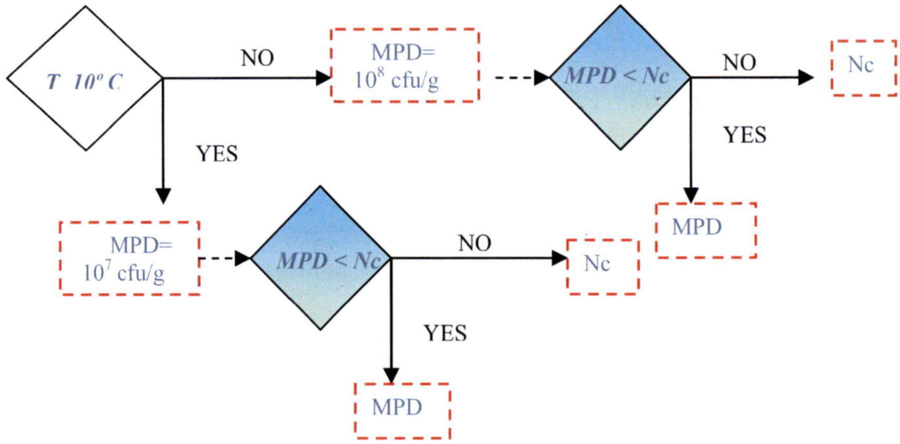

Figure 9. Representation of the temperature-dependent maximum population density (MPD) model used in the QMRA study of *Staphylococcus aureus* in cooked meat in Spain (unpublished). The rhombuses represent "If..." conditioning sentences in the model. Nc: concentration after applying the growth model. MPD: maximum temperature-dependent growth limit.

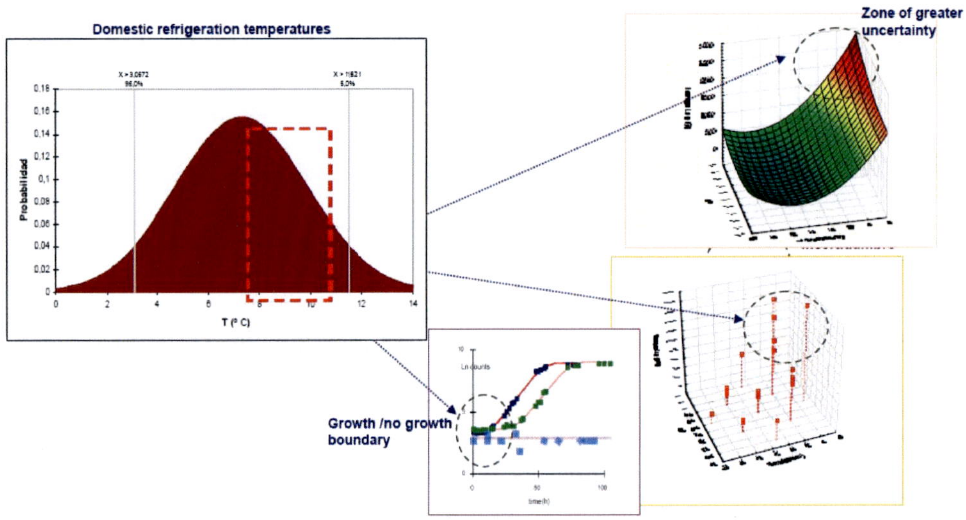

Figure 10. Graphic representation of the impact of refrigeration temperatures on the accuracy of growth models.

Growth/non-growth models are complementary to kinetic models, since they help to determine which conditions must be included in the design of the latter. The combination of both types of models can be understood more clearly when taking practical approach to QMRA. For example, a growth/non-growth model would be used to determine whether the conditions of a certain storage phase (temperature, storage time, pH, etc) permit growth or not. If the result is growth, the kinetic model would then be applied in order to estimate the concentration following storage time. In the case of non-growth, the kinetic model would not be applied, and therefore, the concentration value following storage would be equal to the initial value.

Growth/non-growth models can also incorporate *probability* when the experimental design implies several repetitions. These models estimate, therefore, the probability of growth (whether the microorganism grows or not) within an interval of time, using logistical, linear and non-linear mathematical functions. Once again, their application to QMRA is linked to kinetic models, providing a probabilistic component to the quantitative assessment of growth: not only do they indicate whether there is growth or not; they also specify the frequency of growth. For example, in the previous case, for certain storage conditions, the model could predict growth probability of 50%. This means that, in 100 cases, 50 would give growth, and therefore, the kinetic model would be applied, and the other 50 would not produce growth, and so the concentration would remain at the initial level (pre-storage level).

Inactivation Models

Death or inactivation models are applied to estimate the ratio of microbial death as a consequence of a process of microbial destruction. Most of these models were developed with a view to applying them in the design of more effective heat processes against pathogenic and spoilage microorganisms; however, inactivation models can also be applied to describe the effect of irradiation treatments or high pressure processes.

Inactivation models can be classified as either linear models, based on first order death kinetics, or non-linear models, which assume population heterogeneity in terms of sensitivity or resistance to the inactivation treatment. More recently, non-linear models have been considered an alternative to linear models, given the deviations from linearity (log) observed in certain inactivation processes. The most common models are the logistic equation and the Gompertz model (van Gerwen and Zwietering, 1998) (Table 5 and 6).

Survival Models

These models describe inactivation under non-extreme situations and often outside of growth boundaries. Processes such as heating the food up to around 50-60°C, drying on surfaces, the presence of toxic-inhibiting substances, etc. are the right kinds of situations to be described using this type of model. These could be classed as specific cases of inactivation when the death process occurs more slowly. They are less common than the previous groups, and although the data are treated in the same way as for inactivation, the selection of models is slightly different (Legan et al. 2002). Their inclusion in QMRA models could be crucial; for example, survival processes on surfaces or in atmospheres (air, aerosols, pools) are important in recontamination or cross-contamination processes. The number of cells transferred will depend on the number that survives the conditions of hydric stress normally present in these scenarios.

2.5.2. Dose-Response Models

Basically, a D-R model is a mathematical function that relates dose measure (number of cells ingested) and the probability of an associated adverse effect (infection, illness, death, etc). The probability value ranges between 0 and 1, where 0 indicates absence of effect and 1 the complete appearance of the effect.

The main reason for using D-R models is that contamination in foods and the adverse effects associated are so infrequent that direct observation methods are not effective for detection or compiling information; hence, in order to assess risk, models are required (generated with high doses) which enable extrapolation towards low doses.

Exposure to a pathogen can lead to different responses in the host, as shown in Figure 11. For example, the ingestion of a microorganism by a population will cause infection in a fraction of the population (P_I). Infection is manifested as a multiplication of the microorganism inside the host, followed by excretion. Analytical determinations of antibodies and the increase in body temperature (fever) could indicate infection. Out of all those infected, $P_{P:I}$ will get sick, and a fraction of these, $P_{M:D}$ will die.

Based on the type of response considered, models can be classified as:

- Dose-infection
- Dose-illness
- Dose-death.

To describe mathematically the relationship between dose and the probability of an adverse effect, we should focus on the following premise, which although logical, is crucial in the choice of the type of function: when the dose increases, the probability of the adverse effect also increases. Therefore, only cumulative probability functions can be used for this type of model.

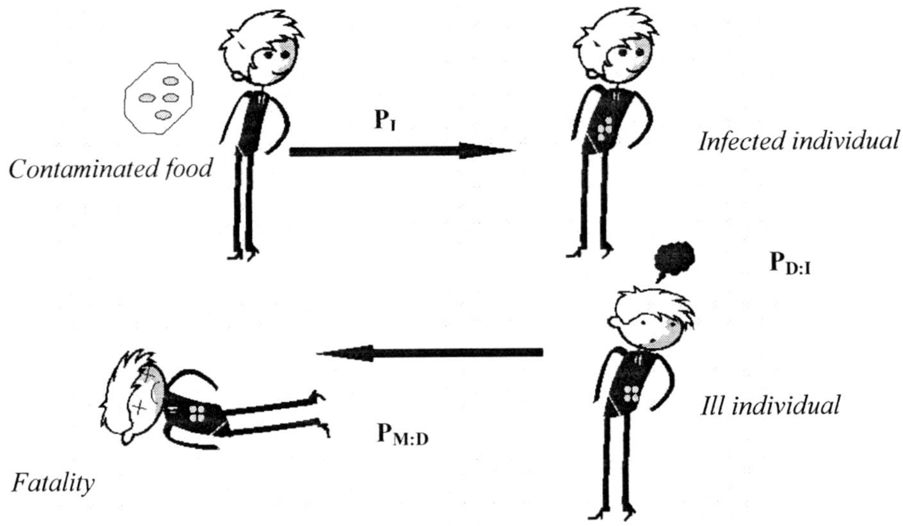

Figure 11. Graphic representation of the final points that could be the result of exposure to a microorganism .

Furthermore, it has been observed that the response of a group of individuals exposed to a dose (frequently expressed in logarithmic units) presents a sigmoidal or S-shape (Figure 12).

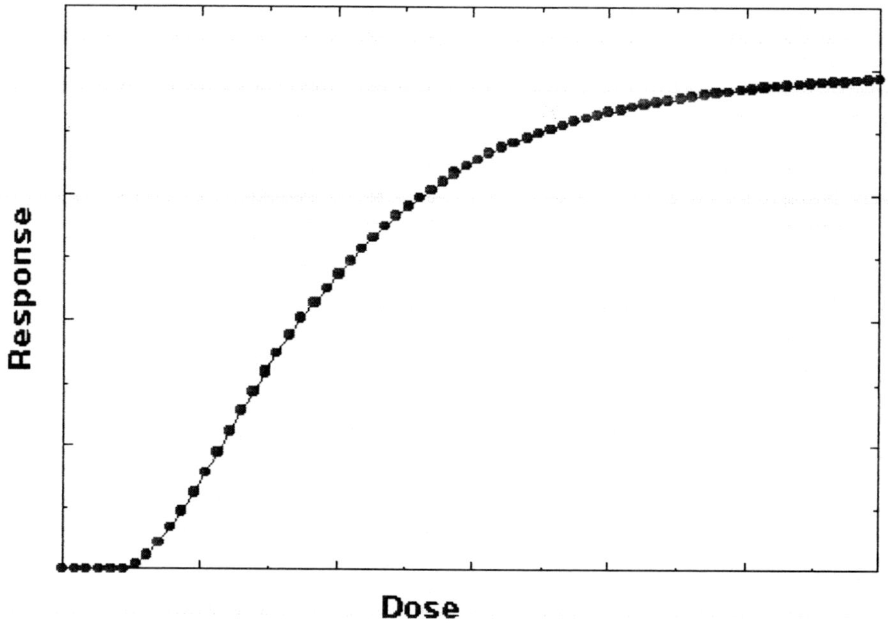

Figure 12. Sigmoidal curve that describes the physiological response (adverse effect) of a living being when exposed to a microbial pathogen.

There are many mathematical functions that meet these conditions and which, therefore, would be appropriate for use in a D-R model. However, when extrapolated out of the range of the experimental data, many of these models give rise to fairly inaccurate predictions (Coleman and Marks, 1998; Holcomb et al., 1999). The most widely accepted models base their mathematical development on theoretical hypotheses about how the microorganism acts in the host; in other words, they consider the interaction between the three factors: food matrix, pathogen and host.

In the case of "Predictive Microbiology", it is important to consider the separation between mechanistic and empirical models. In this case, there appear to be certain differences in the classification of the functions into one or the other group. According to Haas et al. (1999), Exponential and Beta-Poisson models are mechanistic, since they are developed on the basis of several hypotheses related with the variability and infectiousness of the microorganism or susceptibility of the host, and functions such as the log-logistic, log-probit and Weibull-gamma are empirical, because they are limited to describing the data range. However, Buchanan et al. (2000) question whether the models mentioned by Haas et al. (1999) are mechanistic, since the main justification for their use is the good fit to data, rather than as a means to uncover more in-depth information about the action mechanism of the pathogen (McLauchlin et al. 2004). Hence, Buchanan et al. (2000) define mechanistic models as those that describe the D-R relationship as the result of the compartmentalised modelling of the infection/illness mechanism: gastric acid barrier, adhesion/infectiousness and morbidity/mortality. The other models, including both the mechanistic and empirical models defined by Haas et al. (1999), are classified by Buchanan et al. (2000) as empirical. In the opinion of Buchanan et al. (2000), mechanistic models are a good alternative to improve prediction by extrapolation where empirical models usually fail. However, to date, very few of these models have been created, partly due to the lack of knowledge about how the three factors in the epidemiological triad interact. This section will treat the models included by Haas et al. (1999) as mechanistic since they are the most commonly used, and will explain the theoretical basis that supports them.

Theoretical Basis and Mechanistic Models

Hypothetically, in order for an infection process to occur, two sub-processes need to occur sequentially:

- The host should ingest one or more microorganisms capable of producing an adverse effect.
- Many of the microorganisms are eliminated or destroyed by the host's defence mechanisms, so only a fraction of the microorganisms ingested will reach the place where infection can begin.

These hypotheses imply, in mathematical terms, that the probability of infection, $P(k)$, is the product of the probability of ingesting j microorganisms per dose (d), $P_1(j/d)$, by the probability that k microorganisms will survive in order to initiate the infection process, $P_2(k/j)$:

$$P(k) = \sum_{j=1}^{\infty} P_1(j/d)P_2(k/j) \tag{5}$$

An important number of D-R functions can be derived from equation 5. The way in which the parameters P_1 ,P_2 , $k_{\min}$ (minimum infective dose) are defined will determine the type of D-R model obtained.

When developing the models, two alternatives can be taken, depending on whether the existence of an infection threshold is considered or not. Based on this consideration, models can be classified as follows:

- Threshold models: Infection can only occur after a certain number of inoculated pathogens have been ingested (minimum infective dose).
- Non-threshold models: It is not necessary to inject a certain number of inoculated pathogens in order for infection to occur.

There are two sets of hypotheses which are decisive for the mathematical development of any model, which are:

- The hypothesis of independent action and cooperative action. The first hypothesis means that the probability of causing infection per pathogen inoculated is different from zero and independent from the total number of pathogens inoculated. The second hypothesis, on the other hand, states that the probability of infection per pathogen is dependent on the number of pathogens inoculated.

- The concept of single-hit and multi-hit. The first states that a single pathogen, by itself, is capable of producing infection, in other words $k_{min} = 1$. The multi-hit concept hypothesis, summed up in mathematical terms, states that $k_{min} > 1$. Non-threshold models are based on single-hit hypotheses and threshold models are based on the multi-hit concept.

Even though threshold models are based on a solid theoretical foundation, to date all efforts to define the minimum infective dose of certain pathogens have proved unsatisfactory. The single-hit theory, on the other hand, the basis of the non-threshold models, has been proven experimentally through the study of outbreaks caused by *E. coli* O157:H7, *Salmonella Entiritidis* and *Shigella* (Buchanan et al. 2000), in which a single cell can be the cause of an outbreak. This fact has cast a shadow on the validity of threshold models and has relaunched the use of single-hit (non-threshold) models. The most commonly used non-threshold model is the exponential model, thanks to its mathematical simplicity and its excellent fit to experimental data. For the primary function in equation 5 we applied:

- The single-hit concept hypothesis ($k_{min} = 1$) and independent action hypothesis.
- d is not a single value but rather a dose distribution, the Poisson distribution (the pathogen is distributed homogeneously among the portions).

Hence, following mathematical development, the exponential D-R equation is as follows:

$$P_I(d) = 1 - \exp(-rd) \tag{6}$$

where r is the probability that a single microorganism produces infection.

The exponential model can approximate a linear model. When $rd << 1$, $\exp(-rd) \sim 1 - rd$, so:

$$P_I(d) \sim rd, \text{ for } rd << 1 \tag{7}$$

The Beta-Poisson D-R model is an exponential model with a variable r parameter. This variation could be due to the diversity of human response, differences in virulence between strains of the same species or both. The Beta-Poisson model describes the r parameter through a Beta distribution (Moran, 1954) and assumes cooperative action between pathogens (Buchanan et al., 2000); hence r is dose-dependent. Given the complexity of the explicit function of the model, a simpler approach has been adopted, as shown below (for further details about its mathematical development, see Hass et al. (1999)):

$$P_I(d) = 1 - \left(1 + \frac{d}{\beta}\right)^{-\alpha} \tag{8}$$

Parameters α and β are derived from the mathematical fit process. It is important to bear in mind that the approximation is less accurate at the low values of β or high values of d. Once again this equation can be approximated linearly for low doses.

Other models can be derived from the previous two functions, by modifying the type of distribution for the doses or even changing the distribution for r (Cassin et al. 1998).

Threshold models can be obtained from the exponential function, by modifying the starting hypothesis in terms of the minimum infective dose. In the exponential model $k_{min}=1$; however, in the infective threshold model, $k_{min} > 1$. On this basis, gamma type threshold models can be generated, which in their compact form can be defined as:

$$P_I(d) = \Gamma\left(k_{min}, dr\right) \quad (9)$$

Just as with non infective threshold models, these can include variability in the hosts' sensitivity, through the r parameter, by using the Beta distribution. It is also possible to construct a variable threshold model in which k_{min} is a distribution. Both types of models present certain difficulties in terms of their mathematical development. For more details see Hass et al. (1999).

2.5.3. Recontamination and/or Cross Contamination Models

Although the phenomena of cross contamination and recontamination have always been understood as an important source of food contamination, it is only over the last few years that researchers have shown interest in studying them in greater depth. The review carried out by Reij et al. (2004) is particularly significant in this aspect and stresses the importance of epidemiological information in order to identify pathways and risk factors associated with cross contamination. It also highlights the importance of including these phenomena in QMRA models and studying their impact on final risk. However, the bibliography reviewed is fairly inadequate and still lacks a unified view in terms of how cross contamination can be understood mathematically and therefore modelled and included in QMRA (Schaffner, 2004).

The few models included in the bibliography are based on the application of transfer rates (TR), which is the percentage of bacteria transferred from a contaminated surface to another (Zhao et al. 1998):

$$TR(\%) = \frac{cells\ on\ recipient}{cells\ on\ donor} \times 100$$

These cases refer to phenomena of cross contamination owing to contact between surfaces (food, utensils, hands, etc); however cross contamination can occur through the transfer of bacteria through liquids or air; in these cases, transfer rates can also be used with certain modifications in relation to the units considered (volume of air and liquid, sedimentation coefficients, etc.).

2.6. Validation: A Necessary Step in the Application of Models

Validation is an essential part of the modelling process. A model should not be applied until it has been validated, which occurs when the predictions are confirmed experimentally, statistically or using other quantitative method (Dym, 2004).

Any model, regardless of its nature (growth model, cross contamination model, etc) should be subjected to a process of validation. The process of validation can refer to four important aspects:

- Conceptual validation: this evaluates whether the model reasonably represents the system being studied.
- Validation of algorithms: this evaluates whether the mathematical functions represent the conceptual model
- Validation of the computer programming code (verification): this checks whether the algorithms or mathematical functions have been implemented correctly.
- Functional validation: this evaluates whether the model simulates the system being studied, in other words the similarity between the model's results and the observations of the system being studied.

In "Predictive Microbiology", functional validation has been developed to a great extent. In practice, most of the models in scientific literature, parallel to the development of the model, present a validation process. Furthermore, text books about "Predictive Microbiology" include very clear and detailed descriptions of the best and the most efficient way to approach the validation process (McKellar and Lu, 2004; Legan et al. 2002).

In "Predictive Microbiology", experimental analysis of growth in food is the basis of model validation (challenge tests): experimental growth data are compared with the model's predictions. This process is called validation in food in order to differentiate it from other types of validation, such as mathematical validation or validation using data from scientific literature. Mathematical validation takes place prior to validation in the food itself, and aims to estimate the goodness of interpolation of the model. Validation using data from the literature has the same purpose as validation in food but instead uses data from the literature. The latter is the most commonly used method since it is currently so easy to access a wide variety of information. An example is the well-known ComBase database (Baranyi and Tamplin, 2004), which can be accessed through the website http://wyndmoore.arserrc.gov/combase. This database provides valuable information about growth parameters (growth rate, lag, MPD) for different microorganisms and environmental factors.

In order to facilitate this comparison, a wide variety of statistical indices has been proposed; the most commonly used examples include the accuracy factor (Af) and the bias factor (Bf) (Ross, 1996). The first factor indicates the confidence level of the model, i.e. the accuracy of the model, and the second factor indicates whether the predictions are above or below the observed values. When they are below, the model's error is qualified as "dangerous", meaning that the model underestimates growth and therefore its application to food safety is not justified. The graphic representation of the values observed in comparison with the corresponding predictions gives a good idea of how well a model predicts and if these predictions over or underestimate growth. The use of several indices at the same time provides better knowledge of the predictive capacity of the model (te Giffel and Zwietering, 1999).

In the case of D-R models, all the information available is usually used to develop the model itself, so data from the validation process are not available. However, when new data are provided, the reliability of the predictions and the suitability of the assumptions made in the modelling process can be evaluated, by comparing the predictions of the model with the new data, which should not have been used for the model (Dennis et al. 2002; FAO/WHO, 2003). An example might be to compare the results of D-R model derived from data on mice with the detailed information about one or several outbreaks in humans. For more practical information about validation processes in D-R models, see Haas et al. 1999 and McCullagh and Nelder, 1989.

Cross contamination models can also be validated by comparing the transfer obtained before applying the model (transfer rate) to the values observed experimentally under similar conditions to those described by the model.

The QMRA model is also a mathematical model and as such should be validated as a whole. Once again, the method is based on the comparison between the model values and those obtained through observing the real system. For example, in the case of EA, predictive values for concentration and prevalence can be compared to data obtained from prospective studies carried out by companies, state institutions and industry associations.

To date, very few assessment studies have incorporated a validation stage. The difficulty in finding quantitative information and the lack of mathematical procedures might explain this gap. The study performed by Giovannini et al. (2004) did include a validation stage, using data obtained from telephone interviews to validate a QMRA of *Salmonella* spp. for consumers of pork products in a region of Italy. The data from the interviews were transformed using a second order Bayesian model into incidence values for cases of diarrhoea caused by the pathogen. The incidence values from the QMRA model (13.7%) and the data for validation, expressed as confidence intervals, were compared and found to be fairly consistent. Rosenquist et al. (2003) compared the number of cases of listeriosis obtained by their model with the real number reported for the year 2000 and found a good level of consistency between the two. In the both cases, the criteria used to make a decision about validity were based more on purely personal and subjective aspects and not on statistical analysis, which would provide greater robustness.

In other cases, when there are no data that can be used for validation, the review process is carried out by other advisers and scientists who are not involved in the assessment itself, which can be a good tool to confirm and determine whether the assumptions, approximations and limits of the model are suitable for its purpose. Voysey et al. (2002) states that it is essential to assure that the results of the risk assessment are in line with common sense and experience. A clear example of this is the QMRA performed by the FDA/USDA (2003) for *L. monocytogenes* in ready to eat products, which included an impressive review process that acted as a validation process.

Furthermore, the separation between variability and uncertainty in second-order models can be used for validation purposes. The way in which uncertainty affects the results of the model determines how valid the results are.

By means of a summary, in general terms, models are incomplete representations of reality (European Commission, 2003). However, be that as it may, models can be useful for our purposes. Quantitative and/or qualitative approaches can be used in order to prove the validity of a model depending on the availability of data or lack thereof, respectively. In the end, regardless of the approach taken to a validation, the important fact is that the model and its simulation provide a more or less accurate method for risk managers to define policies and make decisions that are effective in the reduction of risk.

REFERENCES

Adams, M.R., C.L. Little and M.C. Easter (1991). Modelling the effect of pH, acidulant and temperature on growth of *Yersinia enterocolitica*. *J. Appl. Bacteriol.* 71, 65 - 71.

Baranyi, J., T.P. Robinson, A. Kaloti, and B.M. Mackey (1995). Predicting growth of *Brochotrix thermosphacta* at changing temperature. *Int. J. Food Microbiol.* 27, 61 - 75.

Baranyi, J., aned M. Tamplin (2004). ComBase: A Common Database on Microbial Responses to Food Environments. *J. Food Prot.* 67, 1834 - 1840.

Baranyi, J., and T.A., Roberts (1994). A dynamic approach to predicting bacterial growth in food. *Int. J. Food Microbiol.* 23, 277 – 294.

Bemrah, N., H. Bergis, C. Colmin, A. Beaufort, Y. Millemann, B. Dufour, J.J. Benet, O. Cerf, and M. Sanaa (2003). Quantitative risk assessment of human salmonellosis from the consumption of a turkey product in collective catering establishments. *Int. J. Food Microbiol.* 80, 17-30.

Bemrah, N., M. Sanaa, M.H. Cassin, M.W. Griffiths, and O. Cerf (1998). Quantitative risk assessment of human listeriosis from consumption of soft cheese made from raw milk. *Prev. Vet. Med.* 37, 129 – 145.

Bernaerts, K., and J.F. Van Impe (2004). Data-driven approaches to the modelling of bioprocesses. *Transactions of the Institute of Measurement and Control* 26, 349 – 372.

Beumer, R., and R., Kusumaningrum (2003). Kitchen hygiene in daily life. *Int. Biodeterior. Biodegrad.* 51, 299 - 302.

Bigelow, W.D., G.S. Bohart, A.C. Richardson, and C.O. Ball (1920). Heat penetration in processing canned foods. *Natl. Canners Assoc. Bull.* 16-L.

Buchanan, R.L., J.L. Smith, and W. Long (2000). Microbial risk assessment: dose-response relations and risk characterization. *Int. J. Food Microbiol.* 58, 159 – 172.

CAC (Codex Alimentarius Commission) (1999). Principles and guidelines for the conduct of a Microbiological Risk Assessment. CAC/GL-30-1999. Secretariat of the Joint FAO/WHO Food Standards Programme, FAO, Rome.

Cassin, M.H., A.M. Lammerding, E.C.D. Todd, W. Ross, and R.S. McColl (1998). Quantitative risk assessment for *Escherichia coli* O157:H7 in ground beef hamburgers. *Int. J. Food Microbiol.* 41, 21 - 44.

Cayley, S., B.A. Lewis, and T.M. Record (1992). Origins of osmoprotective properties of betaine and proline in *Escherichia coli* K2. *J. Bacteriol.* 174, 1586 - 1595.

CFSAN/FDA (Center for Food Safety and Applied Nutrition/Food and Drug Administration Quantitative) (2005). Risk Assessment on the Public Health Impact of Pathogenic *Vibrio parahaemolyticus* in Raw Oysters. <http://www.cfsan.fda.gov/~acrobat/vpra.pdf>

Chen, Y., W.H. Ross, V.N. Scott, and D.E. Gombas (2003). *Listeria monocytogenes*: Low Levels Equal Low Risk. *J. Food. Prot.* 66, 570 – 577.

Christensen, B.B., H. Rosenquist, H.M. Sommer, N.L. Nielsen, S.F. Andersen, and B. Nørrung (2005). A Model of Hygiene Practices and Consumption Patterns in the Consumer Phase. *Risk Anal.* 25, 49 – 60.

Coleman, M and H. Marks (1998). Topics in dose-response modelling. *J. Food Prot.* 61, 1550-1559.

Cullen, A.C. and Frey, H.C. (1999) *Probabilistic Techniques in Exposure Assessment.* Plenum Press, NY.

Dennis, S.B., M.D. Milotis and R.L. Buchanan (2002). *Microbiological risk assessment in food processing*. Chapter 5: Hazard characterization/dose-response assessment. Ed. M. Brown and M. Stringer. Woodhead Publishing Limited, Cambridge, England.

Dym, C.L. (2004). *Principles of Mathematical Modeling*. Elsevier Academic Press Academic, London, UK. p. 4.

ECR (Economic Research Service). (2000). Economics of Foodborne Disease: Estimating the Benefits of Reducing Foodborne <http://www.ers.usda.gov/data/Foodborneillness/>

European Commission (2003). Risk assessment of food borne bacterial pathogens: Quantitative methodology relevant for human exposure assessment (final report). <http://ec.europa.eu/food/fs/sc/ssc/out308_en.pdf>

FAO/WHO (Food Agriculture Organization/ World Health Organization) (1995). Application of risk analysis to food standards. *Report of the Joint FAO/WHO Expert Consultation*. FAO/WHO, Geneva.

FAO/WHO (Food Agriculture Organization/ World Health Organization) (2002). Risk assessments of *Salmonella* in eggs and broiler chickens: technical report. <http://www.who.int/foodsafety/publications/micro/en/salmonella.pdf>

FAO/WHO (Food Agriculture Organization/ World Health Organization) (2003). Guidelines on Hazard characterization for pathogens in food and water for foods modeling approach.

FAO/WHO (Food Agriculture Organization/ World Health Organization) (2004). Risk assessment of *Listeria monocytogenes* in ready-to-eat foods: technical report <http://www.fao.org/es/ESN/jemra/listeria_en.stm>

FAO/WHO (Food Agriculture Organization/ World Health Organization) (2005). Risk assessment of *Vibrio vulnificus* in raw oysters: interpretative summary and technical report. <http://www.who.int/foodsafety/publications/micro/mra8.pdf>

FDA/USDA/CDC (Food Safety and Inspection Service/ U.S. Department of Agriculture/Center for Diseases Control) (2003). Quantitative Assessment of Relative Risk to Public Health from Foodborne *Listeria monocytogenes* among selected categories of Ready-to-Eat Foods. <http://www.foodsafety.gov/~dms/lmr2-toc.html>

FSIS/USDA (Food Safety and Inspection Service/ U.S. Department of Agriculture) (2001). Draft Risk Assessment of the Public Health Impact of *Escherichia coli* O157:H7 in Ground Beef. <http://www.fsis.usda.gov/OPPDE/rdad/FRPubs/00-023N/00-023NReport.pdf>

FSIS/USDA (Food Safety and Inspection Service/ U.S. Department of Agriculture) (2003). FSIS Risk Assessment for *Listeria monocytogenes* in Deli Meats. <http://www.fsis.usda.gov/OPPDE/rdad/FRPubs/97-013F/ListeriaReport.pdf>

FSIS/USDA (Food Safety and Inspection Service/ U.S. Department of Agriculture) (2005a). Risk Assessment for the Impact of Lethality Standards on Salmonellosis from Ready-to-Eat Meat and Poultry Products. <http://www.fsis.usda.gov/PDF/Salm_RTE_Risk_Assess_Sep2005.pdf>

FSIS/USDA (Food Safety and Inspection Service/ U.S. Department of Agriculture) (2005b). Risk Assessments of *Salmonella Enteritidis* in Shell Eggs and *Salmonella* spp. in Egg Products. <http://www.fsis.usda.gov/PDF/SE_Risk_Assess_Oct2005.pdf>

FSIS/USDA (Food Safety and Inspection Service/ U.S. Department of Agriculture) (2005c). Risk Assessment for *Clostridium perfringens* in Ready-to-Eat and Partially Cooked Meat and Poultry Products. <http://www.fsis.usda.gov/PDF/CPerfringens_Risk_Assess_Exec Summ_Sep2005.pdf>

ILSI (International Life Science Institute) (2000). Revised framework for Microbial Risk Assessment. ILSI press, Washington.

Gill, C.O., T. Jones, K. Rahn, A. Houde, J.C. McGinnis, S. Campbell, R.A. Holley, and D.I. LeBlanc (2002). Control of Product Temperatures during the Distribution of Retail Ready Beef to Stores and Vacuum Packaged Beef to Restaturants. *Dairy, Food and Enviornmental Sanitation* 22, 422 – 428.

Giovannini, A., V. Prencipe, A. Conte, L. Mario, A. Petrini, F. Pomilio, V. Rizzi, and G. Migliorati (2004). Quantitative risk assessment of *Salmonella* spp. infection for the consumer of pork products in an Italian region. *Food Control* 15, 139 - 144.

Haas, C.H., J.B. Rose, C.P. and Gerba (1999). Quantitative Microbial Risk Assessment. Ed. John Wiley and Sons, Inc. Nueva York, USA.

Hald, T., D. Vose, H.C. Wegener, and T. Koupeev (2004). A Bayesian approach to quantify the contribution of animal-food sources to human salmonellosis. *Risk Anal.* 24, 255 - 269.

Hartnett, E., L.A. Kelly, G. Gettinby, and M. Wooldridge (2002). A quantitative risk assessment for campylobacters in broilers: work in progress. *Int. Biodeterior. Biodegrad.* 50, 161 - 165.

Holcomb, D.L., M.A. Smith, G.O. Ware, Y.C. Hung, R.E. Brackett, and M.P. Doyle (1999). Comparison of six dose-response models for use with food-borne pathogens. *Risk Anal.* 19, 1091 - 1100.

ICMSF (International Commission on the Microbiological Specifications of Foods) (2002). Microorganisms in Foods 7.Microbiological testing in food safety management. Kluwer Academic/Plenum Publisher, NY.

Jouve, J.L. (2002). Microbiological risk assessment in food processing. Chapter 2: Microbiological risk assessment (MRA): an introduction. Woodhead Publishing Limited, Cambridge, England. p. 35.

Kennedy, J., I.S. Blair, D.A. McDowell and D.J. Bolton (2005). The Microbiological Status of Non/Food Contact Surfaces in Domestic Kitchens and the Growth of *Staphylococcus aureus* in Domestic Refrigerators. *Food Prot. Trends* 35, 974 – 980.

Lammerding, A.M. and A. Fazil (2000). Hazard identification and exposure assessment for microbial food safety risk assessment. *Int. J. Food Microbiol.* 58, 147 - 157.

Lammerding, A.M. and G.M. Paoli (1997). Quantiative risk assessmen: an emerging tool for emerging foodborne pathogens. *Emerg. Infect. Dis.* 3(4), 483 - 487.

Legan, D., M. Vandeven, C. Stewart and M. Cole (2002) Modelling the growth, survival and death of bacterial pathogens in foods. In Foodborne pathogens ed. Blackburn,C.d.W. and McClure, P.J. pp. 53 - 95. Cambridge, UK: Woodhead Publishing Ltd.

Lindqvist, R., S. Sylvén and I. Vagsholm (2002). Quantitative microbial risk assessment exemplified by *Staphylococcus aureus* in unripened cheese made from raw milk. *Int. J. Food Microbiol.* 78, 155 - 170.

Malakar, P.K., G.C. Barkera, M.H. Zwietering, and K. van't Riet (2003). Relevance of microbial interactions to predictive microbiology. *Int. J. Food Microbiol.* 84, 263 - 272.

McCullagh, P., and J.A. Nelder (1989). *Generalized linear models*. Chapter 1 – Introduction. 2 ed. Chapman and Hall, London.

McKellar, R.C. (2001). Development of a dynamic continuous-discrete-continuous model describing the lag phase of individual bacterial cells. *J. Appl. Microbiol.* 90, 407 – 413.

McKellar, R.C. and X. Lu (2004). *Modeling microbial responses in foods*. CRC Press. Boca Raton, Florida.

McLauchlin J, R.T. Mitchell, W.J. Smerdon, and K. Jewell (2004). *Listeria monocytogenes* and listeriosis: a review of hazard characterisation for use in microbiological risk assessment of foods. *Int. J. Food Microbiol.* 92, 15 - 33.

McMeekin, T.A., J.N. Olley, T. Ross, and D.A. Ratkowsky (1993). *Predictive Microbiology: Theory and Application*, Wiley, NY.

McMeekin, T.A., R.E. Chandler, P.E. Doe, C.D. Garland, J. Olley, S. Putro, and D.A. Ratkowsky (1987). Model for combined effect of temperature and salt concentration/water activity on the growth rate of *Staphylococcus xylosus*. *J. Appl. Bacteriol.* 62, 543 - 550.

Mellefont, L.A. and T. Ross (2003). The effect of abrupt shifts in temperature on the lag phase duration of *Escherichia coli* and *Klebsiella oxytoca*. *Int. J. Food Microbiol.* 83, 295-305.

Mètris, A., Y. Le Marc, A. Elfwing, A. Ballagi, and J. Baranyi (2005). Modelling the variability of lag times and the first generation times of single cells of *E. coli*. *Int. J. Food Microbiol.* 100, 13 - 19.

Moran, P.A. (1954). The dilution assay of viruses. *J. Hyg.* 52, 189 – 193.

Morgan, M.G. and M. Henrion (1990). Uncertainty: a guide dealing with uncertainty in quantitative risk assessment and policy analysis. Cambridge University Press, NY.

Nauta, M.J. (2001). A modular process risk model structure for quantitative microbiological risk assessment and its application in an exposure assessment of *Bacillus cereus* in REPFED. RIVM report 149106007. Bilthoven, the Netherlands.

Nauta, M.J. (2005). Microbiological risk assessment models for partioning and mixing during food handling. *Int. J. Food Microbiol.* 100, 311 - 322.

Nauta, M.J., E.G. Evers, K. Takumi, and A.H. Havelaar (2001). Risk assessment of Shiga-toxin producing *Escherichia coli* O157 in steak tartare in the Netherlands in a REPFED. RIVM-rerport 257851003. Bilthoven, the Netherlands.

Oscar, P. (2004). A quantitative risk assessment model for *Salmonella* and whole chickens. *Int. J. Food Microbiol.* 93, 231 - 247.

Ratkowsky, D.A, J. Olley, T.A. McMeekin, and A. Ball (1982). Relation between temperature and growth rate of bacterial cultures. *J. Bacteriol.* 149, 1-5.

Reij, M.W., W.D. Aantrekker, W.D., and ILSI Europe Risk Analysis in Microbiology Task Force (2004). Recontamination as a source of pathogens in processed foods. *Int. J. Food Microbiol.* 91, 1 - 11.

Rosenquist, H., N.L. Nielsen, H.M. Sommer, B. Norrung, and B.B. Christensen (2003). Quantitative risk assessment of human campylobateriosis associated with thermophilic *Campylobacter* species in chickens. *Int. J. Food Microbiol.* 83, 87 - 103.

Ross, T. (1996). Indices for performance evaluation of predictive models in food microbiology. *J. Appl. Bacteriol.* 81, 501 - 508.

Sanaa, M., L. Coroller, and O. Cerf (2004). Risk assessment of listeriosis linked to the consumption of two soft cheeses made from raw milk: camembert of Normandy and brie of meaux. *Risk Anal.* 24, 389 – 399.

Schaffner, D.W. (2004). *Modeling microbial responses in foods*. Chapter 10: Models-What comes after the next generation? CRC Press. Boca Raton, Florida.

Swinnen, I.A.M., K. Bernaerts, E.J.J. Dens, A.H. Geeraerd, and J.F. Van Impe (2004). Predictive modelling of the microbial lag phase: a review. *Int. J. Food Microbiol.* 94, 137 - 159 te Giffel, M.C., and M.H. Zwietering (1999). Validation of predictive models describing the growth of *Listeria monocytogenes*. *Int. J. Food Microbiol.* 46, 135 - 149.

van Gerwen, S.J. and M.H. Zwietering (1998). Growth and inactivation models to be used in quantitative risk assessments. *J. Food Prot.* 61, 1541 - 1549.

Vose, D. (2000a). *Risk Analysis. A Quantitative Guide*. John Wile and Sons Inc., NY.

Voysey, P., K. Jewell, and M. Stringer (2002). *Microbiological risk assessment in food processing.* Chapter 7: Risk Characterization. Ed. M. Brown and M. Stringer. Woodhead Publishing Limited, Cambridge, England.

Whiting, R.C., and R.L. Buchanan (1997). Development of a quantitative risk assessment model for Salmonella enteritidis in pasteurized liquid eggs. *Int. J. Food Microbiol.* 36, 111 – 125.

Zhao, P., T. Zhao, M.P. Doyle, J.R. Rubino, and J. Meng (1998). Development of a model for evaluation of microbial cross-contamination in the kitchen. *J. Food Protect.* 61, 960 - 963.

Zwietering M.H., and F. Pérez-Rodríguez (2006). *Exposure to microorganisms: the devil is in the tail.* Food Micro 2006, 20th International Symposium of the International Committee on Food Microbiology and Hygiene (ICFMH), Bologna, Italy, 29 August - 2 September.

Zwietering, M.H and S.J. van Gerwen (2000). Sensitivity analysis in quantitative microbial risk assessment. *Int. J. Food Microbiol.* 58, 213–221.

Zwietering, M.H., T. Wijtzes, J.C. De Wit and K. van't Riet (1992). A decision support system for prediction of the microbial spoilage in foods. *J. Food Prot.* 55, 973 - 979.

In: Food Microbiology Research Trends
Editor: Marta V. Palino, pp. 131-154
ISBN 978-1-60021-723-4

Chapter 4

Current Status on the Etiology, Epidemiology, Food Safety Implications and Control Measures in *Escherichia coli* O157:H7 Infections

G. Normanno*

Department of Health and Animal Welfare, Faculty of Veterinary Medicine
Str. Prov. per Casamassima Km 3, 70010 Valenzano (Ba), Italy

Abstract

Escherichia coli O157:H7 is an important foodborne pathogen. It was recognized as a cause of severe human illness only in the early 1980's. Since then it has been implicated in foodborne disease outbreaks in many countries throughout the world. Cattle are the main reservoir of this microrganism, and transmission of the infection to humans occurs primarily through the consumption of contaminated food. The pathogenicity of *E. coli* O157:H7 is associated to genes encoding for a number of virulence factors, especially the verotoxin-encoding genes. The infective dose is extremely low - in the order of a few bacterial units. *E. coli* O157:H7 is acid-tolerant and survives in acidic environments such as the gastric barrier and acidic foods. The most severe illnesses induced by *E. coli* O157:H7 are Hemorrhagic Colitis and Hemolytic Uremic Syndrome. Both these clinical forms are characterized by severe morbidity and might sometimes be lethal. Children are the most commonly affected age group and they are affected by the highest death rate. *E. coli* O157:H7 is one of the main food safety hazards having important implications for human health worldwide, since outbreaks lead sometimes to cases of mortality. This review examines the etiological and epidemiological aspects of *E. coli* O157:H7 infection and focuses on the food safety concerns raised by *E. coli* O157:H7 and on control methods for the prevention of food poisoning.

Keywords: Escherichia coli O157:H7, food safety.

---

*Correspondence. Tel:+39 080 4679895; fax:+39 0804679854; E-mail address: g.normanno@veterinaria.uniba.it

1. Introduction

Escherichia coli was firstly described by Theodor Escherich in 1885 and it is considered a typical inhabitant of the intestinal tract of humans and warm-blooded animals. *E. coli* colonizes the intestine during the first few hours after birth and stays on indefinitely as a commensal (Drasar and Hill, 1974). It was used as an indicator of faecal contamination of food and water as early as 1890.

A few clones causing human disease developed within the species. Most of them induce variably severe gastrointestinal symptoms while others cause serious systemic diseases with a high mortality rate. The main pathogenic variants of *E. coli* are categorized according to the virulence factors they produce or based on their pathogenetic mechanisms during infection. Enterotoxigenic strains (Enterotoxigenic *E. coli*, ETEC) synthesize heat-labile and heat-stable (LT and ST) enterotoxins; enteropathogenic strains (Enteropathogenic *E. coli*, EPEC) synthesize adherence factors that generate lesions in the enterocytes; enteroaggregative strains have either diffuse or aggregate patterns of adherence (Enteroaggregative *E. coli*, DAEC, EAggEC, or EAEC) and form microcolonies on the enterocytes; enteroinvasive strains (Enteroinvasive *E. coli*, EIEC) penetrate the enterocytes; cytolethal distending toxin producing *E. coli* (CDTEC) and, finally, enterohemorrhagic strains (Enterohemorrhagic *E. coli* EHEC) produce Verotoxins (VTs) or Shiga toxins (Stx) and are thus also called verotoxin producing-*E. coli* (VTEC or STEC) (Doyle and Padhye, 1989; Batt, 2000; Clarke, 2001). *E. coli* O157:H7 is considered to be the prototype of the EHEC strains (Nataro and Kaper, 1998) and is a genuine emerging foodborne pathogen that probably evolved from the Enteropathogenic *E. coli* O55:H7 (Feng et al., 1998). It is the predominant VTEC serotype in many parts of the world and it has been most commonly associated with large outbreaks (Karmali, 1989). *E. coli* O157:H7 was firstly isolated in 1975 in a woman with grossly bloody diarrhea and recognized as a major foodborne pathogen in 1983 after Riley reported two outbreaks of Hemorrhagic Colitis (HC) associated with the consumption of contaminated hamburgers from a fast food chain in Michigan and Oregon (Riley et al., 1983). In both places the *E. coli* serotype isolated from the patients and the food was O157:H7. Since then, *E. coli* O157:H7 has been considered a serious public health concern.

1.1. Metabolic and Growth Characteristics

E. coli O157:H7 presents many biochemical reactions typical of other strains of *E. coli* although there are some important differences. First, most *E. coli* O157:H7 isolates are slow- or non-fermenters of sorbitol and this feature distinguishes this serotype from the other *E. coli* strains that ferment sorbitol within 24h (Doyle and Schoeni, 1984). Second, *E. coli* O157:H7 is negative with the MUG assay and thus lacks β-glucuronidase activity (Tortorello, 2000). Both features are helpful in the isolation and presumptive identification of the microorganism although sorbitol-fermenting and β-glucuronidase-producing strains of *E. coli* O157:H7 have been isolated from patients with HUS (Gunzer et al., 1992; Karch and Bielaszewsca, 2001). Other important biochemical and metabolic characteristics of *E. coli* and *E. coli* O157:H7 are summarized in Table 1.

Table 1. Main biochemical differential parameters, parameters of development and of survival of *E. coli* and *E. coli* O157

	E. coli	*E. coli* O157
Gram	-	-
Cytochrome Oxidase	-	-
Indole Production	+	+
Citrate and Urease	-	-
β-Glucuronidase	+	-
Enterohemolysin	-	+
Lactose Fermentation	+	+
Sorbitol Fermentation (24h)	+	- (+)
Grow 44°-44.5°C	Well	Poorly
Min. Grow T°C	5	5
Chilling Survival	Yes	Yes
Minimum pH Value	4-4.5	2.5
a_w	0.97	0.97
Max [NaCl]	6.5	6.5
Milk Pasteurization Survival	No	No
Inactivation T°		>63.3 (160°F)

2. Virulence Factors

The following steps are required for an active *E. coli* O157:H7 infection to develop: colonization of the intestine, *in situ* replication of the microrganism, evasion of the host immune response, and production of toxins. *E. coli* O157:H7 is considered to be one of the most formidable biological agents that induce disease in humans because of its considerable range of virulence factors, which has been extensively reviewed elsewhere (Law, 2000b; Caprioli et al., 2005). *E. coli* O157:H7 exerts its pathogenic action through the synthesis of adherence factors, haemolytic factors, iron subtraction factors, host immune response evasion factors, and verocytotoxins (VTs). The most important virulent factor of *E. coli* O157:H7 is the synthesis of VTs and the adhesin *intimin*. Virulence factors are encoded on phage, plasmid, or chromosomal genes, but novel or unknown virulence factors may also exist. These pathogenicity determinants have been acquired by *E. coli* O157:H7 via other organisms which are *E. coli* and non- *E. coli* species (Law, 2000a). In the chromosome genes encoding for the VTs are present together with a pathogenicity island called Locus of Enterocyte Effacing (LEE) that contains the information encoding the set of proteins needed for adherence and for the *attaching and effacing lesions* to occur (A/E lesions) (McDaniel and Kaper, 1997). The information required to synthesize enterohemolysin and the fimbriae potentially involved in the colonization of the mucosal gut is present in the characteristic 60 MDa plasmid of *E. coli* O157:H7, designated as the pO157 plasmid (Karch et al., 1987; Schmidt et al., 1995).

2.1. Adherence Factors

Several adherence factors have been shown to exist in *E. coli* O157:H7, but the main pathogenetic role is played by an outer membrane protein (OMP) called *intimin*. This is a 94-97-kDa OMP, made up of 939 aminoacids, encoded by the *eae* gene (*E. coli* attaching and effacing). Besides the gene coding for intimin, LEE contains several genes, such as *EspA, EspB,* and *EspD*, and a type III secretion apparatus. All these encode for a few secretory proteins needed for A/E activity, as well as the *Tir* gene that codes for the intimin receptor (Tir: translocated intimin receptor) (Kenny et al., 1997; Sinclair et al., 2006). Intimin produces a characteristic, close adherence of the microorganism to the enterocyte surface known as "attaching and effacing adherence". This mechanism is also typical of EPEC strains in their action on the enterocytes of the small intestine whereas *E. coli* O157:H7 targets the large intestine (Yu and Kaper, 1992). The bacterium/host cell interaction leads to an exchange of biochemical signals resulting in the localized destruction of microvilli (Mead and Griffin, 1998) and the formation of a cup-like structure on the enterocyte surface. The cup-like structure is the result of a rearrangement of the cytoskeletal components of the cytoskeleton and is the place where *E. coli* becomes anchored and causes the A/E lesions (Knutton et al., 1989; Donnenberg et al., 1997; Paton and Paton, 1998). The *E. coli* O157:H7 strains carrying the *eae* gene isolated in HC and HUS cases were more numerous than the *eae* negative strains isolated during these diseases (Paton, and Paton, 1998). It appears that the presence of *eae* is not essential to cause human disease and that *eae* strains may synthesize other yet non-identified virulence factors (Barret et al., 1992), such as fimbriae, other OMPs, and lipopolysaccharides (LPS), which may be involved in enteric cell adhesion (Karch et al., 1987; Fratamico et al., 1993; McKee and O'Brien, 1995). Further studies are needed to shed light on the role of the fimbriae, LPSs, and other structures that may be involved in the complex bacterial cell/enterocyte interaction, especially in the *eae* negative strains.

2.2. Verotoxins (VTs)

The most potent virulence factor of *E. coli* O157:H7 is associated with the synthesis of VTs or Shiga-like toxins (STXs). Although VTs alone cannot induce disease, they are responsible for the severe symptoms of HC and HUS (Mead and Griffin, 1998). Verotoxins are one of the most potent bacterial cytotoxins known and their name comes from the irreversible cytopathic effect they have on monolayers of VERO (African green monkey kidney) cells (Konowalchuk et al., 1977). Verotoxins are also referred to as *Shiga-like toxins* because of the similarity between VT1 and the Shiga toxin produced by Type 1 *Shigella dysenteriae.* The only difference between the two proteins is in one amino acid of the A subunit (Weeratna and Doyle, 1991). VTs are exoproteins with a molecular weight of about 70,000. Antigenically they are distinguished as VT1 and VT2 and are encoded by genes, namely *stx1* and *stx2,* of temperate bacteriophages inserted into the microrganism chromosome (O'Brien and Holmes, 1987). The compositions of VT1 and VT2 are extremely different since VT2 presents only 56% amino acid homology with VT1. VT1 is a relatively homogeneous protein (5 genetic variants have been described so far), while 12 variants of VT2 have been observed - VT2c, VT2d, VT2e, VT2f, etc. (Ramachandran et al., 2001; Scheutz et al., 2001). Not all the variants of VT2 are produced by *E. coli* O157 or are associated to disease in humans. VT2e has been found in strains of *E. coli* O138, O139, and O140, the agents of piglet edema disease (Konowalchuk et al., 1977), and VT2f is produced by pigeon-adapted *E. coli* strains (Schmidt et al., 2000). VTs have a common structure, made up of the two subunits, A and B. The B subunit has a pentameric structure (MW of each monomer is 5,000-7,000) and it is involved in the binding between the toxin and the cell receptor. The A subunit (MW of 29,000-31,000) is the active part of the toxin. After being proteolytically nicked and reduced to the A_1 fragment, the A subunit inhibits protein synthesis through catalytic interaction with the 60S ribosomal subunit and the inhibition of the peptide chain elongation step of protein synthesis causes cell death (Sandving and van Deurs, 1996). VTs interact with eukaryotic cells through binding of B subunit with its specific receptor, globotryaosilceramide (Gb_3). This receptor is commonly present on the surface of colonic cells, on the surface of the endothelium of renal microvessels, and in VERO cells (Nataro and Kaper, 1998). VT2c is a particular VT subgroup, since it presents poor affinity with Gb_3 and has a smaller effect on the VERO cell line (Lindgren et al., 1994). The VT2e subgroup, instead, binds to the cell receptor, Gb_4 (Nataro and Kaper, 1998).

The biological effects of VTs on laboratory animals vary depending on the species and amount of toxin used. The most sensitive species are rabbits that present with pooling of fluid in the ileal loop, followed by paralysis and death (Keusch et al., 1972). Besides the VERO cells, other cell lines sensitive to the cytopathic effect of VTs are HeLa, Daubi, KB, human liver cells, and human foreskin fibroblasts. By contrast WI-38, Henle 407, CHO, L, BHK, and several other cells of human neoplastic origin are resistant to the action of VTs (Karmali, 1989).

VTEC strains are able in producing one or both VTs simultaneously. While most of the *E. coli* O157:H7 strains produce both VT1 and VT2, having a cytotoxic action on the renal microvessels, only few strains produce VT1. VT2 is known to be about 1000 times more effective than VT1 (Law, 2000b).

To date, few studies have been carried out on the impact food containing preformed VTs may have on human health. VT1 is known to be a relatively heat-stable toxin, hence VT1 produced by VTEC strains may remain active if the food undergoes mild heat treatment, such as pasteurization. VT1 heated at 45-70°C for up to 60 min has been shown to maintain its toxicity while it becomes completely inactivated when heated at 80°C for 60 min or at 85°C for 5 min (Kittel et al., 1991; Weeratna and Doyle, 1991). This issue requires further investigations.

2.3. Enterohemolysin

Most *E. coli* O157:H7 strains synthesize a particular hemolysin called *enterohemolysin* (Ehx). Ehx is encoded by the *hlyA* gene present in the 60-Mda virulence plasmid pO157 which is highly conserved in EHEC strains (Beutin et al., 1989). Unlike *E. coli* alpha-hemolysin, which produces large clear zones of haemolysis on blood agar after few hours of incubation, Ehx produces a small turbid haemolytic zone on washed sheep's blood agar after an incubation of 24 hours. The specific role of enterohaemolysin in the pathogenesis of HUS is still unclear. Enterohaemolysin has been found in most *E. coli* O157:H7 strains isolated from clinical cases but other experiments have shown that this factor is not essential for the development of HC and HUS (Law, 2000b). Ehx may be involved in *E. coli*'s ability to survive in the host organism since it lyses erythrocytes leading to a release of haemoglobin which provides a source of iron for the microorganism to grow in the gut (Law and Kelly, 1995).

2.4. Acid Resistance

The acid resistance of *E. coli* O157:H7 has very important effects both on human disease and food safety. By virtue of its acid tolerance *E. coli* O157:H7 can easily pass the acid barrier of the stomach and reach areas of the intestine where it can bind to the sites of enteric adherence: this biological feature of the microorganism is associated with the extraordinarily low infective dose required to produce illness in humans which ranges from tens to a few hundreds of bacterial cells (Tuttle et al., 1999). Three induction mechanisms of acid resistance have been reported to exist in strains of enterohemorrhagic *E. coli*: oxidative, arginine-dependent, and glutamine-dependent (Lin et al., 1996). Due to these mechanisms the microorganism can survive at pH 2.5 for 2-7 h at 37°C (Buchanan and Doyle, 1997). Nevertheless, the ability of *E. coli* O157:H7 to survive in an acid environment depends on the other growth parameters, the type of acid it is exposed to, temperature and on the growth phase of the microbial population (cells in the stationary phase are more than 1000 times more acid resistant than cells in the exponential phase). The possible contamination of foods with acid resistant bacteria calls for more careful surveillance of foodstuffs. Foods considered to be microbiologically safe because of their low pH value, such as mayonnaise, yoghurt, cured meats, and apple juice can be vehicles of *E. coli* O157:H7 infection (Meng and Doyle, 1998). Tests have shown that *E. coli* O157:H7 can survive for 5-7 weeks in mayonnaise (pH 3.6-3.9), for weeks in yoghurt kept at 4°C, and for 10-31 days in unpasteurized apple juice (pH 3.6-4.0) kept at 8°C (Caprioli et al., 2001).

2.5. Other Virulence Factors

Several other protein factors have been considered to be putative virulence factors synthesized by *E. coli* O157:H7. They include Extracellular Serin Protease (EspP), the *Clostridium-difficile*-like toxin, the enteroaggregative heat-stable toxin (EAST1) the production of Catalase/Peroxidase and the Cytolethal Distending Toxin (Law, 2000b; Janka et al., 2003). The involvement of these factors in the pathogenesis of *E. coli* O157:H7 infections is still to be clarified.

3. Epidemiology

The epidemiology of *E. coli* O157:H7 infections is very complex and still obscure. It is still unclear whether certain animal species are involved in spreading the microrganism. Transmission is known to occur mainly through contaminated foods but, several other routes of infection have been acknowledged. Victims may become infected after swimming in lakes and other recreational waters, and animal-to-person infection via the faecal-oral route may occur especially with school children during farm visits (Table 2). Since reporting the infection is not obligatory in many countries, the data available are an underestimation of the real incidence of *E. coli* illnesses. The epidemiological picture of the infection is far from being complete because no exhaustive information is available about the presence of *E. coli* O157:H7 in animal reservoirs or in foods of animal origin, nor is there any data on the ability of the microrganism to survive in several foodstuffs. Since 1982, when the first outbreak of HC and HUS ascribable to *E. coli* O157:H7 infection was reported (Riley et al., 1983), several outbreaks have been described (Remis et al., 1984). The most significant of these are reported in Table 2.

3.1. Reservoirs

The association between *E. coli* O157:H7 infection and the consumption of bovine meat was evident from the first epidemic outbreak. Several studies subsequently established that cattle are the main reservoir of *E. coli* O157:H7 (Montenegro et al., 1990; Armstrong et al., 1996). Carriers are usually young animals with no clinical signs that intermittently shed *E. coli* O157:H7 into the environment through their faeces, at about 6.8×10^5 CFU/g (Fegan et al., 2004) with a peak during the spring-summer months (Hancock et al., 1997). This seasonality in shedding the germ is consistently reflected in disease presenting in humans during the same months, probably because of an increase in the consumption of barbecued meat during outdoor activities (Chapman, 1995). The absence of clinical signs and symptoms in carriers is another factor enhancing the risk of meat contamination. Since they have no visible signs of infection these animals are regularly slaughtered. Their meat may become contaminated at the slaughterhouse due to spillage from the intestine or because of contact between carcass surface and hide and may thus become the gateway through which the microrganism enters the human food chain. A number of surveys have been carried out to assess the prevalence of cattle carrying and shedding *E. coli* O157:H7 and the percentage of contaminated carcasses. These studies were carried out in many countries and the results were extremely variable. In the US the percentage of bovine carriers ranged from 2 to 45% (Karch et al., 1987; Faith et al., 1998; Elder et al., 2000); in the United Kingdom the prevalence of the microrganism in animals was 15.7% (Chapman et al., 1997), while in other European countries the percentage of animal carriers and contaminated carcasses was 1% to about 37% (Wells et al., 1991; Blanco et al., 1996; Albihn et al., 1997; Heuvelink et al., 1999; Bonardi et al., 1999; Chapman et al., 2001; Bonardi et al., 2001; Guyon et al., 2001). This variability may be due to the differences in sampling criteria, geographic origin, the number of samples analyzed and survey protocols used in the surveys. *E. coli* O157:H7 has also been isolated from carcasses in Australia, Canada and Japan (Sekiya, 1997; Power et al., 2000; Philips et al., 2001). Cattle are not the only source of *E. coli* O157:H7 infection for humans since the microrganism has also been isolated from bison, water-buffalo, sheep, goats, pigs, deer, horses, dogs, birds, flies, and other animals (Griffin and Tauxe, 1991; Caprioli et al., 1993; Chapman et al., 2001; Johnsen et al., 2001; Bouvet et al., 2002; Bonardi et al., 2003; Li et al., 2004; Kaufmann et al., 2006). Non-ruminant species are believed to be transient carriers of *E. coli* O157:H7 (Caprioli et al., 2005).

Table 2. Listing of prior *E. coli* O157 outbreaks occurs from 1982 to 2006

Year	Country	Persons Affected	Age	Cases of Hus	Hospital-izations	Deaths	Serotype	Implicated Food; Source of Infection (confirmed by laboratory or by epidemiological evidences)
1985	Canada	73	/	12	19	19	*E. coli* O157:H7	Sandwich meal
1988	Wisconsin	61	/	/	/	/	*E. coli* O157:H7	Roast Beef
1989/90	Missouri (U.S.)	243	/	2	32	4	*E. coli* O157:H7	Municipal water
1991	Massachusett	18	/	/	/	/	*E. coli* O157:H7	Unpasteurized Apple juice
1992/93	California	34	/	/	/	4	*E. coli* O157:H7	Hamburger
1992/93	Washington (U.S.)	501	/	45	151	3	*E. coli* O157:H7	Hamburger
1993	Nevada	58	/	/	/	/	*E. coli* O157:H7	Hamburger
1992/96	England and Wales (U.K.)	381	/	59	120	14	*E. coli* O157	Cold cooked meat; Milk; Raw vegetables; Cooked minced beef dishes
1994	Virginia (U.S.)	156	9/22 years	1	3	0	*E. coli* O157:H7	Raw (red or pink) minced beef
1994	Washington, California (U.S.)	23	23 months/ 77 years	1	6	0	*E. coli* O157:H7	Dry cured salami
1994	Wisconsin	193	Young to middle-aged adults	0	2	0	*E. coli* O157:H7	Undercooked roast beef and salad cross-contaminated
1995	Illinois (U.S.)	12	2/12 years	3	3	0	*E. coli* O157:H7	Lake water
1995	Ontario (Canada)	21	/	0	21	0	*E. coli* O157:H7	Green salad
1995	Bohemia (Czech Republic)	4	Children	4	4	0	*E. coli* O157	Raw goat's milk
1995	Montana (U.S)	92	/	/	/	/	*E. coli* O157	Leaf Lettuce
1995	Oregon (U.S)	11	/	/	/	/	*E. coli* O157	Jan Deer Jerky
1996	Scotland (U.K.)	496	/	/	151	21	*E. coli* O157	Cold cooked meat; Steak pie

1996	Connecticut	14	2/73 years	3	7	0	*E. coli* O157:H7	Unpasteurized apple juice
1996	British Columbia, California, Colorado, Washington (U.S.)	45	1/41 years	12	/	0	*E. coli* O157:H7	Unpasteurized apple juice
1996	Connecticut and Illinois (U.S.)	/	/	/	/	/	*E. coli* O157:H7	Mesclun mix lettuce
1996	Sakai City (Japan)	9000	/	0	/	/	*E. coli* O157:H7	White radish sprouts
1997	Michigan (U.S.)	60	2/79 years	2	25	0	*E. coli* O157:H7	Alfalfa sprouts
1997	Virginia (U.S.)	48	6/67 years	0	11	0	*E. coli* O157:H7	Alfalfa sprouts
1997	Colorado (U.S.)	15	/	/	/	/	*E. coli* O157:H7	Hamburger
1997	Finland	14	3/8 years	0	5	0	*E. coli* O157:H7	Lake water
1997	Fuerteventura (Spain)	14	/	3	/	0	*E. coli* O157:H7	Raw vegetables
1998	England (U.K.)	7	/	/	4	0	*E. coli* O157:H7	Unpasteurized cream
1998	Wyoming (U.S.)	157	9/60 years	4	4	0	*E. coli* O157:H7	Municipal water
1998	Georgia (U.S.)	32	/	/	/	/	*E. coli* O157:H7	Swimming pool
1998	Ontario (Canada)	39	18 months/ 69 years	2	14	0	*E. coli* O157:H7	Genoa salami
1998	Washington (U.S.)	12	Children	/	/	/	*E. coli* O157:H7	School lunch
1999	Ohio (U.S.)	/	/	/	/	/	*E. coli* O157:H7	Coleslaw
1999	Oklahoma (U.S.)	7	/	/	/	/	*E. coli* O157:H7	Apple Cider
2000	Barcelona (Spain)	158	Children <5 years old	6	/	0	*E. coli* O157:H7	Sausages made with pork
2000	New York (U.S.)	1000	/	2	/	2	*E. coli* O157:H7	Water well located on the fairgrounds
2000	Oregon (U.S.)	171	/	/	/	/	*E. coli* O157:H7	Fast food
2000	Pennsylvania and Washington (U.S.)	75	1/52 years	9	19	0	*E. coli* O157:H7	Direct contact with farm animals

Table 2. (Continued)

Year	Country	Persons Affected	Age	Cases of Hus	Hospital-izations	Deaths	Serotype	Implicated Food; Source of Infection (confirmed by laboratory or by epidemiological evidences)
2001	British Columbia (U.S.)	5	1/7 years	2	3	0	*E. coli* O157:H7	Non-pasteurized goat's milk
2002	Connecticut (U.S.)	45	10/90 years	7	24	1	*E. coli* O157:H7; *E. coli* O157:NM	Not identified
2002	Japan	43	1/77 years	0	/	0	*E. coli* O157:H7	Grilled beef
2002	Texas (U.S.)	45	/	0	/	0	*E. coli* O157:H7	Cajun potatoes; Gravy; Rice; Chili.
2002	Colorado California; Iowa; Michigan; South Dakota; Washington; Wyoming (U.S.)	28	1/72 years	5	7	0	*E. coli* O157:H7	Minced beef
2004	Denmark	25	18 children and 7 adults	0	/	0	*E. coli* O157	Milk
2005	Wales (U.K.)	160	Children	/	/	/	*E. coli* O157	Cooked meat at School meals service
2005	Sweden	120	Adults	7	/	7	*E. coli* O157	Iceberg lettuce
2005	Scotland	15	Adults	/	/	/	*E. coli* O157	Not identified
2005	France	26	/	13	20	0	*E. coli* O157:H7	Beefburgers
2006	Norway	7	2/8 years	7	7	0	*E. coli* O157	Minced beef
2006	Scotland	13	4 children and 9 adults	4	4	0	*E. coli* O157	/
2006	Scotland	3	/	0	3	0	*E. coli* O157	/

Whether chicken is a potential reservoir of *E. coli* O157:H7 is still unclear. *E. coli* O157:H7 has been shown to colonize the intestine of chickens rapidly as a result of experimental infection, but investigations carried out in order to isolate the microorganism from farm animals provided negative results (Chapman et al., 1997; Heuvelink et al., 1999). Nevertheless, the results of a survey conducted in Slovakia showed that, out of 216 cloacal swabs analyzed, 20 (9.2%) were positive at the *E. coli* O157 detection, but all the isolates tested negative for H7 antigen (Pilipčinec et al., 1999). *E. coli* O157:H7 was isolated from chicken carcasses on the retail market (Doyle and Schoeni, 1987) suggesting that cross-contamination may occur during storage and marketing. Pathogenic strains of *E. coli* O157:H7 have been isolated from specimens of turkey faeces (Heuvelink et al., 1999).

The environment is a continuous source of contamination along the beef chain. *E. coli* O157:H7 can survive and replicate in animal feeds and in sedimented water on farms. It remains infective in water for at least six months and may thus survive and spread on farms during periods when it cannot be detected in cattle, especially during cold months (Hancock et al., 2001). *E. coli* O157:H7 has been frequently isolated from surfaces (walls, floors, ceilings, etc.) on the farm (24%), from the surfaces of the animal transport vehicles (7.3%), and from surfaces at slaughterhouses (7.8%) (Buncic et al., 2004). Finally, humans may also potentially spread *E. coli* O157:H7 since person-to-person infection and asymptomatic carriers have been identified in nursing home, hospitals daycares and among abattoir workers and others working in close contact with cattle and beef carcasses. When proper personal hygiene is not maintained, these individuals may be a source of food contamination and may cause secondary person-to-person spread.

3.2. Foods

Various foods have been implicated as potential vehicles in incidents or outbreaks of *E. coli* O157:H7 infections: burgers, minced beef, sausages, wurstels, turkey rolls, unpasteurized cow, sheep, and goat milk, cheese made from raw milk, vegetables, unchlorinated drinking water and others (Tab. 2). The fact that *E. coli* O157:H7 may be present in foods of animal origin and in the environment is predictable. Carcasses are mainly contaminated during slaughtering (skinning, evisceration, equipment tools), while meat products (ground meat, etc.) may be cross-contaminated during storage, processing, and marketing. Similarly, milk may become contaminated on the farm or cross-contaminated at processing facilities (McKee et al., 2003). Vegetables, such as coleslaw, alfalfa sprouts, and lettuce have been associated with outbreaks of *E. coli* O157:H7 infections (Sanz et al., 2003), and may become contaminated by manure from infected animals or by contaminated irrigation water. Sea foods may probably become contaminated by *E. coli* O157:H7 present in water from human or animal waste. Food safety issues are raised by the ability of *E. coli* O157:H7 to survive in ready-to-eat foods that have been processed to become microbiologically safe for their low pH (yoghurt, mayonnaise, and dry fermented salami) and have however caused infection outbreaks. After an important outbreak involving 20 laboratory-confirmed cases of *E. coli* O157:H7 diarrhoea associated with commercially distributed dry-cured salami in Washington and California (Centers for Diseases Control and Prevention, 1995), several studies have been carried out on the ability of *E. coli* O157:H7 to survive and produce toxins after maturation. These studies showed that the microbial load of *E. coli* O157:H7 gradually decreases by about a logarithm a week during maturation of cured meats (Normanno et al., 2002). Hence, the safety of these foods depends on the bacterial load contaminating the raw material at the time of sausage making and on the diameter of the salami as larger diameters require longer curing.

Many surveys have been performed to assess the presence of *E. coli* O157:H7 in several foodstuffs such as ground meat, cured meat, milk, cheese, and vegetables (Pradel et al., 2000; Coia et al., 2001; Normanno et al., 2002; Stampi et al., 2004; Normanno et al., 2004). Hussein and Bollinger (2005) reported a global prevalence rates of *E. coli* O157:H7 in beef (whole carcasses, ground, sausage, etc.) ranges from 0.01 to 54.2%.

The results differed greatly with values ranging from very low to very high detection rates probably because the sampling methods used, foodstuffs considered and detection protocols were broadly different.

3.3. Prevalence and Economic Burden

The incidence of *E. coli* O157:H7 infections is substantially low compared to other foodborne pathogens. In 2000, the incidence of human *E. coli* O157:H7 infection per 100,000 persons was 4.10 for Scotland and 2.36 for Northern Ireland (McKee et al., 2003).

It is estimated that in the US there are annually over 73,000 infections and about 60 deaths annually due to *E. coli* O157:H7 infections (Mead et al., 1999). However, the data for 2003 showed that the incidence of *E. coli* O157 infections was 42% lower in the US than in previous years, with 443 laboratory-diagnosed cases (Center For Diseases Control and Prevention, 2004). In 2002 more than 2500 human infections or HUS were reported in the European Union (Kasbohrer, 2004).

The target risk groups are children and the elderly. The highest isolation rate is in the 0-4 year-old group followed by the 5-14 year-old group and the group over 65 (Simmons, 1997).

In a recent study it has been estimated that the yearly cost due to the illnesses caused by O157 infections is about $405 millions of which $370 millions for premature deaths, $30 millions for medical treatments, and $5 millions in productivity losses (Frenzen et al., 2005).

4. Pathology

E. coli O157:H7 infections present a wide spectrum of clinical manifestations ranging from asymptomatic carriage and uncomplicated non-bloody diarrhea to severe illness, including bloody diarrhea and renal failure. Papers of clinical interest present an extensive overview of the pathological features of the infection (Riley, 1987; Karmali, 1989; Nataro and Kaper, 1998; Tkalcic et al., 2003), while this review outlines the main pathogenetic and clinical aspects of the most important diseases associated with the infection: Hemorrhagic Colitis (HC) and the Hemolytic Uraemic Syndrome (HUS). After oral ingestion of *E. coli* O157:H7 and its passage through the acid barrier of the stomach, the microrganism adheres to enterocytes causing A/E lesions and non-bloody diarrhea. The average incubation period is 3 days ranging from 1 to 8 days. At this stage the infection may resolve or continue through replication of *E. coli* O157:H7 and VT production with toxemia. Infection progress to HUS is about 2 to 15% of cases (Dundas et al., 2001) and some risk factors for HUS have been individuated in sex (i.e. females), age (i.e. <5 years), elevated leukocyte count (> 13,000/μL), vomiting, fever, >3 days diarrhea and proteinuria (Rowe et al., 1991; Tserenpuntsag, 2005). The VTs induce microangiopathic lesions in the microvessels of the bowel (HC) and kidneys (HUS), with fibrin/platelet thrombi, capillary occlusion and swelling of the endothelial cells (Karmali, 1989). HC is characterized by severe bloody diarrhea and abdominal pains. In most cases the symptoms will resolve without sequelae. However, in about 10% of young patients (≤ 10 years) the illness will progress to HUS, a syndrome characterized by microangiopathic haemolytic anaemia, thrombocytopenia, renal failure, and central nervous system symptoms. The mortality rate is 3-5% (Nataro and Kaper, 1998). Many survivors (12 to 30%) present severe life-threatening sequelae, including chronic renal failure, hypertension, and neurological symptoms (headache, lethargy, and convulsions). Antibiotic therapy for the management of acute illness may lead to further complications. The case fatality rate ranges from 3 to 7% (Griffin and Tauxe, 1991). The most important clinical features of *E. coli* O157:H7 infection are summarized in Table 3 (Todd and Dundas, 2001).

Table 3. Clinical features of *E. coli* O157 infection (Todd and Dundas, 2001 modified)

Clinical features	Percentage incidence
Vomiting	30-60
Fever	< 30
Abdominal pain	70-80
Diarrhea	95
Diarrhea with blood	>70
Complication rate (HUS)	2-15 in sporadic, 20 in some outbreaks
Asymptomatic infection	10-15
Fatality rate	3-7

5. Detection from Foods

Over the course of time, comprehensive reviews on the methods used for detecting *E. coli* O157:H7 in foods and clinical samples have appeared in the literature (Remis et al., 1984; Padhye and Doyle, 1992; Meng et al., 1994; Phillips, 1999; Chapman, 2000; Bettelheim and Beutin, 2003). Traditional procedures for detecting *E. coli* in foods are unlikely to detect *E. coli* O157:H7 because of specific physiological characteristics of this serotype (Tab. 1). Since the infective dose of *E. coli* O157:H7 is low, sensitive and specific techniques for its detection in food are needed. Rapid and sensitive detection of viable *E. coli* O157 cells, in food is a critical step in the development of effective strategies for the control of the infections. Several methods have been developed for the detection of *E. coli* O157:H7 in food and clinical samples, but to date no universally accepted protocol or "gold standard" method exists for the detection, isolation, and characterization of these bacteria. One obstacle to establishing a method for the detection of *E. coli* O157:H7 in foods is that the indigenous microflora in samples is generally present in greater amounts than few bacterial units of the germ sought for. Another constraint is represented by the correct identification and characterization of isolates, which require serological, biological, and molecular investigations. In terms of risk analysis, the isolation of the strain is not enough and the presence of virulence factors needs to be determined because they might be absent in some *E. coli* O157:H7 strains. Moreover many selective isolation media are based on *E. coli* O157's inability to ferment sorbitol in 24 hours while infections from sorbitol-fermenting strains are becoming more frequent, especially in Europe (Karch and Bielaszewsca, 2001). The method providing the most comprehensive results is based on a selective culture of *E. coli* O157:H7 from which the isolated strains can be typed. One protocol is commonly used in several laboratories and requires neither specialized personnel nor particularly sophisticated equipment while providing satisfactory results. It includes a phase of selective enrichment in modified Tryptone Soy Broth or EC broth added with novobiocin (20mg/l) and incubated at 41±0.5°C for 6-18h, followed by immunomagnetic separation (IMS) and isolation on a selective and differential solid medium, such as Cefixime (0.05mg/l)- Potassium Tellurite (2.5mg/l) (CT)-SMAC, or in a chromogenic medium (ISO 16654, 2001). IMS may be used to capture the target bacteria cells from interfering foodstuffs and concentrate them in a small volume. This step remarkably improves the sensitivity of detection by cultural methods and reduces total analysis time (Wright et al., 1994, Islam et al., 2006). Suspect colonies can be screened for O157 LPS using serological methods (ELISA, latex reagents, etc.) and then identified by biochemical reactions. Once the strain is identified, the presence of virulence factors can be assessed using molecular methods, such as PCR or hybridization with DNA probes for the detection of *eae*, *stx*1 and *stx*2 genes. Biological or serological methods (VERO cell assay, ELISA, RPLA, etc) are useful tools for the evaluation of VT production.

Another approach to detect *E. coli* O157:H7 in foods consists in the screening of the samples (generally after a selective enrichment step) using specific markers for *E. coli* O157 and isolating the microrganism by a selective plating. In fact, molecular techniques (such as PCR, Multiplex-PCR, Real Time-PCR, IMS-PCR, Quantitative Competitive-PCR, etc.) for the detection of genetic markers of virulence, immunoassays for the detection of the O157 antigen, as well as biological assays such as the VERO cell test, are usually employed to directly detect *E. coli* O157:H7 from foods. Nonetheless, each of these methods may have limitations and the results they provide should always be confirmed by isolation and characterization of the strain.

6. Preventing and Control Measures

E. coli O157:H7 infections are a serious public health issue since they can cause epidemic outbreaks affecting many persons with cases of mortality. The epidemiology and control of the infection are rather complex. Since several animals may act as reservoirs, many foods of animal origin (meat and milk) may be contaminated. Faecal contamination of the environment may lead to the contamination of vegetables and water. Moreover, because of its high acid resistance, *E. coli* O157:H7 may persist in processed foods such as cheeses and cured meat. Nevertheless, the infection may be controlled by acting at three levels: the farm, the slaughterhouse (with proper sanitary practices), and the public at large (with information about the safe use of foods).

6.1. Farm

Eradication of the infection on farms where cattle are reared is difficult. The strategies proposed include immunization with different types of vaccines, the administration of probiotics, competitive exclusion treatment or phage therapy (Zhao et al., 1998; Kudva et al., 1999; Tkalcic et al., 2003). These approaches cannot fully ensure eradication of the infection on a farm, and require further investigations. Proper sanitary practices and good hygiene are much more effective as infection control measures. They include changing beverage water frequently, cleaning troughs, reducing faecal contamination of feeds, and the frequent removal and correct disposal of manure (storage and spreading as a soil amendment). Manure is an important source of infection on the farm and for the contamination of fertilized vegetable products, since it has been demonstrated that *E. coli* O157:H7 can survive in bovine faeces from 1 to 18 weeks and for 231 days at 21°C in manure-amended autoclaved soil (Fukushima et al., 1999; Jiang et al., 2002). Furthermore, cattle drinking water represents a source of *E. coli* O157:H7 in farms (Zhao et al., 2006). In addition, in a recent study conducted in order to determine the survival timing of the pathogen in different biological matters, it has been showed that the microrganism survives 97 and 109 days in bovine faeces and water respectively (Scott et al., 2006).

Precautions to be taken during school visits to livestock farms are extremely important. Several cases of infection in school-age children have been ascribed to contact with animals or tasting raw milk on these occasions (Tozzi et al., 2001).

6.2. Slaughterhouse

In abattoirs, a critical point in the meat production chain, the consistent use of Good Hygienic Practices at slaughtering, especially during skinning and evisceration, may considerably reduce the percentages of contaminated carcasses. Particular care should be taken to ensure cleanliness of the hide and sanitation of the stunning box. Recent studies have pointed out that stunning boxes may be an important source of contamination for the hide given that the on-hide median prevalence of positive cattle at slaughter units is 23.6% (Buncic et al., 2004; Small et al., 2005). Pre-skinning decontamination, proper removal of the gastrointestinal tract, avoidance of direct carcass-carcass contact, and frequent disinfection of utensils and surfaces can reduce the percentages of contaminated carcasses at slaughterhouses. Finally, several methods have been proposed to decontaminate carcasses at the end of the processing cycle (washing with hot water, the use of steam, organic acids, gamma radiation, ozone, etc.) (Castillo et al., 2003). Each has limits in terms of applicability and efficiency, at times for technical and/or prescriptive reasons.

6.3. Food Industry and Home

Health education and information on safe food handling and preparation practices may effectively reduce the infection rates. Cooking ground beef to the point that typical strains of salmonellae are killed will also kill *E. coli* O157:H7. Pasteurization of milk (72°C for 16.2 s) will kill more than 10^4 *E. coli* O157:H7 per ml (Doyle, 1991). *E. coli* O157 is more heat-sensitive than salmonellae and is killed by heating to an internal temperature of 68.3°C. Cooking raw meats to an internal temperature of 70°C for two minutes (until there are no pink areas left inside) or an equivalent treatment will provide effective means of preventing infection (Advisory Committee on the microbiological Safety of Food, 1995).

Special care should be taken in the production of ready-to-eat foods such as fermented salami, especially in countries where manufacturing regulations do not prescribe a technological procedure for the reduction of the bacterial load contaminating the mixture before sausage making. It would be appropriate to constantly check the microbiological quality of the raw materials used and perform challenge tests by experimentally inoculating the mixtures to check whether the processing procedure used reduces the load of *E. coli* O157:H7 in the finished product by at least 5 log (Faith, 1998).

Strict observance of hygienic and sanitary practices during food preparation is very important, especially in mass catering services. Accurate washing of vegetables, cleansing of work surfaces, uninterrupted refrigeration of stored foods (4°C), and especially care taken to avoid cross-contamination of raw, cooked and ready-to-eat foods can dramatically reduce the risk of infection.

7. Conclusion

The microbiological safety of food needs to be carefully assessed. The evaluation of potential risks derived from the presence of *E. coli* O157:H7 in foods is an important aspect of assessment of food safety that has not received much attention in the past. To provide the scientific basis for decisions regarding human health, new methods and policies to assess such food safety need to be developed and agreed upon internationally. A this proposal, since 1st January 2006, in European Community is come into force a new body of legislation on the hygiene, safety and control of foods. This regulation includes several Regulations (European Regulation n. 178/2002, 852/2004, 853/2004, 854/2004, 2073/2005, 2074/2005 and others) and aimed to guarantee a high level of safety of foodstuff produced and marketed in EC. Surprisingly, the Reg. (CE) 2073/2005, dealing with the microbiological criteria applicable on foods, considers that the application of microbiological criteria for *E. coli* O157 for the control of the microbiological safety of foods is unlikely to reduce significantly the infection risks for consumers. Nevertheless, it underlines that the application of Good Manufacturing Practices in order to reduce the fecal contamination of foods during the producing chain may constitute a helpful tool for reducing the number of infections due to *E. coli* O157.

In conclusion, an efficient risk analysis for preventing *E. coli* O157:H7 infections requires further investigations. Investigations are needed to shed light on the actual prevalence of *E. coli* O157:H7 in animals acting as reservoirs and in foods of animal origin (using standardized protocols), on its ability to survive in several types of processed foods, and on the likelihood of intoxications due to the presence of preformed VTs in food. Reduction and control of this dangerous infection are predicated on the realization of effective surveillance systems, infection case-studies, and proper health education of food industry personnel handling products "from farm to fork". Finally, efficient food labelling can ensure the provision of adequate information to consumers on safe food preparation practices.

In this perspective, further efforts are needed to integrate medical and veterinary expertise in order to improve competencies and scientific knowledge about the epidemiology of the disease, the ecology of the pathogen along with its behaviour both in the food chain and in complex foods (i.e. fermented and ready to eat foods) and detection methods of the microrganism from foods.

Finally, clear communication of the basis for safety assessment in this area is generally lacking at national and international levels. For this reason, governments all over the world should intensify their efforts to improve food safety in response to an increasing number of food safety problems and rising consumer concerns.

Acknowledgments

The author is grateful to Dr. N.C. Quaglia for her assistance, suggestions, criticism and for her invaluable help in consulting literature cited and table draft.

References

Advisory Committee on the microbiological Safety of Food. 1995. Report on Vero Cytotoxin-producing *Escherichia coli*. *HMSO*, London.

Albihn, A., Zimmernan, U., Rehbinder, V., Jansson, V., Tysén, E., Engvall, A., 1997. Enterohemorrhagic *E. coli* (EHEC)- a nation wide Swedish survey of bovine faeces. *Epidémiol.* Santé Anim. 4, 31-32.

Armstrong, G.L., Hollingsworth, J., Morris, J.G. Jr, 1996. Emerging foodborne pathogens: *Escherichia coli* O157:H7 as a model of entry of a new pathogen into the food supply of the developed countries. *Epidemiol. Rev.* 18, 29-51.

Barret, T.J., Kaper, J.B., Jerse, A.E., Wachsmut, I.K., 1992. Virulence factors in Shiga Toxin-producing *Escherichia coli* isolated from humans and cattle. *J. Infect. Dis.* 165, 979-980.

Batt, C.A., 2000. *Escherichia coli*. In: Robinson, R., Batt, C., and Patel, P. (Ed.), *Encyclopedia of Food Microbiology*, vol. 1. Academic Press, New York, pp. 633-640.

Bettelheim, K.A., and Beutin, L., 2003. Rapid laboratory identification and characterization of verocytotoxigenic (Shiga toxin producing) *Escherichia coli* (VTEC/STEC). *J. Appl. Microbiol.* 95, 205-217.

Beutin, L., Montenegro, M.A., Orskov, I., Prada, J., Zimmermann, S., Stephan, R., 1989. Close association of verotoxin (Shiga-like toxin) production with enterohemolysin production in strains of *Escherichia coli*. *J. Clin. Microbiol.* 27, 2559-2564.

Blanco, M., Blanco, J.E., Blanco, J., Gonzales, E.A., Mora, A., Prado, C., Fernandez, L., Rio, M., Ramos, J., Alonso, M.P., 1996. Prevalence and characteristics of *Escherichia coli* serotypes O157:H7 and other verocytotoxin-producing *Escherichia coli* in healthy cattle. *Epidemiol. Infect.* 117, 251-257.

Bonardi, S., Brindani, V., Pizzin, G., Lucidi, L., D'Incau, M., Liebana, E., Morabito, S., 2003. Detection of *Salmonella* spp., *Yersinia enterocolitica* and verocytotoxin-producing *E. coli* O157 in pigs at slaughter in Italy. *Int. J. Food Microbiol.* 85, 101-110.

Bonardi, S., Maggi, E., Bottarelli, A., Pacciarini, M.L., Ansuini, A., Vellini, G., Morabito, S., Caprioli, A., 1999. Isolation of Verocytotoxin-producing *Escherichia coli* O157:H7 from cattle at slaughter in Italy. *Vet. Microbiol.* 67, 203-211.

Bonardi, S., Maggi, E., Pizzin, G., Morabito, S., Caprioli, A., 2001. Faecal carriage of verocytotoxin-producing *Escherichia coli* O157 and carcass contamination in cattle at slaughter in northern Italy. *Int. J. Food Microbiol.* 66, 47-53.

Bouvet, J., Montet, M.P., Rossel, R., Le Roux, A., Bavai, C., Ray-Gueniot, S., Mazuy, C., Atrache, V., Vernozy-Rozand, C., 2002. Prevalence of verocytotoxin producing *Escherichia coli* (VTEC) and *E. coli* O157 in French pork. *J. Appl. Microbiol.* 93, 7-14.

Buchanan, R.L., and Doyle, M.P., 1997. Foodborne significance of *Escherichia coli* O157:H7 and other enterohemorrhagic *E. coli*. *Food Technology* 51, 69-76.

Buncic, S., Avery, S.M., and De Zutter, L., 2004. Epidemiology of *Escherichia coli* O157 in cattle from farm to fork. pp. 67-84. *Proceedings of International EU-RAIN Conference*, Padua, Italy, 2-3 Dec.

Caprioli, A., Lucangeli, C., Severini, M., 2001. *Escherichia coli* O157. In: De Felip, G. (Ed.), Recenti sviluppi di igiene e microbiologia degli alimenti. *Tecniche nuove*, Italy, pp. 625-641.

Caprioli, A., Morabito, S., Brugère, H., Oswald, E., 2005. Enterohaemorragic *Escherichia coli*: emerging issues on virulence and modes of transmission. *Vet. Res.* 36, 289-311.

Caprioli, A., Nigrelli, A., Gatti, R., Zavanella, M., Blando, A.M., Minelli, F., and Donelli, G., 1993. Characterization of verocytotoxin- producing *Escherichia coli* isolated from pigs and cattle in northern Italy. *Vet. Rec.* 133, 323-324.

Castillo, A., McKenzie, K.S., Lucia, L.M., Acuff, G.R., 2003. Ozone treatment for reduction of *Escherichia coli* O157:H7 and *Salmonella* serotype *Typhimurium* on beef carcasses surfaces. *J. Food Prot.* 66, 775-779.

Center For Diseases Control and Prevention. 2004. Preliminary FoodNet data on the incidence of infection with pathogens transmitted commonly through food- Selected sites, US, 2003. *Morbidity and Mortality Weekly Report* 2004, 53, 338-342.

Centers for Diseases Control and Prevention. 1995. *Escherichia coli* O157:H7 outbreak linked to commercially distributed dry-cured salami- Washington and California, 1994. *Morbidity and Mortality Weekly Report*, 44.

Chapman, P.A., 1995. Verocytotoxin-producing *Escherichia coli*: an overview with emphasis on the epidemiology and prospects for control of *E. coli* O157. *Food Control* 4, 187-193.

Chapman, P.A., 2000. Methods available for the detection of *Escherichia coli* O157 in clinical, food and environmental samples. *W. J. Microbiol. Biotechnol.* 16, 733-740.

Chapman, P.A., Cerdàn Malo, A.T., Ellin, M., Ashton, R., Harkin, M.A., 2001. *Escherichia coli* O157 in cattle and sheep at slaughter, on beef and lamb carcasses, and in raw beef and lamb products in South Yorkshire, UK. *Int. J. Food Microbiol.* 64, 139-150.

Chapman, P.A., Siddons, C.A., Malo, A.T.C., Harkin, M.A., 1997. A one year study of *Escherichia coli* O157 in cattle, sheep, pigs, and poultry. Epidemol. Infect. 119, 245-250.

Clarke, S.C., 2001. Diarrhoegenic *Escherichia coli* – an emerging problem? *Diagn. Microbiol. Infect. Dis.* 41, 93-98.

Coia, J.E., Jhonston, Y., Steers, N.J., Halson, M.F., 2001. A survey of *Escherichia coli* O157 in raw meats, raw cow's milk and raw milk cheeses in south-east Scotland. *Int. J. Food Microbiol.* 66, 63-69.

Commission Regulation (EC) No 2073/2005 of 15 November 2005 on microbiological criteria for foodstuffs. *Official Journal of the European Union L.* 338 of 22 December 2005.

Donnenberg, M.S., Kaper, J.B., Finlay, B.B., 1997. Interactions between enteropathogenic *Escherichia coli* and host epithelial cells. *Trends Microbiol.* 5, 109-114.

Doyle, M.P., 1991. *Escherichia coli* O157:H7 and its significance in foods. *Int. J. Food Microbiol.* 12, 289-302.

Doyle, M.P., and Schoeni, J.L., 1984. Survival and growth characteristics of *Escherichia coli* associated with hemorrhagic colitis. *Appl. Environ. Microbiol.* 48, 855-856.

Doyle, M.P., and Schoeni, J.L., 1987. Isolation of *Escherichia coli* O157:H7 in meats and poultry. *Appl. Environ. Microbiol.* 53, 2394-2396.

Doyle, M.P., Padhye, V.V., 1989. *Escherichia coli.* p. In: Doyle, M.P. (Ed.), *Foodborne Bacterial Pathogens*. Marcel Dekker, New York, pp. 235-271.

Drasar, B.S., and Hill, M.J., 1974. *Human intestinal flora.* Academic Press, Ltd., London, UK.

Dundas, S., Todd W.T., Stewart, A.I., Murdoc, P.S., Chaudhuri, A.K.R., Hutchinson, S.J. 2001. The central Scotland *Escherichia coli* O157:H7 outbreak: risk factor for hemolytic uremic syndrome and death among hospitalized patients. *Clin. Infect. Dis.* 33, 923-931.

Elder, R.O., Keen, J.E., Siragusa, G.R., Barkocy-Gallagher, G.A., Koohmaraie, M., Laegreid, W., 2000. Correlation of enterohemorrhagic *Escherichia coli* O157 prevalence in faeces, hides, and carcasses of beef cattle during processing. *Proc. Natl. Acad. Sci. USA* 97, 2999-3003.

Faith, N.G., Parniere, N., Larson, T., Lorang, T.D., Kaspar, C.W., Luchansky, J.B., 1998. Viability of *Escherichia coli* O157:H7 in salami following conditioning of batter, fermentation, and drying of sticks, and storage of slices. *J. Food Prot.* 61, 377-382.

Fegan, N., Higgs, G., Vanderlinde, P., and Desmarchelier, P., 2004. Enumeration of *Escherichia coli* O157 in cattle faeces using the most probable number technique and automated immunomagnetic separation. *Lett. Appl. Microbiol.* 38, 56-59.

Feng, P., Lampel, K.A., Karch, H., and Whittam, T.S., 1998. Genotypic and phenotypic changes in the emergence of *Escherichia coli* O157:H7. *J. Infect. Dis.* 177, 1750-1753.

Fratamico, P.M., Bhaduri, S., Buchanan, R.L., 1993. Studies on *E. coli* serotype O157:H7 strains containing a 60-Mda plasmid and on 60-Mda plasmid-cured derivatives. *J. Med. Microbiol.* 39, 371-381.

Frenzen, P.D., Drake, A., Angulo, F.J., 2005. Emerging Infections Program FoodNet Working Group. Economic cost of illness due to *Escherichia coli* O157 infection in the United States. *J. Food Prot.* 68, 2623-2630.

Fukushima, H., Hoshina, K., Gomyoda, M., 1999. Long-term survival of Shiga toxin-producing *Escherichia coli* O26, O111, and O157 in bovine faeces. *Appl. Environ. Microbiol.* 65, 5177-5181.

Griffin, P.M., Tauxe, R.V., 1991. The epidemiology of infections caused by *Escherichia coli* O157:H7, other enterohemorrhagic *E. coli*, and the associated hemolytic uraemic syndrome. *Epidemiol. Rev.* 13, 60-98.

Gunzer, F., Bohm, H., Russman, H., Bitzan, M., Aleksic, S., Karch, H., 1992. Molecular detection of sorbitol fermenting *Escherichia coli* O157 in patients with hemolytic-uraemic syndrome. *J. Clin. Microbiol.* 30, 1807-1810.

Guyon, R., Dorey, F., Malas, J.P., Grimont, F., Foret, J., Rouviere, B., Collobert, J.F., 2001. Superficial contamination of bovine carcasses by *Escherichia coli* O157:H7 in a slaughterhouse in Normandy (France). *Meat Sci.* 58, 329-331.

Hancock D.D., Besser, T.E., Rice, D.H., Herriott, D.E., Tar, P.I., 1997. A longitudinal study of *Escherichia coli* O157 in fourteen cattle herds. *Epidemiol. Infect.* 118, 193-195.

Hancock, D., Besser, T., Lejeune, J., Davis, M., Rice, D., 2001. The control of VTEC in animal reservoir. *Int. J. Food Microb.* 66, 71-78.

Heuvelink, A.E., Zwartkruis, J.T.M., van den Biggelaar, F.L.A.M., van Leeuwen, W.J., de Boer, E., 1999. Isolation and characterization of verocytotox-producing *Escherichia coli* O157 from slaughter pigs and poultry. *Int. J. Food Microbiol.* 52, 67-75.

Hussein H.S., Bollinger, L.M., 2005. Prevalence of shiga toxin-producing *Escherichia coli* in beef. *J. Food Prot.* 68, 2224-2241.

Islam, A.M., Heuvelink, A.E., Talukder, K.A., de Boer, E., 2006. Immunoconcentration of Shiga toxin-producing *Escherichia coli* O157 from animal faeces and raw meats by using Dynabeads anti-*E. coli* O157 and the VIDAS system. *Int. J. Food Microbiol.* 109,151-156.

ISO 16654, 2001. Microbiology of food and animal feeding stuffs. Horizontal method for the detection of *Escherichia coli* O157.

Janka, A., Bielaszewsca, M., Dobrindt, U., Greune, L., Schimdt, M.A., and Karch, H., 2003. Cytolethal distending toxin gene cluster in enterohemorragic *Escherichia coli* O157:H⁻ and *Escherichia coli* O157:H7: characterization and evolutionary considerations. *Infect. Immun.* 71, 3634-3638.

Jiang, X., Morgan, J., Doyle, M.P., 2002. Fate of *Escherichia coli* O157:H7 in manure-amended soil. *Appl. Environ. Microbiol.* 68, 2605-2609.

Johnsen, G., Wasteson, Y., Heir, E., I. Berget, O., Herikstad, H., 2001. *Escherichia coli* O157:H7 in faeces from cattle, sheep and pigs in the southwest part of Norway during 1998 and 1999. *Int. J. Food Microbiol.* 65, 193-200.

Karch, H., and Bielaszewsca, M., 2001. Sorbitol-fermenting Shiga toxin producing *Escherichia coli* O157:H⁻ strains: epidemiology, phenotypic and molecular characteristics, and microbiological diagnosis. *J. Clin. Microbiol.* 39, 2043-2049.

Karch, H., Heesemann, J., Lauls, R., O'Brien, A.D., Tacket, C.O., Levine, M.M., 1987. A plasmid of enterohemorrhagic *Escherichia coli* O157:H7 is required for expression of a new fimbrial antigen and for adhesion to epithelial cells. *Infect. Immun.* 55, 455-461.

Karmali, M.A., 1989. Infection by verotoxin-producing *Escherichia coli. Clin. Microbiol.* Rev. 2, 15-38.

Kasbohrer, A., 2004. Current European trends in foodborne zoonoses. 1-3. *Proceedings of International EU-RAIN Conference*, Padua, Italy, 2-3 Dec.

Kaufmann, M., Zweifel, C., Blanco, M., Blanco, J.E., Blanco, J., Beutin, L., Stephan, R. 2006. *Escherichia coli* O157 and non-O157 Shiga toxin-producing *Escherichia coli* in fecal samples of finished pigs at slaughter in Switzerland. *J. Food Prot.* 69, 260-266.

Kenny, B., DeVinney, R., Stein, M., Reinscheid, D.J., Frey, E.A., Finlay, B.B., 1997. Enteropathogenic *E. coli* (EPEC) transfers its receptor for intimate adherence into mammalian cells. *Cell* 91, 511-520.

Keusch, G.T., Grady, G.F., Mata, L.J., McIver, J., 1972. The pathogenesis of *Shigella diarrhea* 1. Enterotoxin production of *Shigella dysenteriae* 1. *J. Clin. Invest.* 51, 1212-1218.

Kittel, F.B., Padhey, N.V., and Doyle, M.P., 1991. Characterization and inactivation of verotoxin-1 produced by *Escherichia coli* O157:H7. *J. Agr. Food Chem.* 39, 141-145.

Knutton, S., Baldwin, T., Williams, P.H., McNeisch, A. S., 1989. Actin accumulation at sites of bacterial adhesion to tissue culture cells: basis of a new diagnostic test for enteropathogenic and enterohemorrhagic *Escherichia coli*. *Infect. Immun.* 57, 1290-1298.

Konowalchuk, J., Speirs, J.I., Stavric, S., 1977. Vero response to a cytotoxin of *Escherichia coli*. *Infect. Immun.* 18, 775-779.

Kudva, I.T., Jelacic, S., Tarr, P.I., Youderian, P., Hovde, C., 1999. Biocontrol of *Escherichia coli* O157 with O157-specific bacteriophages. Appl. Environ. Microbiol. 65, 3767-3773.

Law, D., 2000a. The history and evolution of *Escherichia coli* O157 and other shiga toxin-producing *E. coli*. *World J. Microbiol. Biotechnol.* 16, 701-709.

Law, D., 2000b. Virulence factors of *Escherichia coli* O157 and other shiga toxin-producing *E. coli*. *J. Appl. Microbiol.* 88, 729-745.

Law, D., Kelly, J., 1995. Use of heme and hemoglobin by *Escherichia coli* O157 and other Shiga-like-toxin-producing *Escherichia coli* serogroups. *Infect. Immun.* 63, 700-702.

Li, Q., Sherwood, S.J., Logue, C.M., 2004. The prevalence of *Listeria*, *Salmonella*, *Escherichia coli* and *E. coli* O157 on bison carcasses during processing. *Food Microbiol.* 21, 791-799.

Lin, J., Smith, M.P., Chapin, K.C., Baik, H.S., Bennet, G.N., Foster, J.W., 1996. Mechanism of acid resistance in enterohemorragic *Escherichia coli*. *Appl. Environ. Microbiol.* 62, 3094-3100.

Lindgren, S.W., Samuel, J.E., Schmitt, C.K., O'Brien, A.D., 1994. The specific activities of Shiga toxin type (SLT-II) and SLT-II related toxins of enterohemorragic *Escherichia coli* differ when measured by Vero cell cytotoxicity but not by mouse lethality. *Infect. Immun.* 62, 623-631.

McDaniel, T.K., Kaper, J.B., 1997. A cloned pathogenicity island from enteropathogenic *E. coli* confers the attaching and effacing phenotype on *E. coli* K-12. *Mol. Microbiol.* 23, 399-407.

McKee, M.L., O'Brien, A.D., 1995. Investigation of enterohemorragic *E. coli* O157:H7 adherence characteristics and invasion potential reveals new attachment pattern shared by intestinal *E. coli*. *Infect. Immun.* 63, 2070-2074.

McKee, R., Madden, R.H., Gilmur, A., 2003. Occurrence of verocytotoxin producing *Escherichia coli* in dairy and meat processing environments. *J. Food Protect.* 66, 1576-1580.

Mead, P.S., and Griffin, P.M., 1998. *Escherichia coli* O157. *Lancet* 352, 1207-1212.

Mead, P.S., Slutscker, L., Dietz, V., McCaig, L.F., Bresee, J.S., Shapiro, C., Griffin, M.P., Tauxe, R.V., 1999. Food-related ilnesses in the United States. *Emerg. Infect. Dis.* 5, 607-625.

Meng, J., and Doyle, M.P., 1998. Microbiology of Shiga toxin-producing *Escherichia coli* in foods. In: Kaper, J.B., and O'Brien, A.D. (Ed.), *Escherichia coli* O157:H7 and other Shiga toxin-producing *E. coli* strains. *American Society for Microbiology*, Washington, D.C., pp. 92-108.

Meng, J., Doyle, M.P., Zhao, T., and Zhao, S., 1994. Detection and control of *Escherichia coli* O157:H7 in foods. *Trends F. Sci. Technol.* 5, 179-185.

Montenegro, M.A., Bulte, M., Trumpf, T., Aleksic, S., Rauter, G., Helmut, R., 1990. Detection and characterization of faecal vero-toxin-producing *Escherichia coli* O157 from healthy cattle. *J. Clin. Microbiol.* 28, 1417-1421.

Nataro, J.P., and Kaper, J.B., 1998. Diarrhagenic *Escherichia coli*. *Clin. Microbiol. Rev.* 11, 142-201.

Normanno, G., Dambrosio, A., Parisi, A., Quaglia, N.C., La Porta, L., Celano, G.V., 2002. Survival of *Escherichia coli* O157:H7 in a short ripened fermented sausage. *Ital. J. Food Sci.* 14, 181-185.

Normanno, G., Parisi, A., Dambrosio, A., Quaglia, N.C., Montagna, C.O., Chiocco, D., and Celano, G.V., 2004. Typing of *Escherichia coli* O157:H7 isolated from fresh sausage. *Food Microbiol.* 21, 79-82.

O'Brien, A.D., Holmes, R.K., 1987. Shiga and Shiga-like toxins. *Microbiol. Rev.* 51, 206-220.

Padhye, N.V., and Doyle, M.P., 1992. *Escherichia coli* O157:H7: epidemiology, pathogenesis, and methods for detection in food. *J. Food Protect.* 55, 555-565.

Paton, J.C., and Paton, A.W., 1998. Pathogenesis and diagnosis of Shiga toxin-producing *Escherichia coli* infection. *Clin. Microbiol. Rev.* 11, 450-479.

Philips, D., Summer, J., Alexander, J.F., Dutton, K.M., 2001. Microbiological quality of Australian beef. *J. Food Prot.* 64, 692-696.

Phillips, C.A., 1999. The epidemiology, detection and control of *Escherichia coli* O157. *J. Sci. Food Agric.* 79, 1367-1381.

Pilipčinec, E., Tkáčiková, L., Naas, H.T., Cabadaj, R., Mikula, I., 1999. Isolation of verotoxigenic *Escherichia coli* O157 from poultry. *Folia Microbiol.* 44, 455-456.

Power, C.A., Johnson, R.P., McEwen, S.A., McNab, W.B., Griffiths, M.W., Usborne, W.R., De Grandis, S.A., 2000. Evaluation of the Reveal and SafePath rapid *Escherichia coli* O157 detection tests for use on bovine faeces and carcasses. *J. Food Prot.* 63, 860-866.

Pradel, N., Livrelli, V., De Champs, C., Palcoux, J.B., Reynaud, A., Scheutz, F., Sirot, J., Joly, B., and Forestier, C., 2000. Prevalence and characterization of Shiga-toxin producing *Escherichia coli* isolated from cattle, food, and children during one year prospective study in France. *J. Clin. Microbiol.* 38, 1023-1031.

Ramachandran, V., Hornitzky, M.A., Bettelheim, K.A., Walker, M.J., Djordjevic, S.P., 2001. The common ovine shiga toxin 2- containing *Escherichia coli* serotypes and human isolates of the same serotypes possess a Stx2d toxin type. *J. Clin. Microbiol.* 39, 1932-1937.

Regulation (EC) No 178/2002 of the European Parliament and of the Council of 28 January 2002 laying down the general principles and requirements of food law, establishing the European Food Safety Authority and laying down procedures in matters of food safety. *Official Journal of the European Communities L.* 31/1 of 1February 2002.

Regulation (EC) No 852/2004 of The European Parliament and of the Council of 29 April 2004 on the hygiene of foodstuffs. *Official Journal of the European Union L.* 139, vol. 47 of 30 april 2004.

Regulation (EC) No 853/2004 of The European Parliament and of the Council of 29 April 2004 laying down specific hygiene rules for on the hygiene of foodstuffs. *Official Journal of the European Union L.* 139, vol. 47 of 30 april 2004.

Regulation (EC) No 854/2004 of The European Parliament and of the Council of 29 April 2004 laying down specific rules for the organization of official controls on products of animal origin intended for human consumption. *Official Journal of the European Union L.* 139, vol. 47 of 30 April 2004.

Remis, R.S., McDonald, K.L., Riley, L.W., Puhr, N.D., Wells, J.G., Davis, B.R., Blake, P.A., Cohen, M.L., 1984. Sporadic cases of hemorrhagic colitis associated with *Escherichia coli* O157:H7. *Ann. Intern. Med.* 101, 738-742.

Riley, L.W., 1987. The epidemiologic, clinical, and microbiologic features of hemorrhagic colitis. *Ann. Rev. Microbiol.* 41, 383-407.

Riley, L.W., Remis, R.S., Helgerson, S.D., McGee, H.B., Wells, J.G., Davis, B.R., Hebert, R.J., Olcott, H.M., Johnson, L.M., Hargrett, N.T., Blake, P.A., and Cohen, M.L., 1983. Hemorrhagic colitis associated with a rare *Escherichia coli* serotype. *N. Engl. J. Med.* 308, 681-685.

Rowe, P.C., Walop, W., Lior, H., Mackenzie, A.M., 1991. Hemolytic anemia after childhood *Escherichia coli* O157:H7 infection: are females at increased risk? *Epidemiol. Infect.* 106, 523-530.

Sandving, K., van Deurs, B., 1996. Endocytosis, intracellular transport, and cytotoxic action of Shiga toxin and ricin. *Physiol. Rev.* 76, 949-966.

Sanz, S., Giménez, M., and Olarte, C., 2003. Survival and growth of *Listeria monocytogenes* and *E. coli* O157:H7 in minimally processed artichokes. *J. Food Prot.* 66, 2203-2209.

Scheutz, F., Beutin, L., Pierard, D., and Smith, H., 2001. Nomenclature of verocytotoxins. In: Duffy, G., Garvey, P., and McDowell, D. (Ed.), Verocytotoxigenic *E. coli. Food and Nutrition* Press Inc., Trumbull, CT, USA, pp. 447-452.

Schmidt, H., Beutin, L., Karch, H., 1995. Molecular analysis of the plasmid-coded hemolysin of *Escherichia coli* O157:H7 strain EDL 933. *Infect. Immun.* 63, 1055-1061.

Schmidt, H., Scheef, J., Morabito, S., Caprioli, A., Wieler, L., Karch, H., 2000. A new shiga toxin variant (Stx2f) from *Escherichia coli* isolated from pigeons. *Appl. Environ. Microbiol.* 66, 1205-1208.

Scott L., McGee, P., Sheridan, J.J., Earley, B., Leonard, N., 2006. A comparison of the survival of *Escherichia coli* O157:H7 grown under laboratory condition or obtained from cattle faeces. *J. Food Prot.* 69, 6-11.

Sekiya, J., 1997. *Escherichia coli* O157:H7 in livestock in Japan. *Rev. Sci. Tech. Off. Int. Epiz.* 16, 391-394.

Simmons, N.A., 1997. Global perspectives on *Escherichia coli* O157:H7 and other verocytotoxic *E. coli* spp.: UK views. *J. Food Prot.* 60, 1463-1465.

Sinclair J.F., Dean-Nystrom, E.A., O'Brien, A.D., 2006. The established intimin receptor Tir and the putative eucaryotic intimin receptors nucleolin and beta1 integrin localize at or near the site of enterohemorrhagic *Escherichia coli* O157:H7 adherence to enterocytes in vivo. *Infect Immun.* 74, 1255-1265.

Small, A., Wells-Burr, B., Buncic, S., 2005. An evaluation of selected methods for the decontamination of cattle hides prior to skinning. *Meat Science* 69, 263-268.

Stampi, S., Caprioli, A., De Luca, G., Quaglio, P., Sacchetti, R., Zanetti, F., 2004. Detection of *Escherichia coli* O157 in bovine meat products in northern Italy. *Int. J. Food Microbiol.* 90, 257-262.

Tkalcic, S., Zhao, T., Harmon, B.G., Doyle, M.P., Brown, C.A., Zaho, P., 2003. Faecal shedding of enterohemorragic *Escherichia coli* in weaned calves following treatment with probiotic *Escherichia coli. J. Food Prot.* 66, 1184-1189.

Todd, W.T.A., and Dundas, S., 2001. The management of VTEC O157 infection. *Int. J. Food Microbiol.* 66, 103-110.

Tortorello, M.L., 2000. *Escherichia coli* O157:H7. In: Robinson, R., Batt, C., and Patel, P. (Ed.), *Encyclopedia of Food Microbiology*. Academic Press, New York, vol. 1 pp. 646-652.

Tozzi, A.E., Gorietti, S., Caprioli, A., 2001. Epidemiology of human infections by *Escherichia coli* O157 and other verocytotoxin-producing *Escherichia coli*. In: Duffy, G., Garvey, P., McDowell, D. (Ed.), Verocytotoxigenic *Escherichia coli. Food and Nutrition* Press Inc., Trumbull, pp. 161-179.

Tserenpuntsag, B., Chang, H.G., Smith, P.F., Morse, D.L., 2005. Hemolytic uremic syndrome risk and *Escherichia coli* O157:H7. *Emerg. Infect. Dis.* vol.11 No 12.

Tuttle, J., Gomez, T., Doyle, M.P., Wells, J.G., Zhao, T., Tauxe, R.V., and Griffin, P.M., 1999. Lessons from a large outbreak of *Escherichia coli* O157:H7 infections: insights into the infective dose and method of widespread contamination of hamburger patties. *Epidemiol. Infect.* 122, 185-192.

Weeratna, R.D., Doyle, M.P., 1991. Detection and production of verotoxin 1 of *Escherichia coli* O157:H7 in food. *Appl. Environ. Microbiol.* 57, 2951-2955.

Wells, J.G., Shipman, L.D., Greene, K.D., Sowers, E.G., Green, J.H., Cameron, D.N., Downes, F.P., Martin, M.L., Griffin, P.M., Ostroff, S.M., Potter, M.E., Tauxe, R.V., and Wachsmuth, I.K., 1991. Isolation of *Escherichia coli* serotype O157:H7 and other Shiga-like-toxin producing *Escherichia coli* from dairy cattle. *J. Clin. Microbiol.* 29, 985-989.

Wright, D.J., Chapman, P.A., and Siddons, C.A., 1994. Immunomagnetic separation as a sensitive method for isolating *Escherichia coli* O157 from food samples. *Epidemiol. Infect.* 113, 31-39.

Yu, J., and Kaper, J.B., 1992. Cloning and characterization of the *eae* gene of enterohemorrhagic *E. coli* O157:H7. *Mol. Microbiol.* 6, 411-417.

Zhao, T., Doyle, M.P., Harmon, B.G., Brown, C.A., Mueller, P.O.E., Parks, A.H., 1998. Reduction of carriage of enterohaemorragic *Escherichia coli* O157 in cattle by inoculation with probiotic bacteria. *J. Clin. Microbiol.* 36, 641-647.

Zhao, T., Zhao, P., West, J.W., Bernard, J.K., Cross, H.G., Doyle, M.P., 2006. Inactivation of enterohemorrhagic *Escherichia coli* in rumen content- or faeces-contaminated drinking water for cattle. *Appl. Environ. Microbiol.* 72, 3268-3273.

In: Food Microbiology Research Trends
Editor: Marta V. Palino, pp. 155-188
ISBN 978-1-60021-723-4

Chapter 5

LACTIC ACID FERMENTATION OF FRUITS AND VEGETABLES - AN OVERVIEW

Ramesh C. Ray* and Smita H. Panda
Central Tuber Crops Research Institute (Regional Centre),
PO: Dumduma Housing Board, Bhubaneswar - 751019, India

ABSTRACT

Lactic acid (LA) fermentation has many benefits including feasible in small scale, inexpensive, does not require additives and confers organoleptic characteristics to food stuffs. It not only produces LA but also imparts taste, aroma and flavour to the fermented products and preserve nutrients, vitamins such as ascorbic acid and coloured pigments (β-carotene and anthocyanin) of fruits and vegetables, which are considered as antioxidants. Fermented fruits and vegetables contribute significantly to the diets of rural populations in a number of developing countries. The main reasons of LA fermentation are nutritional, physiological and hygienic aspects of the process. Fermentation of fruits and vegetables can occur 'spontaneously' by the natural lactic bacterial surface microflora, i.e. *Lactobacillus* spp., *Leuconostoc* spp., *Pediococcus* spp., etc.; however, the use of starter culture such as *Lactobacillus plantarum* and *Lb. acidophilus* provides consistency and reliability of performance. LA bacteria are considered to be powerful probiotics that improves human intestinal microbial balance and enhance health by inhibiting the growth of pathogens such as *Escherichia coli*, *Salmonella* and *Staphylococcus*. The probiotic species include *Lb. plantarum*, *Lb. acidophilus*, *Lb. rhamnosus* and *Lb. gasseri*.

Many products like *kimchi*, *sauerkraut* and pickles are produced by LA fermentation of fruits and vegetables, but most economically relevant are fermentations of olives, cucumbers and cabbage. Fermented vegetables commonly produced in Asia are *gundruk* (fermentation of mustard, radish, cauliflower and other leafy vegetables in Nepal and Bhutan), *kimchi* (fermented Korean cabbage and radish), *sinki* (fermented radish tap roots in Nepal and Bhutan), *kanji* (fermented carrots in India and Pakistan), *dhamuoi* (fermented cabbage and other vegetables in Vietnam), *dak-gua-dong* (fermented mustard leaves in Thailand), *hua-chai po* (fermentation of carrots and turnips in Thailand) and *khalpi* (fermented cucumbers in Nepal). The other examples of LA bacterial and yeast

* Corresponding author: Fax: +91-674-2470528; e- mail: rc_ray@rediffmaill.com

fermentations are *gari, fufu, lafun, dawa dawa* and *agbelima* which are produced in African continent from cassava, a tropical root crop. Likewise, sweet potatoes are fermented by LA bacteria into lacto-pickle and lacto-juice.

INTRODUCTION

LA fermented fruits and vegetables (Table 1) have an important role in feeding the world's population on every continent today (Battcock and Azam- Ali, 2001; Panda *et al.*, 2005). They play an important role in preservation, production of wholesome nutritious foods in a wide variety of flavours, aromas and textures which enrich the human diet and remove antinutritional factors to make the food safe to eat (Steinkraus, 1997; Bernardeau *et al.*, 2006; Demir *et al.*, 2006). Fermentation serves many benefits, i.e. provides food security, improves nutrition and uplifts social well being of the people living in marginalized and vulnerable society (Ross *et al.*, 2002; Montet *et al.*, 2006). Fermentation based industries are important source of income and employment in Asia, Africa and Latin America (Ogungobi *et al.*, 2005). Fermentation of fruits and vegetables can occur 'spontaneously' by the natural lactic bacterial surface microflora, i.e. *Lactobacillus* spp., *Leuconosto*c spp., *Pediococcu*s spp., etc.; however, the use of starter culture such as *Lb. plantarum*, *Lb. rhamnosus*, *Lb. gasseri* and *Lb. acidophilus* provides consistency and reliability of performance (Karovicova *et al.*, 1999). Fruits are commonly processed for alcoholic fermentation of wine and other beverages as they are rich in sugars, vitamins, minerals and juices are slightly acidic. They are therefore a suitable medium for the growth of yeasts, and fruit sugars are rapidly converted into ethanol. Fruits can also be fermented by lactic acid bacteria (LAB). Vegetables have low sugar content but are rich in minerals, vitamins, have neutral pH and thus provide a natural medium for LA fermentation (Daeschel, 1987; Buckenhuskes, 1993, 1997). LA fermentation of vegetable products applied as a preservation method for the production of finished and half-finished products is considered as an important technology and is further investigated because of the growing amount of raw materials processed in the food industry (Cheigh and Park, 1994; Montet *et al.*, 2006), and these foods are well suited to promoting the positive health image of probiotics (Heller, 2001).

Table 1. Common and botanical names of the fruit and vegetables mentioned in the text

Common Name	*Botanical Name*
African locust bean	*Parkia filicoidea* L.
Apple	*Malus sylvestris* L.
Bamboo	*Bambusa glaucescens* L.
Banana	*Mus* *a pardisiaca* L.
Beet root	*Beta vulgaris* L.
Cabbage	*Brassica oleracea var. capitata* L.
Capsicum	*Capsicum annum* L
Carrot	*Daucus carota* L.
Cassava	*Manihot esculenta Crantz*
Cauliflower	*Brassica oleracea var. botrytis* L.
Celeriac	*Apium graveolens var. rapaceum* L.

Table 1. (continued)

Common Name	*Botanical Name*
Chinese cabbage	*Brasicca chinensis* L.
Cloves	*Syzygium aromaticum* L.
Cucumber	*Cucumis sativus* L.
False banana	*Ensete ventricosum*
Garlic	*Allium sativum* L.
Ginger	*Zingiber officinale* L.
Gourd	*Momordica charantia* L., *Lagenaria siceraria* L., *Luffa cylindrica* L.
Green pea	*Pisum sativum* L.
Horse-radish	*Lactuca sativa* L.
Immature palm	*Phoenix dactylifera* L.
Juniper berries	*Juniperus communis* L.
Lemon	*Citrus lemon* L.
Lime	*Citrus aurantifolia* L.
Mango	*Mangifera indica* L.
Mustard	*Brassica juncea* L., *B. campestris* L.
Okra	*Abelmoschus esculentus* L.
Olives	*Olea europea* L.
Pak-sian	*Gynadropsis pentaphylla*
Parsley	*Petroselinum crispum* L.
Pear	*Pyrus communis* L.
Pepper	*Piper nigrum* L.
Radish	*Raphanus sativus* L.
Red chilli	*Capsicum frutescens* L.
Spinach	*Spinacea oleracea*
Sweet potato	*Ipomoea batatas* L.
Taro	*Colocasia esculenta* var. *antiquorum* L.
Tomato	*Lycopersicum esculentum* L
Turnip	*Brassica rapa* L.

The other reasons for this interest are the nutritional, physiological and hygienic aspects of the process and their corresponding implementation and production costs (Karovicova *et al.*, 2001,2002b). The consumption of LA fermented fruits and vegetables helps in human nutrition in several ways such as attainment of balanced nutrition, providing vitamins, minerals and carbohydrates and preventing several diseases such as diarrhoea and cerosis of liver because of probiotic properties (Gardner *et al.*, 2001; Rakin *et al.*, 2004; Yamano *et al.*, 2006). This review provides an overview on the current research and prospects of LA fermentation of fruits and vegetables in regards to human nutrition and health.

Lactic Acid Bacteria (LAB)

LAB are a group of bacteria that ferment sugar (i.e. glucose) predominantly to LA (Campbell –Platt, 1987; Liu, 2003). They are gram positive, non- sporulating rods and cocci having low GC (guanine- cytosine) content. Most of the LAB grow anaerobically but they are also aerotolerant. This group of bacteria is divided into two sub- groups (Table 2)

(i) Homofermentative

This sub- group of bacteria produces a single fermentation product i.e. LA *via* the glycolytic (Embden- Meyerhof) pathway. Members of the genera are *Pediococcus*, *Streptococcus* and *Lactococcus* (Axelsson, 1998).

$$\underset{\text{Glucose}}{C_6H_{12}O_6} \longrightarrow \underset{\text{Lactic acid}}{2CH_3CHOHCOOH}$$

The fermentation of one mole of glucose yields two moles of LA.

(ii) Heterofermentative

This sub-group of bacteria produces LA plus appreciable amount of ethanol, acetate and CO_2 *via* the 6- phosphogluconate/ phosphoketose pathway (Steinkraus, 1997). Bacteria involved in this group belong to genera *Leuconostoc and Lactobacillus*. The biochemical pathway is as follows.

$$\underset{\text{Glucose}}{C_6H_{12}O_6} \longrightarrow \underset{\text{Lactic acid}}{2CH_3CHOHCOOH} + \underset{\text{Ethanol}}{C_2H_5OH} + \underset{\text{Carbon dioxide}}{CO_2}$$

Stiles and Holzapfel (1997) and Axelsson (1998) have given an in-depth account of the biochemical pathways for homo and hetero- fermentative LAB.

Table 2. Major lactic acid bacteria in fermented vegetables

Homofermenter	*Facultative homofermenter*	*Obligate heterofermenter*
Enterococcus faecium	*Lactobacillus bavaricus*	*Lactobacillus brevis*
Lactobacillus acidophilus	*Lactobacillus casei (syn. Lb. rhamnosus)*	*Lactobacillus buchneri*
Lactobacillus delbrueckii	*Lactobacillus coryniformis*	*Lactobacillus cellobiosus*
Lactobacillus lactis	*Lactobacillus curvatus*	*Lactobacillus confusus*
Lactobacillus leichmannii	*Lactobacillus plantarum*	*Lactobacillus coprophilus*
Lactobacillus salivarius	*Lactobacillus sake*	*Lactobacillus fermentum*
Pediococcus acidilactici	*Lactobacillus manihotivorans*	*Lactobacillus sanfrancisco*
Pediococcus cerevisiae		*Leuconostoc dextranicum*
Pediococcus damnosus		*Leuconostoc lactis*
Pediococcus pentosaceus		*Leuconostoc mesenteroides*
Streptococcus agalactiae		*Leuconostoc paramesenteroides*
Streptococcus bovis		
Streptococcus faecalis		
Streptococcus mutans		
Streptococcus salivarius		
Streptococcus thermophilus		

Source: Romano *et al.* (1979); Guyot *et al.* (2000); Montet *et al.* (2006).

Beneficial Effects of LAB

LAB have strong inhibitory effects on the growth and toxin production of other bacteria. This antagonistic activity can be the result of:

- Competition for available nutrients
- Decrease in redox potential (Ouvrya *et al.*, 2002)
- Production of LA and acetic acids and the resulting decrease in pH
- Production of other inhibitory primary metabolites, such as H_2O_2, CO_2 or diacetyl (Ray and Daeschel, 1992)
- Production of special antimicrobial compounds such as bacteriocins and antibiotics

Each of these properties and especially the combination of some of them can be used to extend the shelf life and safety of food products (Kalantzopoulos, 1997). The characteristic products of LA fermentation are lactic and acetic acids (Eriksson, 1991; Adams and Nicolaides, 1997). The antimicrobial action of these acids is related to the ability of the undissociated acid molecules to penetrate through the bacterial plasma membrane as a function of their diffusion constant. In the cytoplasm, the acid dissociates to release protons and conjugate bases with higher pH; this disrupts the membrane proton-motive force, thus disabling the energy yielding and transport process dependent upon it (Savard *et al.*, 2002). LA also forms the natural protection of the body against various infections, increases immunity and acts as a physiological disinfection agent, improves digestion and acts at liver diseases (Sulc, 1984). Acetic acid is more inhibitory than LA and can inhibit yeasts, moulds and bacteria (Caplice and Fitzerald, 1999). Although lactic and acetic acids in fermented products have antimicrobial effects, high acetic acid concentrations are not typically desirable from sensory and quality aspects (Lee *et al.*, 1999). The combination of acids and alcohol is conductive to the formation of esters that impart desirable flavours (Lee, 1997). CO_2 can directly be produced in an anaerobic environment and is toxic to some aerobic food microorganisms through its action on cell membranes and its ability to reduce internal and external pH (Caplice and Fitzerald, 1999). Produced CO_2 replaces the air and provides anaerobic conditions favourable for the stability of ascorbic acid and the natural colour of vegetables (Lee, 1997). LAB lack true catalase to break down H_2O_2 generated in the presence of oxygen. It is argued that H_2O_2 can accumulate and be inhibitory to some microorganisms. Inhibition is mediated through the strong oxidizing effect on membrane lipids and cell proteins (Caplice and Fitzerald, 1999).

Bacteriocins

Bacteriocins produced by LAB are peptides or small proteins that are frequently inhibitory towards many undesirable bacteria, including food-borne pathogens (e.g. *Listeria monocytogenes*, *Salmonella*, *Staphylococcus*, *Escherichia coli* and *Clostridium botulinum*) (Leroy *et al.*, 2002). Bacteriocins from LAB can be subdivided into four classes (Klaenhammer, 1993; Abee *et al.*, 1995):

- Class I of bacteriocins consists of lanthibiotics. These are small and heat stable peptides that contain thioether amino acids such as lanthionine (van Reenen *et al.*, 1998; Messens and De Vuyst, 2002; Hernandez *et al.*, 2005)
- Class II is divided into three sub-groups of which Class IIa is the most common. This group is composed of pediocins like bacteriocins with antilisterial activity. Pediocins are produced by *Pediococcus* spp. and while they are not very effective against spores, they are more effective than nisin in some food systems (O'Sullivan *et al.*, 2002)
- Class III comprises large heat labile proteins (Eijsink *et al.*, 2002), and
- Class IV a complex of bacteriocins with glyco-and/or lipid moieties (Rodriguez *et al.*, 2003)

An advantage of bacteriocins over classical antibiotics is that digestive enzymes destroy them. Bacteriocins producing strains can be used as part of or adjuncts to starter cultures for fermented foods in order to improve safety and quality (Caplice and Fitzerald, 1999). LAB are considered to have several beneficial physiological effects such as antimicrobial activity enhancing the immune potency (Kullisaar *et al.*, 2002) and to prevent cancer and lower serum cholesterol levels (Kaur *et al.*, 2002). Proposed health and nutritional benefits of *Lactobacillus* spp. are presented in Table 3 (Klaenhammer, 1995).

Because LAB prohibit colonization by the invaders and control the intestinal pH through the release of acetic and lactic acids, these bacteria could effectively prevent constipation and diarrhoea caused by lactose intolerance or pathogenic bacteria (Mohapatra *et al.*, 2007). A synergistic effect of dietary fibre and LAB for the improvement of the large intestinal health of the host may be achieved by providing fermented fibre rich natural plant materials to the host (Kim *et al.*, 2000). Exopolysaccharides produced by LAB are sugar polymers that improve the rheological and textural properties of food products (Leroy *et al.*, 2002). Several research studies confirm the ability of LAB to reduce the mutagenicity of intestinal contents by suppressing the levels of specific bacterial enzymes that promote the activation of procarcinogenic compounds (Daly and Davis, 1998).

Table 3. Proposed health and nutritional benefits of *Lactobacillus* species to human beings

Enzyme (lactase) presentation
Colonization and maintenance of normal intestinal microflora
Competitive exclusion of undesirable microorganisms
Microbial interference and antimicrobial activities
Pathogen clearance
Immunostimulation and modulation
Cholesterol reduction/removal
Deconjugation of bile acids
Anticarcinogenic and antimutagenic activities
Reduction of endotoxaemia from alcoholic liver disease

Source: Karovicova and Kohajdova (2003).

Lactobacilli have been periodically associated with anticarcinogenic, antimutagenic and antitumorigenic activities (Kaur and Kapoor, 2001; Bernardeau *et al.*, 2006). These activities may occur *via* the following mechanisms:

- Binding, inhibition or inactivation of mutagens *in vitro*
- Reductions in carcinogen-generating faecal enzymes *in vivo*
- Stimulation of the immune system
- Suppression of tumour formation (Klaenhammer, 1995)

LAB As Powerful Probiotics

Lactobacillus organisms are commonly considered as probiotics, which are defined as live bacteria that improve intestinal microbial balance and enhance health (Heller, 2001; Molin, 2001). Bifidobacteria first observed and reported in 1990 as probiotic, which is strictly anaerobic and gram positive rods (Cross, 2001; Soomro *et al.*, 2002). These probiotics are normal inhabitants of the gastrointestinal tract. Some of the species are *Bifidobacterium lactis, B. infants, B. longum* and *B. animalis*. Other identified probiotics species include *Lactobacillus casei, Lb. acidophilus, Lb. rhamnosus, Lb. delbrueckii* and *Streptococcus thermophilus*. Probiotics bacteria have a positive effect on the human immunological system, when used in children with innate HIV (Havenaar, 1992; Soomro *et al.*, 2002). Benefits of probiotics to human health include:

- Gastrointestinal well being i.e. resistance to infection, prevention or fast recovery from diarrhoea
- Reduce cancer risk, particularly colon and breast cancer
- Improve tolerance to milk proteins and lactose
- Reduce LDL (low density lipid) cholesterol
- Antitumourgenic activities *in vitro*

LA Fermented Fruits and Vegetables

Fruits and vegetables are rich in nutrients, vitamins and minerals and some of them contain coloured pigments such as flavonoids, lycopene, anthocyanin, β-carotene and glucosinolates, which act as antioxidant in the body by scavenging harmful free radicals implicated in degenerative diseases like cancer, arthritis and ageing (Kaur and Kapoor, 2001). LA fermentation enhances the organoleptic and nutritional quality of the fermented fruits and vegetables as well as retains the nutrients and coloured pigments (Steinkraus, 1997; Dahal *et al.*, 2005). In eastern European and in North American countries, *sauerkraut*, olives and cucumbers are often fermented, while other vegetable are sterilized or frozen (Steinkraus, 1997; Montet *et al.*, 2006). The most reported fermented fruits and vegetables are listed below.

- Root vegetables – such as carrots, turnips, beetroot, radishes, celeriac and sweet potato (Rakin *et al.*, 2004; Panda *et al.*, 2007)
- Vegetable fruits such as cucumbers, olives, tomatoes, peppers, okra and green-peas (Dahal *et al.*, 2005)
- Vegetables juices such as carrot, turnips, tomato pulp, onion, sweet potato, beet and horseradish (Karovicova *et al.*, 1999, 2002b; Klewicka *et al.*, 2004; Panda and Ray, 2007)
- Fruits such as apples, pears, immature mangoes, immature palms, lemons and fruit pulps such as banana (Zhang *et al.*, 2000; Montet *et al.*, 2006)

Table 4 shows some examples of typical LA fermented vegetables produced in different regions of the world and the microorganisms associated with these products. According to Kim *et al.* (2000) the Chinese cabbage, cabbage, tomato, carrot and spinach gave relatively higher fermentability than other vegetables (okra and gourds) because they have more fermentable saccharides. The classic LA vegetable fermentation is *sauerkraut* (Steinkraus, 1997). Fresh cabbage is shredded and 2.3 - 3.0 % salt is added before allowing for natural fermentation. *Sauerkraut* production typically relies on a sequential microbial process that involves heterofermentative and homofermentative LAB, generally involving *Leuconostoc* spp. in the initial phase and *Lactobacillus* and *Pediococcus* spp. in the subsequent phases (Lee *et al.*, 2005). The pH of final product varies from 3.5 - 3.8 (Gardner *et al.*, 2001). At this pH the cabbage or other vegetables will be preserved for a long period of time (Steinkraus, 1997). *Sauerkraut* brine is an important by-product of the cabbage fermentation industry and can be used as a substance for the production of carotenoids by *Rhodotorula rubra* (Shih and Hang, 1996) or for ß-glucosidase production by *Candida wickerhamii* for commercial applications (Sim and Hang, 1996). As a group of traditional vegetable foods in Korea, *kimchi* is known to be the product of natural mixed fermentation carried out principally by LAB (Cheigh and Park, 1994; Montet *et al.*, 2006). *Kimchi* is characterized particularly by its sour, sweet and carbonated taste and differs in flavour from *sauerkraut* and pickles that are popular fermented vegetables in the West (Hong *et al.*, 1999). Optimum taste of *kimchi* is attained when the pH and acidity reach approximately 4.0–4.5 and 0.5–0.6, respectively. Vitamin C content is maximal at this point. As the fermentation temperature increases, the ripening time decreases. *Kimchi* ripens in one week at 15°C and three days at 25°C. Low temperature is preferred in *kimchi* fermentation to prevent the production of strong acid and over-ripening and to extend the period of optimum taste (Lee, 1997).

Pickling cucumbers are made in Africa, Asia, Europe and Latin America. In Egypt cucumbers are pickled by soaking in brine to produce *torshi khiar* (Mahmoud *et al.*, 1972; Steinkraus, 2002). *Khalpi* is a cucumber pickle popular during the summer months in Nepal (Dahal *et al.*, 2005). Although a variety of methods are used, placing the cucumbers in 5% salt brine is a satisfactory method. The cucumbers absorb salt until there is equilibrium between the salt in the cucumbers and the brine (about 3% salt in the brine) (Reina *et al.*, 2005). When the pH attains at about 4.7 – 5.7, the brine is inoculated with either *Lb. plantarum* or *Pediococcus pentosaceus* or a combination of these organisms for a total cell count of 1- 4 billion cells/ gallon of brined cucumbers (Daeschel and Fleming, 1984; McDonald *et al.*, 1991, 1993). The final product has an acidity of 0.6- 1.0% (as LA) and pH of 3.4- 3.6 in about two weeks, depending upon the temperature (Lu *et al.*, 2001).

Table 4. Microorganisms isolated from some lactic acid fermented fruits and vegetables

Product name	*Country*	*Fruit and vegetables*	*Microorganisms*	*References*
Burong mustala	Philippines	Mustard	*Lactobacillus brevis* *Pediococcus cerevisiae*	Karovicova *et al.* (2002a; 2002b)
Dakguadong	Thailand	Mustard leaf, salt	*Lactobacillus plantarum*	Tanasupawat and Komagata (1994)
Dhamuoi	Vietnam	Cabbage, various vegetables	*Leuconostoc mesenteroides* *Lactobacillus plantarum*	Steinkraus (1997)
Gundruk	Nepal	Cabbage, radish	*Pediococcus and Lactobacillus spp.*	Steinkraus *et al.* (2002); Dahal *et al.* (2005); Tamang *et al.* (2005)
Kanji	India and Pakistan	Carrots	*Lactobacillus plantarum* *Lactobacillus brevis*	Berry (1998); Shah (1986)
Khalpi	Nepal	Cucumber	*Lactobacillus plantarum* *Pediococcus pentosaceus*	Dahal *et al.* (2005); Tamang *et al.* (2005)
Kimchi	Korea	Cabbage, radish, various vegetables	*Leuconostoc mesenteroides* *Lactobacillus brevis* *Lactobacillus plantarum*	Cheigh and Park (1994); Choi *et al.* (1994); Lee *et al.* (2005)
Olive	Spain, Italy	Olive	*Lactobacillus plantarum* *Lactobacillus brevis* *Pediococcus cerevisiae* *Leuconostoc mesenteroides*	Etchells *et al.* (1966); Borcakli *et al.* (1993); Ozay and Borcakli (1996); Duran- Quintana *et al.* (1999); Nychas *et al.* (2002)
Pak-sian-dong	Thailand	Leaves of Pak-sian	*Lactobacillus brevis* *Pediococcus cerevisiae* *Lactobacillus plantarum*	Steinkraus (1996); Tanasupawat and Komagata (1994)
Sauerkraut	International	Cabbage	*Leuconostoc mesenteroides* *Lactobacillus plantarum* *Lactobacillus brevis*	Sim and Hang (1996); Kalac *et al.* (2000a,b); Karovicova *et al.* (2002a; 2002b); Viander *et al.* (2003)
Sinki	India, Nepal and Bhutan	Radish	*Lactobacillus plantarum* *Lactobacillus brevis* *Lactobacillus fermentum* *Leuconostoc fallax* *Pediococcus pentosaceus*	Tamang *et al.* (2005)
Sunki	Japan	Leaves of otaki-turnip	*Lactobacillus plantarum* *Lactobacillus brevis* *Pediococcus pentosaceus* *Bacillus coagulans*	Karki (1996); Battcock and Azam-Ali (2001)
Suan-tsai	Taiwan	Mustard leaves	*Pediococcus pentosaceus* *Tetragenococcus halophilus*	Chen *et al.* (2006)

The consumption of LA fermented vegetable juices (lacto- juice) has increased in many countries (Kopec, 2000). Lacto-juices are produced mainly from cabbage, red beet, carrot, celery and tomato (Karovicova *et al.*, 2002b; Demir *et al.*, 2006). They can be produced by either of the following procedures:

- The vegetable is fermented in a usual way and then it is processed by pressing the juice (manufacture from *sauerkraut*)
- The vegetable is processed to mash or raw juice at first and it is consecutively fermented

There are three types of lactic fermentation of vegetable juices:

- Spontaneous fermentation by natural microflora
- Fermentation by starter cultures that are added into raw materials
- Fermentation of heat-treated materials by starter cultures (Hammes, 1990)

At the manufacture of lacto- juices, the pressed juice can be pasteurized at first and consecutively it is inoculated by a culture of selected LAB at a concentration varying from 2 x 10^5 to 5 × 10^6 Colony Forming Unit (CFU)/ml (Šulc, 1984; Demir *et al.*, 2006; Panda and Ray, 2007). For fermentation of juices of highest quality, it is imperative to use commercially supplied starter cultures such as *Lb. plantarum, Lb. bavaricus, Lb. xylosus, Lb. bifidus, Lb. brevis* (McFeeters, 2004). The criteria used for finding out suitability of a strain are as follows (Karovicova *et al.*, 1994, 1999; McFeeters, 2004):

- The rate and total production of LA, change in pH, loss of nutritionally important substances
- Decrease in nitrate concentration and production of biogenic amines (BAs)
- Ability of substrate to accept a starter culture
- Type of metabolism and ability of culture to create desirable sensory properties of fermented products (Holzapfel, 2002)

Bacteriocin producing starter cultures were suggested as a mean to obtain more controlled and reproducible vegetable fermentations (Aukrust *et al.*, 1994; Ausco *et al.*, 1998). Enzyme (pectinase) mash treatment is a well-known modern process to gain more juice from vegetable (Demir *et al.*, 2001; Angelova, 2007; Panda and Ray, 2007). The enzyme retains in mash most of the mineral compounds including calcium, phosphorus and magnesium of the fruits and vegetables (Kopec, 2000). Fermentation takes place at the temperature of about 20 – 30°C (Adams and Nicolaides, 1997). For an optimal course of LA fermentation, the content of sugars in raw materials must be sufficiently high (at least 40 g/kg) and the content of proteins that neutralize emergent acids must be minimal (Kopec, 2000). During fermentation, the pH of juices decreases from 6.0 – 6.5 to 3.8 – 4.5 (Adams and Nicolaides, 1997; Panda and Ray, 2007). A rapid decrease in pH at the beginning of fermentation is of great importance for the quality of the end product (Viander *et al.*, 2003). A rapid increase in acidity minimizes the influence of spoilage bacteria. The acidity below pH 3.6 is undesirable from the sensory aspect (Lopatnikova, 1992; Mohapatra *et al.*, 2007). By fermentation, the juice imparts pleasant acid taste and characteristic aroma. After fermentation, the juices can be filled into bottles and pasteurized (Ray and Ward, 2006).

DIFFERENT PROCESS OF LA FERMENTATION OF FRUITS AND VEGETABLES

Spontaneous Fermentation

Spontaneous fermentation leads to variations in the sensory properties of the products which vary according to the quality of raw material, temperature and harvesting conditions (Gardner *et al.*, 2001). Fresh vegetables and fruits may present high microbial loads (around 10^5 to 10^7 microorganisms/g) after harvesting, most of which are gram (-) ve bacteria and gram (+) ve bacilli, yeasts and moulds. LAB are least prevalent accounting for less than 0.1% of the total microbial population as these bacteria require high nutrients such as amino acids, fatty acids, vitamins and certain minerals. Therefore, the plant environment is not suitable for their growth and development. But the plant medium can be enriched by salting or addition of certain proteinaceous ingredients (whey, bran, etc) for their growth (Rao *et al.*, 2004). During the fermentation process lactic and acetic acids are formed and pH decreases thus inhibiting the gram (-) ve bacteria and sporulating bacteria. In these conditions, only LAB are able to grow.

Some of the LAB isolated from naturally fermented vegetables are *Lb. plantarum*, *Lb. brevis*, *P. cerevisiae*, *Leuconostoc mesenteroides* and *Lactococcus lactis* (Karovicova *et al.*, 1999; Dahal, *et al.*, 2005). Lactobacilli not only produce LA but also H_2O_2 and antibiotics which inhibit the growth of pathogenic bacteria (Steinkraus, 1997; Battcock and Azam- Ali, 2001). Along with organic acids, the heterofermentative lactobacilli produce CO_2 which also has a preservative effect on foods. The dominant bacteria at outset of fermentation are resistant to high salt concentrations. A dominant flora sequence over the course of the LA fermentation has been identified i.e. *Pediococcus*, *Leuconostoc* and *Lactobacillus*, in that order. The bacterial count can reach 10^9 cells/ ml at the end of fermentation (Lee *et al.*, 2005). Fermentation is halted when all the sugars are consumed with an increase in LA and decrease in pH. Residual sugars can induce the growth of acidophilic yeasts which can be observed on the surface of open vats containing fermented products. They modify the products through production of gas or pigments. The amount of BAs (discussed in the section: anti- nutritional aspects of LA fermentation) increases in the fermented products during spontaneous fermentation which are very lethal to human body (Kirschbaum *et al.*, 2000; Karovicova *et al.*, 2002 a, b).

Controlled Fermentation

Irregularities in the quality of products fermented on a small scale are unacceptable on an industrial scale, where product quality has to be consistent (Panda *et al.*, 2005). Quality control is therefore essential for the industrialization of fermentation process. For controlled LA fermentation, conditions must be created which favours the growth of commensal and/or inoculated LAB while excluding other microorganisms (Gardner *et al.*, 2001). Although suppliers of lactic cultures have numerous starter cultures for the dairy and meat sectors, there are few cultures designed for vegetable fermentations (Hansen, 2002). There are seven factors that influence the growth and activity of LAB in fermenting fruits and vegetables. Those are pH, moisture (water activity), O_2 concentration, temperature, nutrients, selected starter culture and inoculum concentration (Lee and Salminen, 1995).

pH

The pH is a critical factor in preservation and developing aroma and flavour of many fermented fruits and vegetables like cabbage, olives, etc (Hubert and Dupuy, 1994; Steinkraus, 1997; Muyanja *et al.*, 2003; Rao *et al.*, 2004)). Most LAB favour conditions with a near neutral pH (Battcock and Azam- Ali, 2001; Adeniyi *et al.*, 2005). Certain bacteria are acid tolerant (i.e., *Lactobacillus* and *Streptococcus* spp.) and can survive at reduced pH levels (3.0- 4.0) (Fleming *et al.*, 1985; McDonald *et al.*, 1990).

O_2 Availability

The O_2 requirements vary from species to species. Unlike many anaerobes, however most LAB are not sensitive to O_2 and can grow in its presence as well as in absence. They are aero-tolerant anaerobes (Molenaar *et al.*, 2005).

Temperature

Temperature is a critical factor for vegetable fermentation. Most LAB have a temperature optimum between 20^0C to 30^0C, there are some (thermophiles) which prefer high temperatures (50 - 55^0C) and those with colder temperatures optima (15 - 20^0 C) (Buckenhuskes, 1993, 1997; Ross *et al.*, 2002). Most LAB work best at temperatures of 18 - 22^0C. The *Leuconostoc* spp. initiate fermentation at an optimum temperature of 18- 22^0C and above this range of temperature, favours the *Lactobacillus* spp. (Battcock and Azam-Ali, 2001).

Salt Concentration

Salting is an important step in vegetable fermentation. NaCl concentration can range from 20 to 80 g/l during fermentation and up to 160g/l in certain stored vegetables. LAB can tolerate high salt concentrations. The salt tolerance gives them an advantage over less tolerant species and allows LA fermentation that inhibits growth of non-desirable organisms and activity of potential pectinolytic and proteolytic enzymes that cause vegetable softening and further putrefaction (Rao *et al.*, 2004). Salt induces plasmolysis in plant cells which releases mineral salts and nutrients from the vacuole and creates anaerobic conditions for proper growth of LAB around the submerged product. *Leuconostoc* spp. are noted for their high salt tolerance and for this reason they initiate the majority of LA fermentation (Gardener *et al.*, 2001; Rakin *et al.*, 2004).

Salting is done either by adding NaCl or crude salt to vegetables or by soaking in brine. In case of dry salting, salt is added regularly to the successive layers of the product, avoiding the high salt concentrations that could promote yeast development and product damage. The optimum salt concentration depends upon the type of vegetable. Anaerobic conditions are more effective in the finely cut and shredded plant material. Whole or non- juicy vegetables are therefore soaked in highly salted brine.

Water Activity

In general, LAB require a fairly high water activity (0.9 or higher) to survive. There are a few species, which can tolerate water activities lower than this, but usually, the yeasts and fungi will predominate on foods with a lower activity.

Nutrients

All bacteria require a source of nutrients for metabolism. The fermentative bacteria require carbohydrates either simple sugars such as glucose and fructose or complex carbohydrates such as starch or cellulose. The energy requirements of microorganisms are very high. Adding a mixture of acetic acid and calcium acetate to the brine can reduce the number of Enterobacteria (McDonald *et al.*, 1990; 1991).

Selected Starter's Fermentation

The selection of starter cultures is based principally on the competitiveness between the starter and the natural flora, as well as on the sensory properties of the resulting products (McFeeters, 2004). Further, using selected culture starters meets overall economic, technological and nutritional requirements. They are selected on the following criteria for fermentation of fruits and vegetables:

- Lack of production of toxic chemicals
- Genetic stability of the species
- Rapid brine acidification
- Production reproducibility between different batch cultures
- Total depletion of fermentable sugars
- Resistance to bacteriocins and bacteriophages from natural strains

Inoculum Concentration

Inoculum concentrations of 10^6- 10^7 CFU/g is normally used to accelerate the process of fermentation in order to prevent spoilage and to obtain products of consistently high and desired quality (Aukrust *et al.*, 1994; Gardner *et al.*, 2001). In a recent study, an inoculum concentration of 3 x 10^5 CFU/g of *Lb. plantarum* was used as starter culture in fermenting carrot mash into juice (Demir *et al.*, 2006)

PRINCIPLES OF LA FERMENTATION

LA fermentations of fruits and vegetables are carried out under three basic types depending upon the salt concentration used for fermentation.

Dry Salted Fermentation of Vegetables and Fruits

In this type dry salt is added to fruits and vegetables. For every 100 kg of vegetables and fruits, approximately 3kg of salt is needed. The vegetables or fruits are properly washed and salts are placed in an alternative layer of about 2.5cm depth in a barrel or keg. Weight is added to compress the biomass and salt mixtures and assist the formation of brine. Fermentation starts as soon as brine formation occurs, which takes about 24 h. Depending upon the ambient temperatures, fermentation takes between one to four weeks. Fermentation is complete when no more bubbles of CO_2 appear, then the pickle can be packaged in jars seasoned with spices, oil and vinegar (Buckenhuskes, 1997). Dry salting is used for pickling many vegetables and fruits like limes, lemons and cucumbers.

Brine Salted Fermentation of Vegetables

Brine is used for those vegetables that inherently contain sufficient water. Fermentation takes place well in brine of about 20^0 salometer (Vaughn, 1985). Most vegetables can be fermented at 12.5 - 20^0 salometer. Under these conditions the sequence of LAB generally follows the classical *sauerkraut* fermentation described by Steinkraus (1997). Once the vegetables have been brined and the container sealed, there is a rapid development of microorganisms in the brine. The rapidly of the fermentation is correlated with the concentration of salt in the brine and its temperature. At high salt level of about 40^0 salometer, the sequence is directed towards the development of a homofermentation, dominated by *Lb. plantarum.* At 60^0 salometer, LA fermentation stops and if any acid is detected, it is acetic acid produced by acid forming yeasts which are still active at this salt concentration (Vaughn, 1985, Gardner *et al.*, 2001).

Non-Salted Fermentation of Vegetables

Some vegetables are fermented by LAB, without prior addition of salt or brine solution. Some of the non-salted products include *gundruk*, *sinki* and *sunki* (discussed under the section: Recipes).

General Processing of Fruits and Vegetables

Raw Materials Pretreatments

Pretreatments can promote growth of lactic flora that can be used depending on the fruit or vegetable to be fermented. Washing fruits and vegetables prior to fermentation reduce the initial microbial count, thus favouring the development of lactic flora (Ana *et al.*, 2002). Vegetables are also macerated with pectinolytic enzymes (Panda and Ray, 2007) to allow their homogenization prior to lactic fermentation, mainly for the production of cocktails and juices (Zetelaki- Horvath and Anderson, 1986; Demir *et al.*, 2001, 2006).

Many vegetables contain glycosides that hamper efficient fermentation (Karovicova *et al.*, 1999; Drewnowski and Gomez- Carneros, 2000). For LA fermentation of tomatoes, choosing very ripe fruit is recommended, since the high solanin content of unripe fruit might inhibit the growth of LAB (Beltran- Edeza and Hernandez- Sachez, 1989).

Role of Ingredients Used in Fermentations of Fruits and Vegetables

Salt

LA fermentation of fruits and vegetables is mostly carried out in a salted medium (Viander *et al.*, 2003). Salting is done by adding common dry salt (NaCl) with high water content or by soaking in brine. The optimum salt concentration depends on the type of vegetable or fruits (Sanchez *et al.*, 2000; Ross *et al.*, 2004). Substituting NaCl by KCl up to 50% in preparation of *kimchi* from cabbage did not affect the sensory qualities (saltiness, bitterness, sourness, hotness and texture) (Choi *et al.*, 1994). The main role of salt is to promote the growth of LAB over spoilage bacteria, and to inhibit of potential pectinolytic and proteolytic enzymes that can cause vegetable softening and further putrefaction. Salt induces plasmolysis in the plant cells and the appearance of a liquid phase, which creates anaerobic conditions around the submerged product. Anaerobic conditions are more effective in the finely cut and shredded cut material.

Ingredients Favouring Bacterial Growth

Some ingredients when added to LA fermented vegetables or fruits seem to enhance the development of lactic flora. They have three major roles:

- They are a source of nutrients such as sugars, vitamins, minerals which initiate fermentation
- They add desirable aroma, flavour and taste to the fermented product
- They help in combating the spoilage bacteria by lowering the pH

For some vegetables with low nutrient contents, such as turnip and cucumber, the addition of sugar promotes bacterial growth, thereby accelerating fermentation. In spanish-style olive fermentation, olives are undergone alkaline treatment to eliminate their bitterness, followed by repeated washings (Sanchez *et al.*, 2000). They are then replaced with glucose on sucrose to improve LA fermentation (Lu *et al.*, 2001; Duran- Quintana *et al.*, 1999). Whey is often recommended for use in traditional LA vegetable fermentation processes as it has high lactose content, which is a potential energy substrate for LAB. It also supplies minerals salts and vitamins necessary for the lactic flora metabolism.

Ingredients with Antiseptic Properties

Spices or aromatic herbs are added to most of the lactic fruits and vegetable fermentation to improve the flavour of the end products (Panda *et al.*, 2007). Certain spices, mainly garlic, cloves, juniper berries and red chilies help to inhibit the growth of spoilage bacteria (Paludan-Muller *et al.*, 1999; Montet *et al.*, 2006). There are many sulphur compounds with antibacterial properties in garlic (Paludan- Muller *et al.*, 1999) which must be combined with other.vegetables at ratios not higher than 150g / kg. of vegetables. Chemical preservatives such as sorbic on benzoic acid salts are sometimes used in industrial production of LA fermented *sauerkraut*, olives or cucumbers (Steinkraus, 2002). The role of essential spice oils such as thyme, sage, lemon and dill is to inhibit the growth of during fermentations of olives (Reina *et al.*, 2005). Mustard seed contains allylisothiocyanate, a volatile aromatic compound with antibacterial and antifungal properties, which inhibits the growth of yeast

(*Saccharomyces cerevisiae*) and promotes growth of LAB (Steinkraus, 2002; Reina *et al.*, 2005).

Ingredients Modifying the pH and Buffers Effect of Brines

To promote the growth of LAB over yeasts, moulds and other pathogenic or unwanted bacterial strains, acids or buffer systems (acid + acid salts) are often added to the fermentation medium. During the fermentation of fruits and vegetables with high fermentable sugar contents, the fermentation medium has to be buffered to slow down acidification, thus allowing the LAB to consume all the sugars. An acetic acid + calcium acetate buffer system has been reported to improve the LA cucumber fermentation process.

NUTRITIONAL ASPECTS OF LA FERMENTED FRUITS AND VEGETABLES

Beneficial Aspects

Food Preservation

The main motive of LA fermentation is preservation of foods, increase its shelf- life, and improve food quality and palatability (Ray and Ravi, 2005; Ward and Ray, 2006). There are several options for preserving fresh fruits and vegetables including drying, freezing, canning and pickling (Karovicova *et al.*, 1999). But, many of these techniques are not suitable for small- scale use in developing countries (Steinkraus, 2002). For instance, the small-scale canning of vegetables has food safety problems and contamination of food- borne pathogens such as *L. monocytogenes* and *E. coli* might occur (Reina *et al.*, 2005). LA fermentation requires very little sophisticated equipments for carrying out the fermentation process (Battcock and Ali, 2001; Steinkraus, 2002). Further, the pathogenic microflora is inhibited by LAB (Reina, 2003; Reina *et al.*, 2005).

Removal of Anti- nutritional Factors

Many vegetables and fruits contain naturally occurring toxins and antinutritional compounds (Drewnowski and Gomez- Carneros, 2000). For example, cassava (a field crop) roots contain two cyanogenic glucosides, linamarin and lotaustralin (Lei *et al.*, 1999; Ray and Ward, 2006). The roots when naturally fermented by a mixed population of yeasts (*S. cereviseae* and *Candid* spp.) and LAB (*Lactobacillus* spp., *Leuconostoc* spp. and *Pediococcus* spp.), the cyanogen level is reduced drastically (Onabolu *et al.*, 2002a, b; Sanni *et al.*, 2002; Kostinek *et al.*, 2005). Likewise, LAB reduce the toxic elements in African locust bean (Antai and Ibrahim, 1986) and in leaves of *Cassava obtusifolia* during preparation of *kawal*, a Sudanese food (Dirar, 1993).

Minerals and Vitamin Preservation

The micronutrient availability is enhanced in LA fermented vegetables because of significant reduction in the enzyme, phytase. Iron bioavailability was also reported to be higher in carrots and sweet potato after LA fermentation (Rakin *et al.*, 2004; Mohapatra *et al.*, 2007).

There are mixed reports on the preservation of vitamin C during LA fermentation. Whole cabbage when fermented in brine its ascorbic acid content decreased during fermentation, leveling off at about 20- 25 mg/ kg (Buckenhuskes *et al.*, 1990). Vaughn (1985), however, reported increase in vitamin C during *sauerkraut* fermentation.

Improvement of Food Digestibility

LAB contain various intracellular and extracellular food digestive enzymes i.e., α-amylase (Guyot *et al.*, 2000; Rao *et al.*, 2004), pectinase (Kunji *et al.*, 1996) and proteinase (Shurkhno *et al.*, 2005). These enzymes aid in improving digestibility of fermented fruits and vegetables. During LA fermentation, certain indigestible compounds such as sulphur compounds in garlic can break down, improving food digestibility (De Castro *et al.*, 1998; Heller, 2001).

Anti- nutritional Aspects

Occurrence of BAs in Fermented Fruits and Vegetable Products

BAs are low- molecular mass amines having a recognized biological activity and which may occur naturally in foods and beverages (Bover-Cid and Holzapfel, 1999; Kirschbaum *et al.*, 2000). They are formed as a result of the microbial decarboxylation of amino acids. These amino compounds can confer an unpleasant flavour to the product and may cause psychoactive and vasoactive effects such as headache, fever and vomiting. Some of the BAs are histamine, tyrosine, putrescine, cadaverine, etc (Kalac *et al.*, 1999; Arena and Manca de Nadra., 2001). The manufacturing of *sauerkraut* takes place in three steps characterized by active microorganisms that produce BAs (Kolesarova, 1995):

- *Lc. mesenteroides*, producing putrescine in a concentration of about 250 mg/kg
- *Lactobacillus* sp., producing putrescine and tyramine
- *P. cerevisiae*, producing histamine in a concentration of about 200 mg/kg

The data in Table 5 show that BAs, especially putrescine, accumulated in *sauerkraut* brine (Halasz *et al.*, 1994). Kalac *et al.* (1999) examined the content of BAs in 121 *sauerkraut* samples. The mean concentrations were 174, 146 and 50 mg/kg for tyramine, putrescine and cadaverine, respectively. Histamine levels were below 2 mg/kg in 44% of samples and above 10 mg/kg in only 19% of samples. Kirschbaum *et al.* (1997, 1999) found extremely high amounts of putrescine in bottled and pasteurized *sauerkraut* juices from Germany (up to 229 or 694 mg/dm^3). Kalac *et al.* (2000a) determined 44.1–387 mg/kg of tyramine, 21.9 to 260 mg/kg of putrescine, 12.4 – 59.6 mg/kg of cadaverine and 2.4 – 22.2 mg/kg of spermidine in different laboratory prepared *sauerkrauts* after six months storage. In lacto- juice prepared from cabbage, Kirschbaum *et al.* (2000) found 38.3– 62.9 mg/dm^3 of histamine, 37.1– 73 mg/dm^3 of tyramine, 83.5 – 366 mg/kg^3 of putrescine and 18.9–59.4 mg/dm^3 of cadaverine. Cadaverine, histamine, putrescine, spermidine and tyramine were found in LA fermented vegetables (such as carrot and red beet) in concentrations ranging from 1 to 15 mg/kg (Hornero-Mendez and Garrido-Fernandez, 1997). Kalac *et al.* (2000b) reported that *sauerkrauts* inoculated by *Lb. plantarum* or a mixed culture of *Lb. plantarum*, *Lb. casei*,

Pediococcus spp. and *Enterococcus faecium* had significantly lower concentrations of tyrosine, putrescine and cadaverine. In conclusion, the amount of BAs increases during controlled or spontaneous microbial fermentation of food or in course of food spoilage.

Table 5. Content of biogenic amines (BAs) in *sauerkraut* and *sauerkraut* brine

BAs	Sauerkraut *(mg/ kg)*	*Brine (mg/ dm^3)*
Cadaverine	53-71	178- 237
Histamine	42- 52	143- 174
Putrescine	100- 200	442- 678
Spermidine	3- 8	10- 76
Spermine	1- 2	3- 5
Tyramine	174	371- 73

Source: Krischbaum *et al.* (2000); Karovicova and Kohajdova (2003).

Synthesis of D (-) LA in Fermented Products

LA normally exists in two isomeric forms i.e. L (+) or D (+) depending on rotation of polarized light to right or left respectively. Both LA isomers are usually present in homemade or small-scale fermented vegetable fermentations. LAs are classified as generally recognized as safe (GRAS) for use in food additives by the Food and Drug Administration, USA (FDA), but D (-) LAs are harmful to human metabolism and can result in acidosis and decalcification (Datta *et al.*, 1995; Wee *et al.*, 2006). L (+) LA isomers are absorbed by the intestinal mucus and are used as energy substrate during the metabolic activity. D (-) forms are not assimilated and are eliminated by the kidneys in form of salt leading to a loss of calcium and magnesium cations (Padmaja, 1995).

Recipes of Some Fermented Fruits and Vegetables

Sauerkraut

This is an example of dry salted fermented vegetables. It results from the natural LA fermentation of salted and shredded cabbage. It is a common way of preserving fresh vegetable in the western world, in China and in Korea (Sim and Hang, 1996). The high nutritive value of *sauerkraut* is mainly due to the increased digestibility in comparison to raw cabbage and relatively low vitamin C losses. For at least 150 years, *sauerkraut* has been made in the home for saving fresh cabbage before spoilage (Steinkraus, 1997; Viander *et al.*, 2003). Other vegetables which are fermented by *sauerkraut* process are carrots, onions and beets (Mrocek-Delclos, 1991; Gardner *et al.*, 2001). Presently, the production of *sauerkraut* has become an important food industry (Steinkraus, 1989, 1996).

Sauerkraut processing at industrial scale consists of trimming the mature and sound heads of cabbage to remove the outer green, broken or dirty leaves (Karovicova and Kohajdova, 2002). The head of cabbage is then sliced by rotary knives into long and fine shreads. Salt is sprinkled on the shreds which are conveyed to the fermentation tanks. Spices could also be added. A salt concentration in the range of 2.25- 2.5% is used for facilitating LA fermentation. Forks are used to uniformly distribute the shreds which are dumped and squeezed into the vat. Once the tank has been filled to the proper level, it is closed. The shredded cabbage should be completely immersed to favour anaerobic conditions and prevent undesirable darkening and flavour changes.

Within a few hours the brine is formed and the fermentation is started, the latter is initiated by *Lc. mesenteroides*, which produces lactic and acetic acids and CO_2. The pH is quickly lowered, thus limiting the activity of undesirable microorganisms and enzymes that might soften the cabbage shreds. In the next phase, homofermentative bacteria such as *Lactobacillus* and *Pediococcus* continue the fermentation to a final pH of 3.5 to 3.8 (Yoon *et al.*, 2006). The CO_2 replaces air and creates anaerobic atmosphere which prevents the oxidation of ascorbic acid and the darkening of the natural cabbage colour. The optimal temperature for *sauerkraut* fermentation is 18 ^{0}C (Viander *et al.*, 2003). At this temperature fermentation is completely performed in three weeks (Hubert and Dupuy, 1994).

Korean *Kimchi*

Korean *kimchi* is made by LA fermentation of Chinese cabbage, radishes and other vegetables and is closely related to *sauerkraut* and pickle fermentation (Cheigh and Park, 1994). It is a main staple in the diet of Korean people. Salting of cabbage can be done as 5 - 7% salt concentration for 12h followed by rinsing and draining. Optimal acidity (as LA) of *kimchi* is 0.4 - 0.8%, with a pH between 4.2 - 4.5 (Lee, 1997). The organism isolated from *kimchi* includes *Lc. mesenteroides, Streptococcus faecalis, Lb. brevis, Lb. plantarum, P. cerevisiae, Flavobacterium* and *Pseudomonas* spp. Korean cabbage and radish are the major substrates; garlic, green onion, ginger, leaf mustard, pepper, parsley and carrot are minor ingredients (Choi *et al.*, 1994).

For the processing of *kimchi* proper cultivars of Chinese cabbage, with light green coloured soft leaves and compact structures are required. The outer leaves are removed and the 'head' is cut into small pieces. The prepared cabbage shreds are is placed in a salt concentration of 5 - 7 % for 12h. It is then rinsed several times with fresh water and drained to remove extra water. After fermentation the product can be left to mature for several weeks in refrigeration condition (Cheigh and Park, 1994).

Kocho

Kocho is a fermented product from false banana (*Ensete ventricosum*) pseudostem. It is produced in Ethiopia (Steinkraus, 1996). False bananas are peeled before placing in the pit, it is left to ferment for three to six weeks, after which it becomes soft, has a strong odour and a paste- like consistency. During fermentation CO_2 builds up in the pit creating an anaerobic atmosphere. As a result of bacterial activity, the temperature rises much higher than the ambient temperature. The pH of the fruit within the pit decreases from 6.7 to 3.7 within about four weeks. Inoculation of the fruit in the pit with LAB greatly speeds up the process. Pit therefore provides a good, reliable, cheap means of storage (Battcock and Azam- Ali, 2001).

Olives

Green olives are LA fermented following NaOH (lye) treatment to remove bitterness (Sanchez *et al.*, 2000). The bacteria involve are *Lb. plantarum, Lb. brevis, P. cerevisiae and Lc. mesenteroides* (Etchells *et al.*, 1966; Borcakli *et al.*, 1993; Duran- Quintana *et al.*, 1999). The pH varied between 4.0- 4.5. The optimum fermentation temperature is 24 ^{0}C. The fermentation period usually takes between two to three months. Once the fermentation is complete, the olives are packed in air-tight jars and sterilized which produces a good quality product with a long storage life. Salting is normally done at 1- 10 % brine solution (Tsapatsaris and Kotzekidou, 2004). With an aim of formulating a probiotic food, the green olive, was used as a vehicle for incorporating probiotic bacterial species such as *Lb. rhamnosus, Lb. paracasei, Bifidobacterium bifidum* and *B. longum*. All these strains showed a good survival rate with a recovery of about 10^6 CFU/g olives after 30 days (Lavermicocca *et al.*, 2005).

Naturally black fermented olives are of equal importance in the international market as green olives (Spyropoulou *et al.*, 2001; Nychas *et al.*, 2002). After harvesting, black olives are transported to the factory, sorted to separate damaged fruits, washed to remove superficial dirt and finally brined in a 6 - 10% salt solution (Balatsouras, 1990; Ozay and Borcakli, 1996). Unlike spanish- style green olives, they do not have a preliminary treatment in NaOH (lye) before fermentation. Nychas *et al.* (2002) reported that yeasts and LAB were the dominant organisms in the brine at the end of the black olive fermentation. Currently both green and black olives are fermented 'naturally' on an industrial scale, without addition of LAB starter cultures.

Cucumbers

Pickled cucumbers are made in Africa, Asia and Latin America. Cucumbers undergo LA fermentation and change from a pale product to a darker green and more transparent product. *Khalpi* is a popular cucumber pickle in Nepal during summer months (McGee, 2004; Dahal *et al.*, 2005). In Nepal, oil is added in preparation of *khalpi*. The *gherkin*, popularly known as pickling 'small cucumber', has emerged as a potential export- oriented delicacy, fetching foreign exchange of US $ 33.3 million from 50,000 tonnes every year from India (Kapur and Singh, 2003).

Fully ripe cucumbers without damage are washed in potable cold water and drained. Then it is sliced into 5-8cm pieces. One kg of salt is added to the cucumbers. As soon as the brine is formed, fermentation starts and bubbles of CO_2 appear. Fermentation takes in between one to four weeks depending on the ambient temperature. Proper salt concentration exerts a selective effect on natural flora, resulting in the growth of LAB. When the pH is about 4.7, the brine is inoculated with either *Lb. plantarum* or *P. pentosaceus* or a combination of these organisms (Steinkraus, 2002). At the end of fermentation, salt (16 % w/w of brine) is traditionally added to the LA fermented cucumbers in order to stop any undesirable bacterial growth during storage (Kapur and Singh, 2003). Cucumber pickle is usually stored in clean capped jars. They keep well if stored in a cool place. Due to high acid level of the final product, the risk of food poisoning is low.

Gundruk

Gundruk is popular pickle consumed in Nepal (Tamang *et al.*, 2005). It is obtained by LA fermentation of leafy vegetables such as cabbage and radish. It is an important source of minerals, particularly when the diet consists of mostly starchy roots, tubers and maize which are low in minerals. This is mainly served as side dish with the main meal and also used as an appetizer.

Shredded leaves are tightly packed in an earthen pot and warm water is added to cover all the leaves. After five to seven days, a mild acidic taste indicates the end of fermentation. The ambient temperature at the time of fermentation is 18 ^{0}C (Steinkraus, 2002). *Pediococcus* and *Lactobacillus* spp. are the predominant microorganisms during fermentation. During fermentation the pH drops slowly to a final value of 4.0 and the amount of acid, i.e. LA increases to about 1% on the sixth day.

Sunki

Sunki is a non- salted fermented vegetable prepared from the leaves of Otaki- turnip in Kiso District of Japan. *Sunki* is eaten with rice or in miso soup. The Otaki- turnip is boiled, mixed with Zumi (a wild small apple) and dried *sunki* from the previous year and allowed to ferment for one to two months. *Sunki* is produced under low temperature (in winter season). Microorganisms involved include *Lb. plantarum*, *Lb. brevis*, *Bacillus coagulans* and *P. pentosaceus* (Karki, 1986; Battcock and Azam- Ali, 2001).

Sinki

Sinki is a sour pickle prepared from radish tap roots. It is consumed traditionally in India, Nepal and parts of Bhutan (Dahal *et al.*, 2005). Fresh radish roots are harvested, washed and wilted by sun–drying for one to two days. They are then shredded, re-washed and packed tightly in glass jars, which are sealed and left to ferment. The optimum fermentation time is 12 days at 3^{0}C. *Sinki* fermentation is initiated by *Lb. fermentum* and *Lb. brevis*, followed by *Lb. plantarum*. The pH drops from 6.7 to 3.3. After fermentation, the radish substrate is sun-dried to a moisture level of about 21%.

Kanji

In Northern India and Pakistan, carrots, especially a variety that is deep purple in colour, are fermented to make a traditional ready to serve drink known as *kanji*. *Kanji* is very popular and considered to have cooling and soothing properties and to be of high nutritional value. After thorough washing the carrots are finely grated. Each kg of grated carrot is mixed with seven litres of water, 200g of salt, 40g of crushed mustard seed and 8g of hot chilli powder. The mixture is then placed in a glazed earthenware vessel, which is almost entirely sealed, leaving only a tiny hole for gases released during fermentation to escape. The mixture is then allowed to ferment naturally for seven to ten days. The final product is slightly acidic in taste and has an attractive purple- red colour. After fermentation the drink is strained through fine muslin and has to be consumed within three to four days after which it goes bad. Each kg of grated carrot yields just over seven litres of *kanji* (Shah, 1986; Berry, 1998).

Lacto- Juice and Pickling of Roots Crops

Sweet potato roots (non-boiled/ fully-boiled) rich in anthocyanin and β-carotene pigments were fermented with *Lb. plantarum* MTCC 1407 at 28± 2^0C for 48h to make lacto- juice. There were no significant variations in biochemical constituents (pH, 2.2-3.3; LA, 1.19-1.27 g/kg root and titratable acidity, 1.23 - 1.46 g/kg root) of lacto- juices prepared from non-boiled and fully-boiled sweet potato roots except β-carotene concentration [130 ± 7.5 mg/kg (fully- boiled roots) and 165± 8.1 mg/kg (non-boiled roots)] (Panda and Ray, 2007).

Sweet potato roots rich in anthocyanin or β-carotene pigments were pickled by LA fermentation by brining the cut and blanched roots in 2-10% NaCl solution and subsequently inoculated with a strain of *Lb. plantarum* MTCC 1407 for 28 days. Treatment with 8-10 % brine solution was found to be organoleptically most acceptable (based on texture, taste, aroma, flavour and after taste) (Panda, 2006; Panda *et al.*, 2007). The pickle prepared from anthocyanin rich sweet potato had a pH (2.5-2.8), titratable acidity (1.5 – 1.7 g/ kg), LA (1.0-1.3 g/ kg), starch (56 - 58 g/ kg) and anthocyanin content (390 mg/ kg) on fresh weight basis. Similarly, β- carotene rich pickle had a pH (2.9-3.0), titratable acidity (2.9 – 3.7 g/ kg), LA (2.6 - 3.2 g/ kg), starch (58 - 68 g/ kg) and β- carotene (163 -165 mg/ kg) (Panda et al., 2007). The pickle prepared from taro roots with 8 - 10% brine solutions had a pH of 5.4- 5.6, ascorbic acid of 3.1- 3.2 g/ kg, LA of 5.2 - 8.3 g/ kg and starch of 90 -106 g/ kg on fresh weight basis (Singh, 2005).

Cassava is another tropical root in which roots are consumed as food in Africa and Latin America (Ray and Ward, 2006). Several fermented foods (*gari, fufu, lafun, dawa dawa, chickwanghe, agbelima, attieke, kivunde* and *peujeum* in Africa, *gaplek* and *putto* in Indonesia and *"cheese" bread* and *"sour" starch* in Latin America) are prepared from cassava roots based on LAB and yeast fermentations (Cereda and Vilpoux, 2006; Ray and Ward, 2006). Fermentation reduces the cyanogenic toxicity and enhances flavour, taste and aroma of the fermented products (Onabolu *et al.*, 2002a, b; Klewicka *et al.*, 2001).

Other Pickled Vegetables and Fruits

- *Naw-mai-dong*: pickled bamboo shoots (*Bambusa glaucescens*) from Thailand (Tanasupawat and Komagata, 1994)
- *Hom- dong*: pickled red onions from Thailand (Tanasupawat and Komagata, 1994)
- *Jeruk*: pickled vegetables including ginger and papaya from Malaysia
- Pickled carrots and turnips are produced in Asia and Africa. They are known as *hua-chai po* in Thailand and *tai tan tsoi* in China
- *Nukamiso- zuke*: vegetables fermented in rice bran, salt and water in Japan
- *Dak-* dua- *dong*: fermented mustard leaves from Thailand (Tanasupawat and Komagata, 1994)
- *Dhamuoi,* fermented cabbage and other vegetables in Vietnam
- Bananas are pickled in the West Indies
- Fermented sweet peppers (*torshi felfel*) are produced in West Asia and Africa
- Cauliflower stalks are fermented to produce *achar tandal* in India
- Aubergines (*torshi bentigen*) are pickled in West Asia

- *Pak-sian-dong*: vegetables Pak Sian (*Gynadropsis pentaphylla*) fermented in Thailand
- *Sayur asin*: fermented wilted mustard cabbage (*Brassica juncea*) from Indonesia
- *Kaktugi*: fermented radish produced in Korea

Current Research in LA Fermented Vegetables or Vegetable Juices

In recent years, several researchers dealt with the LA fermentation of vegetables and vegetable juices De Castro *et al.* (1998) performed the controlled fermentation of peeled, blanched garlic using a starter culture of *Lb. plantarum* and compared it with that of unblanched garlic. Blanching was carried out in hot water (90°C) for 15 min. The starter grew abundantly in the case of blanched garlic, producing mainly LA and reaching pH of 3.8 after seven days. The fermented blanched garlic was microbiologically stable during storage at 30°C in acidified brine. Aukrust *et al.* (1994) used *Lb. sakei* for fermentation of carrot slices that produces the bacteriocins 'sakacin A'. Carrot contained proteinase which was eluted into the brine during fermentation. The results indicated that 'sakacin A' was inactivated by the carrot proteinase. Montano *et al.* (1997) used a mixed culture of *Lb. plantarum* and *S. cerevisiae* and a single culture of *Lb. plantarum* for fermentation of lye-treated carrots. After fermentation, the samples were preserved by addition of sorbic acid, benzoic acid or pasteurized at 80°C for 10 min. The controlled fermentation with the mixed culture resulted in higher utilization of fermentable material than with the single culture. The flavour of carrots fermented by the mixed culture was significantly preferred to that of carrots fermented with the single culture (probably due to the formation of more flavour compounds). The pasteurized samples were microbiologically stable during the storage period (nine months and storage at 30°C) while LAB grew in the samples with preservatives. Kuchta *et al.* (1994) fermented vegetables (gourd, cabbage and celery) by *Lb. plantarum, Lb. brevis* and *Lb. pentosus*. The gourds were fermented for four days, cabbage for seven days and celery for nine days. Sanchez *et al.* (2000) performed fermentation of green olives. It was found that in olives fermented *Lactobacillus pentosus* the highest amount of LA was produced in comparison with spontaneously fermented olives. Karovicova *et al.* (1994) fermented carrot and capsicum by different LAB. It was shown that the best producers of LA were *Lactobacillus* spp. and *Lb. fermentum*.

Lu *et al.* (2001) carried out fermentation of cucumber juices inoculated by *Lb. plantarum*. The juices were fortified with glucose, fructose or a mixture of glucose and fructose. When cucumber juice was supplemented with fructose and/or glucose, the starter culture continued to ferment fructose, but not glucose, resulting in an increase in LA production and decrease in terminal pH. Gardner *et al.* (2001) and Klewicka *et al.* (2004) studied various LAB for the fermentation of cabbage, carrot and beet-based vegetable products. It was found that a starter culture consisting of *Lb. plantarum* NK312, *P. acidilactici* AFERM772 and *Lc. mesenteroides* BLAC accelerated the fermentation process and prevented deterioration of fermented products for up to 90 days. Karovicova *et al.* (1993) used green pea, carrot, celery and onion for the preparation of LA fermented vegetable salads. The salads were inoculated by various *Lactobacillus* strains. The highest production of LA was observed in the salad inoculated by *Lactobacillus* sp. S. Viander *et al.* (2003) examined the impact of low NaCl and mineral salt concentrations on the spontaneous fermentation process of white cabbage into *sauerkraut* and *sauerkraut* juice. It was shown that *sauerkraut* and *sauerkraut* juice could be produced with a very low NaCl concentration as well as with a low mineral salt percentage. The *sauerkraut* juice fermented with 0.5% mineral salt was considered to have the best taste. Karovicova *et al.* (2002a) performed spontaneous fermentation of cabbage juices, fermentation by *Lb. plantarum* 92H and fermentation by a mixed culture of *Lb. plantarum* and *S. cerevisiae* C11-3. It was found that the highest amount of LA was produced in juice inoculated by *Lb. plantarum* 92H and the highest decrease in pH was observed in juice inoculated by the mixed culture. The spontaneously fermented juice had the highest intensity of harmonic taste, acceptance of taste, acceptance of odour and flavour. Drdak *et al.* (1994) tested 16 strains of the genus *Lactobacillus* on samples of white fresh cabbage and of a mixture of sterilized cabbage and carrot juice. Fermentation lasted for 7 days at 27 - 30°C. Karovicova *et al.* (2001, 2002b) prepared cabbage-carrot juices fermented by *Lb. plantarum*92H during 150 h at 24°C and cabbage juices with various concentrations of *Lb. plantarum* 92H during 120 h at 28°C. The juices had the most harmonic taste and flavour in 72–96 h of fermentation. Yoon *et al.* (2004) studied the probiotication of tomato juice by LAB (*Lb. acidophilus* LA39, *Lb. plantarum* C3, *Lb. casei* A4 and *Lb. delbrueckii* D7). Tomato juice was inoculated with a 24-h-old culture and incubated at 30^0 C. The LA fermentation reduced the pH to 4.1 and increased the acidity to 0.65% and the viable cell count (CFU) reached nearly 1.0- 9.0 x 10^9 CFU/ml after 72h fermentation. Similar results were obtained for probiotic- enriched cabbage juice and olive juice fermented by LAB (Tsapatsaris and Kotzekidou, 2004; Yoon *et al.*, 2006).

Modern Techniques Used for Analyzing Microflora of Fermented Fruits / Vegetables

In addition to traditional methods (microscopy, plate count,etc), several modern techniques like RAPD (Random Amplified Polymorphic DNA)-PCR (Polymerase Chain Reaction), species- specific PCR, multiplex PCR, 16s rDNA sequencing, gradient gel electrophoresis, RFLPs(Restriction Fragment Length Polymorphism), cluster analysis of TTGE (Temporal Temperature Gradient Electrophoresis) are employed to isolate and characterize different type LAB strains of fermented fruits and vegetables (Elegado *et al.*, 2004). RFLPs and 16s rDNA were employed to isolate and characterize lactic acid bacteria from *dochi* (fermented black beans), *suan-tsai* (fermented mustard), a traditional fermented food in Taiwan (Chen *et al.*, 2006a, 2006b). The isolated strains are *Lb. plantarum*, *Salmonella enterica*, *E. coli*, *P. pentosaceus*, *Tetragenococcus halophilus*, *Bacillus lichenoformis*, etc. Tamang *et al.* (2005) isolated 269 strains of LAB from *gundruk*, *sinki*, *inziangsang* (a fermented leafy vegetable) and *khalpi* samples and studied the phenotypic characteristics of these strains by genotyping using RAPD- PCR, repetitive element PCR and species- specific PCR techniques. The major representatives of LAB involved in these fermentations were *Lb. plantarum*, *Lb. brevis*, *P. acidilactici* and *Lc. fallax*. RAPD- PCR and gradient gel electrophoresis were used to isolate *Lb. plantarum* strains from *ben saalga*, a traditional fermented gruel from *Burkina Faso* (Omar *et al.*, 2006).

Conclusion

LAB perform an essential role in the preservation and production of wholesome foods ranging from fermented fresh vegetables such as cabbage, cucumber to fermented cereals, yoghurt, sour dough bread, fermented milk, fish and meat products (Bernardeau *et al.*, 2006). These bacteria grow rapidly on most food substrates and lower the pH to a point where other competing organisms are no longer able to grow. Due to LA fermentation, foods become resistant to microbial spoilage and to development of toxins. Lacto- fermented fruits and vegetables would be a good prospect for commercialization in homestead and small- scale industries. Regular consumption of LA fermented food products enriched with coloured pigments such as anthocyanin and β- carotene would be helpful in combating night blindness, liver injury, aging and related ailments because of antioxidant principles.

References

Abee, T., Krockel, L., and Hill, C. (1995). Bacteriocins: modes of action and potentials in food preservation and control of food poisioning. *Int. J. Food Microbiol.* 28(2): 169- 185.

Adams, M.R., and Nicolaides, L. (1997) Review of the sensitivity of different food borne pathogens to fermentation. *Food Control* 8: 227 - 239.

Ana, A.S., Azeredo, D.P., Costa, M. Da., and Macedo, V. (2002). Analysis of risks of minimal processing of vegetables. *Higiene Alimentar.* 16: 80 - 84.

Angelova, M. B. (2007). Microbial pectinases: application in horticultural industries. In: R.C. Ray and O.P. Ward (Eds.), *Microbial Biotechnology in Horticulture, Volume 2*, Science Publishers Inc., Enfield, NH, USA, In press.

Antai, S.P., and Ibrahim, M.H. (1986). Microorganisms associated with African locust bean (*Parkia filicoidea*) fermentation for "dawadawa" production. *J. Appl. Bacteriol.* 61: 145-148.

Arena, M.E., and Manca de Nadra, N.C. (2001). Biogenic amine production by *Lactobacillus*. *J. Appl. Microbiol.* 90: 158- 162.

Aukrust, T.W., Blom, A., Sandtorv, B. F., and Slinde, E. (1994). Interactions between starter cultures and raw material in lactic acid fermentation of sliced carrot. *Lebensm Wiss. Technol.* 27: 337- 341.

Ausco, M., Leal, V., Baras, M., Ruiz-Barba, J.L., Floriano, B., and Jimenez, R. (1998). Bacteriocin production and competitiveness of *Lactobacillus plantarum* LPC010 in olive juice broth, a culture medium obtained from olives. *Int. J. Food Microbiol.* 43: 129- 134.

Axelsson, L. (1998). Lactic acid bacteria: classification and physiology. In: S. Salminen and A.von Wright (Eds.), *Lactic acid bacteria, Microbiology and Functional Aspects*, Marcel Decker Inc., New York, pp.1-72.

Balatsouras, G.D. (1990). Edible olive cultivars, chemical composition of fruit, harvesting, transportation, processing, sorting and packaging, styles of black olives, deterioration, quality standards, chemical analysis, nutritional and biological value of the end product. In: Olio d'Oliva e Olive da Tavola:Ttecnologia e Qualita. Istituto Spermentale per la Elaiotecnica, Pescara, Italy, 25- 28 April, 1990, pp. 291- 330.

Battcock, M., and Azam-Ali, S. (2001). *Fermented Fruits and Vegetables: A Global Perspective*. FAO Agricultural Services Bull., No. 134, Rome, 96 pp.

Beltran-Edeza, L. M., and Hernandez- Sanchez, H. (1989). Preservation of ripe tomatoes by lactic acid fermentation. *Lebensm. Wiss. Technol.* 22: 65-67.

Bernardeau, M., Guguen, M., and Vernoux, J.P. (2006). Beneficial lactobacilli in food and feed: long – term use, biodiversity and proposals for specific and realistic safety assessments. *FEMS Microbiol. Rev.* 30: 487- 513.

Berry, S.N. (1998). *Kanji*, Food Chain 23, Intermediate Technology, London, UK [cited in *Fermented Fruits and Vegetables: A Global Perspective (*M. Battcock and S. Azam- Ali), FAO Agric. Services Bull., No.134, Rome, 2001, 96 pp.]

Borcakli, M., Ozay, G., Alperden, I., Ozsan, E., and Erdek, Y. (1993). Changes in chemical and microbiological composition of olive during fermentation. *Grasa Y Aceites* 44: 253-258.

Bover-Cid, S., and Holzapfel, W.H. (1999). Improved screening procedure for biogenic amine production by lactic acid bacteria. *Int. J. Food Microbiol.* 53(1): 33- 41.

Buckenhuskes, H., Omran, H., and Gierschner, K. (1990). Lactic acid fermentation of whole white cabbage. *Acta Aliment.* 19 (2): 157-163.

Buckenhuskes, H.J. (1993). Selection criteria for lactic acid bacteria to be used as starter cultures for various food commodities. *FEMS Microbiol. Rev.* 12: 253- 272.

Buckenhuskes, H.J. (1997). Fermented vegetables. In: M.P. Doyle, L.R. Beuchat and T.J. Montville (Eds.), *Food Microbiology: Fundamentals and Frontiers*, ASM Press, Washington DC, pp. 665- 680.

Campbell-Platt, G. (1987). *Fermented Foods of the World- A Dictionary and Guide.* Butterworth Publishing Co., London.

Caplice, E., and Fitzerald, G.F. (1999). Food fermentations: role of microorganisms in food production and preservation. *Int. J. Food Microbiol.* 50: 131- 149.

Cereda, M.P., and Vilpoux, O.E. (2006). Cassava fermentation in Latin America: fermented starch – and overview. In: R.C. Ray and O.P. Ward (Eds.), *Microbial Biotechnology in Horticulture, Volume 1*, Science Publishers Inc., Enfield, New Hampshire, pp. 397- 420.

Cheigh, H.S., and Park, K.Y. (1994). Biochemical, microbiological and nutritional aspects of *kimchi* (Korean fermented vegetable products). *Crit. Rev. Food Sci. Nutr.* 34(2): 175-203.

Chen, Y.S., Yanagida, F., and Hsu, J.S. (2006 a). Isolation and characterization of lactic acid bacteria from *suan-tsai* (fermented mustard), a traditional fermented food in Taiwan. *J. Appl. Microbiol.* 101(1): 125- 130.

Chen, Y.S., Yanagida, F., and Hsu, J.S. (2006 b). Isolation and characterization of lactic acid bacteria from *dochi* (fermented black beans), a traditional fermented food in Taiwan. *Lett. Appl. Microbiol.* 43(2): 229- 235.

Choi, S.Y., Beuchat, L.R., Perkins, L.M., and Nakayama, T. (1994). Fermentation and sensory characteristics of *kimchi* containing potassium chloride as a partial replacement for sodium chloride. *Int. J. Food Microbiol.* 21(4): 335- 340.

Cross, M.L., Stevenson, L.M., and Gill, H.S. (2001). Anti-allergy properties of fermented foods: an important immunoregulatory mechanism of lactic acid bacteria. *Int. Pharmacol.* 1: 891- 901.

Daeschel, M.A. (1987). Microbial ecology of fermenting plant materials. *FEMS Microbiol. Rev.* 46: 357- 367.

Daeschel, M.A., and Fleming, H.P. (1984). Selection of lactic acid bacteria for use in vegetable fermentations. *Food Microbiol.* 1: 303- 313.

Dahal, N.R., Karki, T.B., Swamylingppa, B., Li, Q., and Gu, G. (2005). Traditional Foods and beverages of Nepal- a review. *Food Rev. Int.* 21 (1): 1- 25.

Daly, C., and Davis, R. (1998). The biotechnology of lactic acid bacteria with emphasis on applications in food safety and human health. *Agric. Food Sci.* 7(2): 251- 264.

Datta, R., Tsai, S.P., Bonsignore, P., Moon, S.H., and Frank, J.R. (1995). Technological and economic potential of poly (lactic acid) and lactic acid derivatives. *FEMS Microbiol. Rev.* 16 (2- 3): 221-231.

De Castro, A., Montano, A., Sanchez, A.H., and Rejano, L. (1998). Lactic acid fermentation and storage of blanched garlic. *Int. J. Food Microbiol.* 39: 205- 211.

Demir, N., Acar, J., Sar, K., and Mutlu, M. (2001) The use of commercial pectinase in fruit industry. *J. Food Eng.* 47: 275-280.

Demir, N., Bahceci, K. S., and Acar, J. (2006). The effects of differential initial *Lactobacillus plantarum* concentrations on some properties of fermented carrot juice. *J. Food Process. Preserv.* 30 (3): 352- 363.

Dirar, M. (1993). *The Indigenous Fermented Foods of the Sudan.* CAB International. Wallingford, UK.

Drdak, M., Karovicova, J., Grief, G., and Rajniakova, A. (1994). Vyber kmenov *Lactobacillus* species na mliecne fermentaciu zeleniny. *Bull. PV.* 13: 195- 203.

Drewnowski, A., and Gomez-Carneros, C. (2000). Bitter taste, phytonutrients and the consumers: a review. *Am. J. Clin. Nutr.* 72 (6):1424- 1435.

Duran- Quintana, M.C., Garcia Garcia, P., and Garrido Fernandez, A. (1999). Establishment of conditions for green table olive fermentation at low temperature. *Int. J. Food Microbiol.* 51: 133-143.

Eijsink, V.G., Axelsson, L., Havarstein, L.S., Holo, H., and Nes, J.F., and Diep, D.B. (2002). Production of class II bacteriocins by lactic acid bacteria: an example of biological warfare and communication. *Antonie Van Leeuwenhoek* 81(1- 4): 639- 654.

Elegado, F.B., Guerra, M.A., Macavan, R.A., Mendoza, H.A., and Lirazan, M.B. (2004). Spectrum of bacteriocin activity of *Lactobacillus plantarum* BS and fingerprinting by RAPD-PCR. *Int. J. Food Microbiol.* 95(1): 11- 18.

Eriksson, C. (1991). Lactic acid bacteria and vegetables, cereal, meat and fish fermentations. *J. Chem. Technol. Biotechnol.* 51: 4 - 25.

Etchells, J.L., Borg, A.F., Kittel, I.D., Bell, T.A., and Fleming, H.P. (1966). Pure culture fermentation of green olives. *Appl. Environ. Microbiol.* 14(6): 1027- 1041.

Fleming, H.P., McFeeters, R.F., Daeschel, M.A. (1985). The lactobacilli, *Pediococcus* and *Leuconostoc*: vegetable products. In: S.E. Gilliland (Ed.), *Bacterial Starters Cultures for Foods*, CRC Press Inc., Boca Raton, Florida, pp. 97- 118.

Gardner, N. J., Savard, T., Obermeier, P., Caldwell, G., and Champagne, C.P. (2001). Selection and characterization of mixed starter cultures for lactic acid fermentation of carrot, cabbage, beet and onion vegetable mixtures. *Int. J. Food Microbiol.* 64: 261-275.

Guyot, J.P., Calderon, M., and Morlon- Guyot, J. (2000). Effect of pH control on lactic acid fermentation of starch by *Lactobacillus manihotivorans* LMG 18010T. *J. Appl. Microbiol.* 88(1): 176- 182.

Halasz, A., Barath, A., and Holzapfel, H. (1994). Biogenic amines and their production by microorganisms in food. *Trends Food Sci. Technol.* 51: 42- 49.

Hammes, W.P. (1990). Bacterial starter cultures in food production. *Food Biotechnol.* 4: 383-397.

Hansen, E.B. (2002). Commercial bacterial starter cultures for fermented foods of the future. *Int. J. Food Microbiol.* 78(1-2): 119- 131.

Havenaar, R., and Veld, J.H. (1992). Probiotics: a general view. In: B.J.B. Wood (Ed.), *The Lactic Acid Bacteria in Health and Disease*, Elsevier Applied Science Publisher, London, pp. 209 - 224.

Heller, K.J. (2001). Probiotic bacteria in fermented foods: product characteristics and starter organisms. *Am. J. Clin. Nutr.* 73: 373S - 379S.

Hernandez, D., Cardell, E., and Zarate, V. (2005). Antimicrobial activity of lactic acid bacteria isolated from Tenerife cheese: initial characterization of plantaricin TF711, a bacteriocin like substance produced by *Lactobacillus plantarum* TF711. *J. Appl. Microbiol.* 99(1): 77- 84.

Holzapfel, W.H. (2002). Appropriate starter culture technologies for small- scale fermentation in developing countries. *Int. J. Food Microbiol.* 75: 197- 212.

Hong, S., Kim, Y.I., and Pyun, Y.J. (1999). Acid tolerance of *Lactobacillus plantarum* from *kimchi*. *Lebensm. Wiss. Technol.* 32: 142- 148.

Hornero-Mendez, D., and Garrido-Fernanadez, A. (1997). Rapid HPLC analysis of biogenic amines in fermented vegetable brines. *J. Food Proteomics* 60: 414- 419.

Hubert, J.C., and Dupuy, P., (1994). Conservation des fruits et des legumes. In: H. Luquet, (Ed.), *Bacteries Lactiques*, De Roissart, Lorica, Uriage, pp. 233-243.

Kalac, P., Spicka, J., Krizek, M., Steidlova, S., and Pelikanova, T. (1999). Concentrations of seven biogenic amine in *sauerkraut*. *Food Chem.* 67: 275- 280.

Kalac, P., Spicka, J., Krizek, M., and Pelikanova, T. (2000a). Changes in biogenic amines concentration during *sauerkraut* storage. *Food Chem.* 69: 309- 314.

Kalac, P., Spicka, J., Krizek, M., and Pelikanova, T. (2000b). The effects of lactic acid bacteria inoculants on biogenic amines formation in *sauerkraut*. *Food Chem.* 70: 355-359.

Kalantzopoulos, G. (1997). Fermented products with probiotic qualities. *Anaerobe* 3: 185 - 190.

Kapur, A., and Singh, H. (2003). Popularizing gherkin: a cent percent export- oriented delicacy. *Ind. Hort.* April: 42- 44.

Karki, T. (1986) Some Nepalese fermented foods and beverages. In: *Traditional Foods: Some Products and Technologies*, Central Food Technological Research Institute, Mysore, India, [cited in *Fermented Fruits and Vegetables: A Global Perspective* (M. Battcock and S. Azam- Ali), FAO Agric. Services Bull., No. 134, Rome, 2001, 96 pp].

Karovicova, J., Drdak, M., Grief, G., and Hybenova, E. (1993). Capillary isotachophoresis of organic acids produces by selected microorganisms during lactic acid fermentation. *J. Chromatogr.* 638: 241- 246.

Karovicova, J., Drdak, M., Grief, G., and Hybenova, E. (1994). Vyber vhodnych kmenov mikroorganizmov pre konzervaciu karotky a papriky mliecnou fermentaciou. *Potrav. Vedy* 12: 105- 113.

Karovicova, J., Drdak, M., Greif, G., and Hybenova, E. (1999). The choice of strains of *Lactobacillus* species for the lactic acid fermentation of vegetable juices. *Eur. Food Res. Technol.* 210: 53-56.

Karovicova, J., Grief, G., Kohajdova, Z., and Hybenova, E. (2001). Vyuzitie multivariacnej analyzy pri hodoteni mliecne fermentovanych zeleninovych stiav. *Bull. PV.* 40: 119-131.

Karovicova, J., Kohajdova, Z., Greifova, M., Lukacova, D., and Grief, G. (2002a). Porovnanie fermentacii zeleninovych stiav. *Bull. PV.* 41: 197- 213.

Karovicova, J., Kohajdova, Z., Greifova, M., Lukacova, D., and Grief, G. (2002b). Using of multivariate analysis for evaluation of lactic acid fermented cabbage juices. *Chem. Pap.* 56: 267-274.

Karovicova, J., and Kohajdova, Z. (2002). The use of PCA, FA, CA for the evaluation of vegetable juices processed by lactic acid fermentation. *Czech. J. Food Sci.* 20: 135-143.

Karovicova, J., and Kohajdova, Z. (2003). Lactic acid fermented vegetable juices. *Hort. Sci.* 30(4): 152- 158.

Kaur, C., and Kapoor, H.C. (2001). Antioxidants in fruits and vegetables- the millennium's health. *Int. J. Food Sci. Technol.* 36 (7): 703- 725.

Kaur, I.P., Chopra, K., and Saini, A. (2002). Probiotics: potential pharmaceutical applications. *Eur. J. Pharma. Sci.* 15: 1-9.

Kim, H.Y., Min, J.H., Lee, J.H., and Ji, G.E. (2000). Growth of lactic acid bacteria and bifidobacteria in natural media using vegetables, seaweeds, grains and potatoes. *Food Sci. Biotechnol.* 9: 322-324.

Kirschbaum, J., Busch, I., and Bruckner, H. (1997). Determination of biogenic amines in food by automated pre- column derivatization with 2-napthyloxycarbonyl chloride (NOC-CI). *Chromatographia* 45: 263- 268.

Kirschbaum, J., Meier, A., and Bruckner, H. (1999). Determination of biogenic amines in fermented beverages and vinegars by pre-column derivatization with para-nitrobenzyloxycarbonyl chloride (PNZ-CI) and reversed-phase LC. *Chromatographia* 49: 117- 124.

Kirschbaum, J., Rebschen, K., and Brucker, H. (2000). Liquid chromatographic determination of biogenic amines in fermented foods after derivatization with 3, 5- dinitrobenzoyl chloride. *J. Chromatogr.* 881: 517- 530.

Klaenhammer, T.R. (1993). Genetics of bacteriocins produced by lactic acid bacteria. *FEMS Microbiol. Rev.* 12: 39- 86.

Klaenhammer, T.R. (1995). Genetics of intestinal lactobacilli. *Int. J. Diary Sci.* 5: 1019-1058.

Klewicka, E., Motyl, I., and Libudzisz, Z. (2004) Fermentation of beet juice by bacteria of genus *Lactobacillus* sp. *Eur. Food Technol.* 218: 178-183.

Kolesarova, E. (1995). Vyskyt a vznik biogennych aminov v potravinach. *Bull. PV.* 34: 109-122.

Kopec, K. (2000). Jakost mlecne kvasene zeleniny. *Vyziva a Potraviny.* 3: 109- 122.

Kostinek, M., Specht, I., Edward, V.A., Schillinger, U., Hertel, C., Holzapfel, W.H., and Franz, C. M. A. P. (2005). Diversity and technological properties of predominant lactic acid bacteria from fermented cassava used for the preparation of *gari*, a traditional African food. *Syst. Appl. Microbiol.* 28 (6): 527-540.

Kullisaar, T., Yilmer, M., Mikelsaar, M., Vihalemn, T., Annuk, H., Kairane, C., and Kilk, A. (2002). Two antioxidative lactobacilli strains as promising probiotics. *Int. Food Microbiol.* 72: 215- 224.

Kunji, E.R.S., Mierau, I., Hagting, A., Poolman, B., and Kinings, W.N. (1996). The proteolytic systems of lactic acid bacteria. *Antonie Van Leeuwenhoek* 70: 181- 221.

Kutchta, T., Radosovska, R., Gloncakova, B., Belicova, A., and Lopasovska, J. (1994). Priprava kvasenych zeleninovych salatov pomocou bakterii mliecneho kvasenia. *Bull. PV.* 33: 85- 90.

Lavermicocca, P., Valerio, F., Longogro, S.L., Angelis, M.D., Morelli, L., Callegari, M.L., Rizzello, C.G., and Visconti, A. (2005). Study of adhension and survival of lactobacilli and bifidobacteria on table olives with the aim of formulating a new probiotic food. *Appl. Environ. Microbiol.* 71(8): 4233- 4240.

Lee, Y.K., and Salminen, S. (1995). The coming age of probiotics. *Trends Foods Sci. Technol.* 6: 241- 245.

Lee, C. H. (1997). Lactic acid fermented foods and their benefits in Asia. *Food Control* 8: 259- 269.

Lee, C.H. (1999). Fermentation of rice using amylolytic *Bifidobacterium. Int. J. Food Microbiol.* 50: 155-161.

Lee, J.S., Heo, G.Y., Lee, J.W., Oh, Y.J., Park, J.A., Park, Y.h., Pyun, Y.R., and Ahn, J.S. (2005). Analysis of *kimchi* microflora gel electrophoresis. *Int. Food Microbiol.* 102(2): 143- 150.

Lei, V., Amoa-Awua, W. K. A., and Brimer, L. (1999). Degradation of cyanogenic glycosides by *Lactobacillus plantarum* strains from spontaneous cassava fermentation and other microorganisms. *Int. J. Food Microbiol.* 53: 169- 184.

Leroy, F., Degeest, B., and De Vuyst, L. (2002). A novel area of predicting modeling: describing the functioning of beneficial microorganisms in foods. *Int. J. Food Microbiol.* 73: 251- 259.

Liu, S. Q. (2003). Practical implications of lactose and pyruvate metabolism by lactic acid bacteria in food and beverage fermentations. *Int. J. Food Microbiol.* 83: 115-131.

Lopatnikova, J. (1992). Mliecne kvasenie ako perspektivny druh konzervovania. In: *Nove Postupy v Konzervovani Rastlinnej Produkcie.* Bratislava, Dom techniky ZSVTS Bratislava, pp. 68- 81.

Lu, Z., Fleming, H.P., and McFeeters, R.F., (2001). Differential glucose and fructose utilization during cucumber juice fermentation. *J. Food Sci.* 66(1): 162-166.

Mahmoud, S. A. Z., Taha, S.M., Ishak, Y.Z., Abdel-Halim, M., and EL-Nakhal, M.N.S. (1972). Cucumber pickling in Eygpt. II. Old brine method. *Egypt J. Microbiol.* 7: 65- 78.

McDonald, L. C., Fleming, H.P., and Hassan, H.M. (1990). Acid tolerance of *Leuconostoc mesenteroides* and *Lactobacillus plantarum. Appl. Environ. Microbiol.* 56: 2120 - 2124.

McDonald, L.C., Fleming, H.P., and Daeschel, M.A. (1991). Acidification effects on microbial populations during initiation of cucumber fermentation. *J. Food Sci.* 56: 1353-1356.

McDonald, L. C., Shieh, D. H., Fleming, H.P., McFeeters, R.F., and Thompson, R. L. (1993). Evaluation of malolactic-deficient strains of *Lactobacillus plantarum* for use in cucumber fermentations. *Food Microbiol.* 10(6): 489 - 499.

McFeeters, R.F. (2004). Fermentation microorganisms and flavor changes in fermented food. *J. Food Sci.* 69 (1): 35-37.

McGee, H. (2004). On Food and Cooking: The Science and Lore of the Kitchen. "http://en.wikipedia.org/wiki/Pickling". Assessed on 20 January, 2007.

Messens, W., and De Vuyst, L. (2002). Inhibitory substances produced by lactobacilli isolated from sourdoughs – a review. *Int. J. Food Microbiol.* 72: 31- 43.

Mohapatra, S., Panda, S. H., Sahoo, S. K., Sivakumar, P. S., and Ray, R. C. (2007). β-Carotene-rich sweet potato curd: production, nutritional and proximate composition. *Int. J. Food Sci. Technol.* 42: In press.

Molenaar, D., Bringel, F., Schuren, F.H., DeVos, W.M., Siezen, R.J., and Kleerebezem, M. (2005). Exploring *Lactobacillus plantarum* genome diversity by using microarrays. *J. Bacteriol.* 187 (17): 6119- 6127.

Molin, G. (2001). Probiotics in foods not containing milk or milk constituents, with special reference to *Lactobacillus plantarum* 299v. *Am. J Clin. Nutr.* 73(2 Suppl): 380S- 385S.

Montano, A., Sanchez, A.H., Rejano, L., and DeCastro, A. (1997). Processing and storage of lye- treated carrots fermented by a mixed starter culture. *Int. J. Food Microbiol.* 35: 83-90.

Montet, D., Loiseau, G., and Zakhia-Rozis, N. (2006). Microbial Technology of fermented vegetables. In: R.C.Ray and O.P.Ward (Eds.), *Microbial Biotechnology in Horticulture, Volume 1,* Science Publishers Inc., Enfield, New Hampshire, pp. 309-343.

Mrocek-Delclos, P.A. (1991). Vegetable preservation by a mixed organic acid fermentation. Ph.D. thesis, University of Surrey, UK.

Muyanja, C. M., Narvhus, J. A., Treimo, J., and Langsrud, T. (2003). Isolation, characterization and identification of lactic acid bacteria from *bushera*: an Ugandan traditionally fermented beverage. *Int. J. Food Microbiol.* 80 (3): 201- 10.

Nychas, G.J.E., Panagou, E.Z., Parker, M.L., Waldron, K.W., and Tassou, C.C. (2002) Microbial colonization of naturally black olives during fermentation and associated biochemical activities in the cover brine. *Lett. Appl. Microbiol.* 34: 173-177.

O'Sullivan, L., Ross, R.P., and Hill, S. (2002). Potential or bacteriocin- producing lactic acid bacteria for improvements in food safety and quality. *Biochemie* 84 (5- 6): 593- 604.

Ogunjobi, A. A., Adebayo-Tayo, B.C., and Ogunshe, A.A. (2005) Microbiological, proximate analysis and sensory evaluation of processed Irish potato fermented in brine solution. *Afr. J. Biotech.* 4(12): 1409-1412.

Omar, N.B., Abriouel, H., Martinez-Canamero, M., Guyot, J.P., and Galvez, A. (2006). Isolation of bacteriogenic *Lactobacillus plantarum* strains from *ben saalga*, a traditional fermented gruel from Burkina Faso. *Int. J. Food Microbiol.* July 14 (online).

Onabolu, A.O., Oluwole, O.S.A., and Bokanga, M. (2002a). Loss of residual cyanogens in a cassava food during short- term storage. *Int. J. Food Sci. Nutr.* 52: 343- 349.

Onabolu, A.O., Oluwole, O.S.A., Rosling, H., and Bokanga, M. (2002b). Processing factors affecting the level of residual cyanohydrins in *gari*. *J. Sci. Food. Agric.* 82 (9): 966- 969.

Ouvrya, A., Wachea, Y., Tourdot- Marechala, R., Diviesa, C., and Cachona, R., (2002). Effects of oxidoreduction potential combined with acetic acid, NaCl and temperature on the growth, acidification, and membrane properties of *Lactobacillus plantarum*. *FEMS Microbiol. Lett.* 214: 257- 261.

Ozay, G., and Borcakli, M. (1996). Effect of brine replacement and salt concentration on the fermentation of naturally black olives. *Food Res. Int.* 28: 533- 559.

Padmaja, G. (1995). Cyanide detoxification in cassava for food and feed uses. *CRC Crit. Rev. Food Sci. Nutr.* 35 (4): 299- 339.

Paludan-Muller C., Huss, H., and Gram, L. (1999). Characterization of lactic acid bacteria isolated from a Thai low-salt fermented fish and the role of garlic as a substrate of fermentation. *Int. J. Food Microbiol.* 46: 219- 229.

Panda, S. (2006). Lactic acid fermentation of sweet potato- brine mixture by *Lactobacillus plantarum*. M.Sc. thesis, P.G. Department of Biotechnology, Revenshaw Autonomous College, Bhubaneswar, Orissa, India. 75 pp.

Panda, S.H., Parmanik, M., Sharma, P., Panda, S., and Ray, R.C. (2005). Microorganisms in food biotechnology: present and future scenario. In: R.C. Mohanty. and P.K. Chand (Eds.), *Microbes in Our Lives*, Department of Botany and Microbiology, Utkal University, Bhubabneswar, Orissa, pp. 47- 54.

Panda, S. H., Naskar, S. K., and Ray, R. C., (2006). Production, proximate and nutritional evaluation of sweet potato curd. *J. Food Agric. Environ.* 4 (1): 124 -127.

Panda, S.H. and Ray, S.H. (2007). β- Carotene- rich sweet potato lacto- juice. *Plant Foods for Human Nutr.*62: In press.

Panda, S.H., Parmanick, M., and Ray, R .C. (2007.) Lactic acid fermentation of sweet potato (*Ipomoea batatas* L.) into pickles. *J. Food Process. Presv.* 31(1): 83- 101.

Rakin, M., Baras, J., Vukasinovic, M., and Milan, M. (2004). The examination of parameters for lactic acid fermentation and nutritive value of fermented juice of beet root, carrot and brewer's yeast autolysate. *J. Serb. Chem. Soc.* 69 (8-9): 625-634.

Rao, M. S., Pintado, J., Stevens, W.F., and Guyot, J. P. (2004). Kinetic growth parameters of different amylolytic and non- amylolytic *Lactobacillus* strains under various salt and pH conditions. *Biores. Technol.* 94(3): 331- 337.

Ray, B., and Daeschel, M. (1992). *Food Biopreservatives of Microbial Origin*. CRC Press. Boca Raton, Florida, pp. 3- 11.

Ray, R.C., and Ravi, V. (2005). Post harvest spoilage of sweet potato in tropics and control measures. *Crit. Rev. Food Sci. Nutr.* 45: 623- 644.

Ray, R.C., and Ward, O.P. (2006). Post harvest microbial biotechnology of tropical root and tuber crops. In: R .C. Ray., and O. P. Ward (Eds), *Microbial Biotechnology in Horticulture, Volume 1,* Science Publishers Inc., Enfield, New Hampshire, pp. 345-395.

Reina, L.D. (2003). Biocontrol of *Listeria monocytogenes* in minimally processed brined refrigerated cucumbers. Ph.D thesis, North Carolina State University, N.C., Raleigh, USA.

Reina, L.D., Breidt, F., Fleming, H.P., and Kathariou, S. (2005). Isolation and selection of lactic acid bacteria as biocontrol agents for non-acidified refrigerated pickles. *J. Food Sci.* 70(1): 7- 11.

Rodriguez, J. M., Martinez, M.I., Horn, N., and Dodd, H. M. (2003). Heterologous production of bacteriocins by lactic acid bacteria. *Int. J. Food Microbiol.* 80 (2): 101-116.

Romano, A.H., Trifone, J.D., and Brustolon, M. (1979). Distribution of the phosphoenolpyruvate and glucosephosphotransferase system in fermentative bacteria. *J. Bacteriol.* 139(1): 93- 97.

Ross, R.P., Morgan, S., and Hill, C. (2002). Preservation and fermentation: past, present and future. *Int. J. Food Microbiol.* 79 (1- 2): 3- 16.

Sanchez, A. H., De Castro A., Rejano, L., and Montano, A. (2000). Comparative study on chemical changes in olive juice and brine during green olive fermentation. *J. Agric. Food Chem.* 48: 5975- 5980.

Sanni, A.I., Morlon- Guyot, J., and Guyot, J.P. (2002). New efficient amylase- producing strains of *Lactobacillus plantarum* and *Lactobacillus fermentum* isolated from different Nigerian traditional fermented foods. *Int. J. Food Microbiol.* 72(1-2): 53- 62.

Savard, T., Beaulieu, C., Gadren, N.J., and Champagne, C.P. (2002). Characterization of spoilage yeasts isolated from fermented vegetables and inhibition by lactic, acetic and propionic acids. *Food Microbiol.* 19: 363- 373.

Shah, F.H. (1986). Fermented foods of Pakistan. In: *Traditional Foods: Some Products and Technologies*, Central Food Technological Research Institute, Mysore, India, [cited in *Fermented Fruits and Vegetables: A Global Perspective* (M. Battcock and S. Azam- Ali), FAO Agric. Services Bull., No.134, Rome, 2001, pp. 96.]

Shih, C.T., and Hang, Y.D. (1996). Production of carotenoids by *Rhodotorula rubra* from *sauerkraut* brine. *Lebesm. Wiss. Technol.* 29: 570- 572.

Shurkhno, R.A., Gareev, R.G., Abul'khanov, A.G., Validov, Sh. Z., Boronin, A.M., and Naumova, R.P. (2005). Fermentation of a high-protein plant biomass by introduction of lactic acid bacteria. *Prikl. Biokhim. Mikrobiol.* 41(1): 79- 89.

Sim, S.L., and Hang, Y.D. (1996). *Sauerktaut* brine: a potential substrate for production of yeast β– glucosidase. *Lebesm. Wiss. Technol.* 29: 365- 367.

Singh, J. (2005). Lactic acid fermentation of taro (*Colocasia antiquorum* L.)- brine mixture by *Lactobacillus plantarum*. M.Sc. thesis, P.G. Department of Biotechnology, Revenshaw Autonomous College, Bhubaneswar, Orissa, India, 26 pp.

Soomro, A.H., Masud, T., and Anwaar, K. (2002). Role of bacteria (LAB) in food preservation and human health- a Review. *Pakistan J. Nutr.* 1(1): 20- 24.

Spyropoulou, K.E., Chorianopoulos, N.G., Skandamis, P.N., and Nychas, G.J.E. (2001) Control of *Escherichia coli* O157: H7 during the fermentation of spanish- style green table olives (conservolea variety) supplemented with different carbon sources. *Int. J. Food Microbiol.* 66: 3- 11.

Steinkraus, K.H. (1989). *Industrialization of Indigenous Fermented Foods*. Marcel Dekker, New York. pp. 323- 344.

Steinkraus, K.H. (1996). *Handbook of Indigenous Fermented Foods, 2nd Edition*, Marcel Dekker. New York. pp. 776.

Steinkraus, K.H. (1997). Classification of fermented foods: worldwide review of household fermentation techniques. *Food Control* 8: 311- 317.

Steinkraus, K.H. (2002). Fermentations in world food processing. *Comp. Rev. Food Sci. Safety* 1: 23- 32.

Stiles, M.E., and Holzapfel, W.H. (1997). Lactic acid bacteria of foods and their current taxonomy. *Int. J. Food Microbiol.* 36: 1- 29.

Sulc, D. (1984). Gemusesafte. *Flussiges Obst.* 1: 17- 24.

Tanasupawat, S., and Komagata, K. (1994) Lactic acid bacteria in fermented foods in Thailand. *World J. Microbiol. Biotechnol.* 11: 253- 256.

Tamang, J. P., Tamang, B., Schillinger, U., Franz, C. M., Gores, M., and Holzapfel, W. H., (2005). Identification of predominant lactic acid bacteria isolated from traditionally fermented vegetable products of the Eastern Himalayas. *Int. J. Food Microbiol.* 105(3): 347 - 356.

Tsapatsaris, S., and Kotzekidou, P. (2004). Application of central composite design and response surface methodology to the fermentation of olive juice by *Lactobacillus plantarum* and *Debaryomyces hansenii*. *Int. J. Food Microbiol.* 95(2): 157- 168.

van Reenen, C.A., Dicks, L.M., and Chikindas, M.L. (1998). Isolation, purification and partial characterization of plantaricin 423, a bacteriocin produced by *Lactobacillus plantarum*. *J. Appl. Microbiol.* 84(6): 1131- 1137.

Vaughn R.H. (1985). The microbiology of vegetable fermentations. In: B.J.B. Wood (Ed.), *Microbiology of Fermented Foods, Volume 1*, Elsevier Applied Science Publishers, london, pp.49- 54.

Viander, B., Maki, M., and Palva, A. (2003). Impact of low salt concentration, salt quality on natural large- scale *sauerkraut* fermentation. *Food Microbiol.* 20: 391- 395.

Ward, O.P., and Ray, R.C. (2006). Microbial biotechnology in horticulture: an overview. In: R .C. Ray., and O. P. Ward (Eds.), *Microbial Biotechnology in Horticulture, Volume 1*, Science Publishers Inc., Enfield, New Hampshire, pp. 1- 22.

Wee, Y.J., Kim, J.N., and Ryu, H.W. (2006). Biotechnological production of lactic acid and its recent applications. *Food Technol. Biotechnol.* 44(2): 163- 172.

Yamano, T., Iino, H., Takada, M., Blum, S., Rochat, F., and Fukushima, Y. (2006). Improvement of the human intestinal flora by ingestion of the probiotic strain *Lactobacillus johnsonii* Lal. *Brit. J. Nutr.* 95(2): 303- 312.

Yoon, K. Y., Woodams, E. E., and Hang, Y. D. (2006). Production of probiotic cabbage juice by lactic acid bacteria. *Biores. Technol.* 97(12): 1427 - 1430.

Yoon, K.Y., Woodams, E.E., and Hang, Y.D. (2004). Probiotication of tomato juice by lactic acid bacteria. *J. Microbiol.* 42(4): 315- 318.

Zetelaki-Horvath, K., and Anderson, R.E. (1986). Fermented vegetable cocktails. *Acta Aliment.* 15: 265- 268.

Zhang, J.H., Hu, F., and Chen, H.Y. (2000). Processing technique of vegetable juice beverage of Sechium edule Swartz and fermentation beverage of *Cucurbita moschata. Duch. J. Shanghai Agric. coll.* 18 (2): 114-117.

In: Food Microbiology Research Trends
Editor: Marta V. Palino, pp. 189-216

ISBN 978-1-60021-723-4

Chapter 6

CONTROL OF HARMFUL FOOD-BORNE GRAM-NEGATIVE BACTERIA BY WEAKENING CELL MEMBRANES

Hanna-Leena Alakomi and Maria Saarela
VTT Technical Research Centre of Finland
Tietotie 2, FI-02044 VTT, Finland

ABSTRACT

Among common food spoilage microbes there are a number of Gram-negative species, some of which are responsible for many serious infections such as various serovars of *Salmonella enterica* and *Escherichia coli*, and others which create various hygienic problems and spoilage of foods by their metabolic products. Gram-negative bacteria are typically inherently resistant to many external agents intended for antimicrobial action, and for this reason they pose a specifically difficult problem in the food protection. This resistance is largely due to the membrane structures of the cell, i.e. outer membrane (OM) present as the outermost layer in all Gram-negative bacteria. However, it is possible to specifically weaken the OM by various agents that disintegrate the lipopolysaccharide layer. Such agents are collectively termed permeabilizers.

Many parameters effect the survival and growth of microbes in food, one of which is the pH. Weak organic acids, like citric and lactic acid, in addition to their antimicrobial property due to the lowering of the pH, also function as permeabilizers of the Gram-negative bacterial outer membrane. Permeabilization of Gram-negative bacteria by substances suitable for food protection, e.g. lactic acid, would offer an optimal mean to prevent the growth of these microbes in food materials and during different stages of food production.

This chapter focuses on the role Gram-negative bacteria as harmful bacteria and gives an overview of the resistance mechanism of Gram-negative bacteria OM as well as clarifies the mechanisms of antimicrobial agents intended to weaken these protective structures.

Introduction

Structures of Bacterial Cell Envelopes

Traditionally first steps in the identification and characterisation of unknown bacteria are based on phenotypic properties of the isolate, like colony and cell morphology. In addition, microscopic methods with various staining methods are applied to monitor specific cell structures and to obtain information about the bacterial membrane structures. The doublestaining method developed by Hans Christian Gram at the late 1800s is still a valuable tool in the phenotypic classification of bacteria (Holt et al. 1994, Beveridge, 2001). This method reveals differences in the membrane structures of cells according to which they can be classified into two groups. Bacteria that retain the initial crystal violet stain (purple) are called Gram-positive and those that are decolorized and stain red with carbol fuchsin or safranin are called Gram-negative. Normally a reliable result is obtained from young cells, although exceptions may occur. Some bacteria, depending of their cell cycle, stain variable or stain faintly (Beveridge, 2001). Likewise, bacteria with high membrane lipid content (e.g. mycobacteria rich in mycolic acid) require specific staining methods (Brennan and Nikaido, 1995). However, advances in the 16S rRNA gene sequencing have changed classification of some bacteria, which have Gram-negative cell wall but are now assigned to Gram-positive phyla (Schleifer et al. 1990).

During their life microbes encounter several harsh and stressful conditions. They have to cope with changing environmental conditions, e.g. pH, temperature, salinity, antimicrobial compounds, water activity and availability of nutrients (Foster, 2004, Young, 2006). In addition, competition between other microbes for space and nutrients as well as other microbial interactions affects their life (Lazdunski et al. 2004, Young, 2006). Hence, cell membranes have to be multifunctional and respond quickly to changing environmental conditions (Rowley et al. 2006).

Gram-Negative vs. Gram-Positive Cell

Cytoplasma of the cell is surrounded by cytoplasmic membrane (CM), which is composed essentially of a phospholipid bilayer with embedded proteins (Matsumoto et al. 2006, Stenberg et al. 2005). CM is selectively permeable and regulates the transfer of solutes and metabolites in and out of the cell cytoplasm (Holland, 2004). It is also associated with several important enzymes involved in various cell metabolic functions and pumps involved in maintenance of membrane potential and ion gradients (Ruiz et al. 2006, Stenberg et al. 2005). Hence it is considered to be the major target site for biocides (Maillard, 2002). In Gram-positive bacteria CM is covered by a thick (20-80 nm) cell wall layer consisting of peptidoglycan (PG, also called murein) and adjoining polysaccharides, teichoic acids, teichuronic acids and lipoteichoic acids (Cabeen and Jacobs-Wagner, 2005, Schäffer and Messner, 2005). PG is composed of alternating *N*-acetylglucosamine and *N*-acetylmuramic disaccharides, the latter having pentapeptide stems (Meroueh et al. 2006, Scheffers and Pinho, 2005). During the cell wall biosynthesis these pentapeptide stems are cross-linked by transpeptidase enzymes creating cell wall polymer that is responsible for the maintenance of cell shape and osmotic stability (Daniel and Errington, 2003, Merouch et al. 2006). PG layer is significantly thinner (1-7 nm) in Gram-negative bacteria compared to Gram-positive bacteria (Cabeen and Jacobs-Wagner, 2005, Vollmer and Höltje, 2001). Gram-negative bacteria possess as their outermost layer a special outer membrane (OM) that covers periplasmic space, the PG layer and the CM (Nikaido, 2001, 2003). In Gram-negative bacteria the OM and PG are linked to each other with lipoproteins and the OM includes porins, which allow the passage of small hydrophilic molecules (Cabeen and Jacobs-Wagner, 2005, Nikaido, 2003). The structure and composition of the outermost layer of the cells differs greatly between bacteria. On the outer envelope cells may have polysaccharide capsules (Sutherland, 1999; Synder et al. 2006) or protein surface layers (S-layers) (Antikainen et al. 2002; Schäffer and Messner, 2005), which protect bacteria in unfavourable conditions.

Bacterial membranes are responsible for many responses of cells to antimicrobial agents and environmental stresses (Raivio, 2005). For example, the outer membrane of Gram-negative bacteria and mycobacteria acts as a permeability barrier and is responsible for the intrinsic resistance of these micro-organisms to antimicrobial compounds (Brennan and Nikaido, 1995, Nikaido 2001).

Some bacteria have additional structures, such as fimbria and pili in their outermost layer of the cell (Scott and Zahner, 2006). Most *Salmonella* strains produce thin aggregative fimbria (Tafi) and *Escherichia coli* strains are reported to produce similar structures, termed curli (Prigent et al. 2000, White et al. 2001, 2003). These structures in association with other components of extra cellular matrix are suggested to create a coating that affects biofilm formation and cell-cell attachment (White et al. 2003, Prigent et al. 2000, de Rezende et al. 2005, Solano et al. 2002). For *E. coli* and *Salmonella* it has been reported that cellulose and thin aggregative fimbriae are produced together and they form a highly inert, hydrophobic extracellular matrix, in which the cells are embedded (Zojac et al. 2001, Solano et al. 2002).

A number of harmful species and potential pathogens belong to the group of Gram-negative bacteria. The family *Enterobacteriaceae* consists of a large group of Gram-negative, non-spore forming, facultatively anaerobic bacteria (Holt, 1994). Some members of *Enterobacteriaceae* family are responsible for serious infections (e.g. *Salmonella, Campylobacter, Escherichia*) in addition to more common, milder gastrointestinal infections, although they also are abundant in the environment contaminated with faeces (Liebana et al. 2003) and part of normal microbiota of many animals and humans (Holt, 1994).

Harmful Gram-Negative Bacteria – Importance in Different Surroundings

Salmonella as a Pathogen

The nomenclature of the genus *Salmonella* has been variable, since two systems of nomenclature have been in use for the genus *Salmonella* (Tindall et al. 2005). Recently it has been agreed that the genus *Salmonella* currently includes only two species, namely *Salmonella enterica* and *S. bongori.* The type species *Salmonella enterica* is divided into six subspecies and most *Salmonella* belong to the subspecies *Salmonella enterica* subsp. *enterica*

(Tindall et al. 2005, www.bacterio.cict.fr). Members of this subspecies have usually been named based on the original isolation location of the serovar or serotype. The genus *Salmonella* is extremely heterogenous, and more than 2 500 serovars of *Salmonella* currently exist (Popoff et al. 2004).

Salmonella, although being intestinal bacteria, are widespread in the environment and commonly found in farm effluents and in any material subjected to faecal contamination (Liebana et al. 2003, Martinez-Urtaza et al. 2004, Le Minor, 1999). Salmonellosis is an infectious disease of humans and animals caused by living cells of the two species of *Salmonella* (*S. enterica* and *S. bongori*). *S. enterica* is a facultatively intracellular pathogen which preferentially resides inside macrophages, yet it requires both antibodies and a cellular immune response for clearance (Kauffman et al. 2001). Depending on the serotype and host, *Salmonella* can cause diseases ranging from mild gastroenteritis to typhoid fever (Humphrey, 2004, Chiu et al. 2004).

Salmonella has been reported to be the most common food-borne bacterial disease in the world (Plym Forshell and Wierup, 2006). In the United States it has been estimated that 1.4 million non-typhoidal *Salmonella* infections with 400 deaths occur annually (Voetsch et al. 2004). In 2005, salmonellosis remained the second most frequent zoonosis with about 180 000 reported human cases in European Union (EFSA, 2006). Eventhough under-reporting of salmonellosis can be high. *S.* Enteritidis and *S.* Typhimurium have been reported to be the serovars most frequently associated with human illness (EFSA, 2006, Olsen et al. 2002). Human *S.* Enteritis cases have often been associated with consumption of contaminated eggs (Guard-Petter, 2001, Cogan et al. 2004) and broiler meat, while *S.* Typhimurium cases have typically been associated with consumption of contaminated pig, poultry and bovine meat (EFSA, 2006, Jorgensen et al. 2000). In 2005 the Finnish National Health Institute (NIH) reported 2503 salmonella cases (Anonymous, 2006). Majority of the Finnish salmonellosis cases were caused by *S.* Enteritidis and they were obtained from abroad (Niskanen et al. 2006). *S.* Typhi remains an important health threat for humankind, with more than 16 million cases and 600 000 deaths annually world-wide (Pang et al. 1998). Typhoid fever is a disease that usually results from overcrowding and poor sanitary conditions. Hence, the incidence of this disease is highest in developing countries with poor hygienic conditions, inadequate clean water supplies and sewage systems (Gasem et al., 2002, Vollaard et al. 2004). The emergence of antibiotic-resistant *Salmonella* strains, e.g. due to previous uncontrolled use of antimicrobials in animal feeds and an increased therapeutic use in other areas, is an increasing problem world-wide (Sørum and L'Abée-Lund, 2002, Su et al. 2004).

The nature of the pathogenic action of *Salmonella* varies with the serovar, the strain, the infectious dose, the nature of the contaminated food, and the host status (Humphrey, 2004). Infants and immunosuppressed patients are more susceptible to salmonella infection than healthy adults (Tauxe, 2002, Le Minor 1999, Voetsch et al. 2004). Infections with *Salmonella* are initiated when the pathogen invades the gastrointestinal epithelium (Merrell and Falkow, 2004, Reynolds et al. 2006). Virulence of *Salmonella* requires multiple factors (Groisman and Ochman, 1997, Marcus et al. 2000). The cell membranes and their stress responses have an important role in virulence as most virulence determinants reside in, or must transit through them (Raivio, 2005). Lipopolysaccharide (LPS) is the major virulence factor of Gram-negative bacteria (Trent et al. 2006, Raetz and Whitfield, 2002). During the past decade, enormous progress has been achieved in the elucidation of LPS recognition and signalling in mammalian phagocytes (reviewed in Trent et al. 2006, Heumann and Roger 2002, Reynolds et al. 2006). Recognition of LPS leads to rapid activation of an intracellular signalling pathway, which results in the release of pro-inflammatory mediators (Heumann and Roger 2002). Pathogenic *Salmonella* bacteria have evolved many strategies in adapting to the hostile environment of the phagosome (Monack et al. 2004, Raupach and Kauffman 2001; Reynolds et al. 2006).

Salmonella infections have been reported to result either in fatal bacteremia, when unrestricted, or in the generation of neutrophil- and mononuclear-rich microabscesses that lead to bacterial clearance, when successfully controlled (Merrel and Falkow, 2004). In some cases bacteria persist in gall bladders of asymptomatic carriers, who contribute substantially to dissemination of disease by continuous shedding of the pathogen into environment (Raupach and Kauffman 2001).

Origin of *Salmonella*

Salmonella can be associated with many kinds of foods (Humphrey, 2004). Contamination of meat (cattle, pigs, poultry) may originate from animal salmonellosis, but most often it results from the contamination of muscles with intestinal contents during evisceration of animals, washing, and transportation of carcasses (al-Saigh et al. 2004). Likewise, vegetables and fruits may carry salmonella when contaminated with fertilizers of faecal origin, or when washed with polluted water (Das et al. 2006, Duffy et al. 2005). Proper heat treatment of the foods will kill salmonella in food, but caution should be taken to avoid cross-contamination (Reij et al. 2004, van Asselt and Zwietering, 2006). Other essential elements in the prophylaxis of salmonellosis are consumer education (in particular improvement of hygiene), correct storage temperatures (preventing multiplication of salmonella in food), and use of pasteurization (e.g. for milk) or sterilization when possible (Humphrey, 2004).

In the European Union, the Zoonoses Directive 92/117/EC requires collection of information on zoonosis and zoonotic agents in humans, animals, foods and feeds as well as requires monitoring of e.g. breeding flocks for Salmonella (EFSA, 2006). Food products of the animal origin are considered to be the major source of human salmonella infections and Salmonella infections of production animals play an important role in the public health and in food safety (Plym Forshell and Wierup, 2006). Therefore monitoring of zoonotic agents should cover the whole food chain. Since feed contaminated with Salmonella is a potential source for contamination of farm livestock investigation of feed and feed raw materials for salmonella is also essential (Plym Forshell and Wierup, 2006, Sauli et al. 2005). Competative exclusion used in poultry industry is an additional way to control salmonella and it has been applied in Finland and Sweden successfully for many years (Nuotio et al. 1992, Schneitz et al. 1992, Schneitz & Renney, 2003).

Other Gram-Negative Pathogens

Along salmonella, camplylobacter and other Gram-negative foodborne pathogens e.g. *Yersinia* have emerged (Tauxe, 2002). Due to its ability to grow at low temperatures *Yersinia* is able to multiply during prolonged cold-storage into high enough levels to cause a disease

(Little and Knochel, 1994). *Yersinia enterocolitica* is an important food-borne pathogen that can cause yersiniosis in humans and animals. Pigs are considered to be an important source of human yersiniosis. Similar *Yersinia* genotypes have been found among canine, feline and wild rodent strains, indicating that these animals are other possible infection sources for humans (Fredriksson-Ahomaa et al. 2006).

In 2005, campylobacteriosis was the most frequently reported zoonotic disease in humans in European Union (EFSA, 2006). Due to its ability to cause water-borne outbreaks campylobacter epidemics normally infect a large number of people (Niskanen et al. 2006). The most important vehicle of *Campylobacter* transmission to humans is poultry meat that has been contaminated during processing (Wagenaar et al. 2006). *Escherichia coli* infections remain one of the most common causes of gastro-enteritidis. Majority of *E. coli* strains are harmless, but a small portion can cause serious gastroenteritidis (Harrington et al. 2006). *E. coli* enteritidis is caused by at least six distinct *E. coli* pathotypes: enterotoxigenic (ETEC), enteropathogenic (EPEC), enterohemorrhagic (EHEC), enteroinvasive (EIEC), diffusively adherent (DAEC) and enteroaggregative (EAEC) *E. coli* (Harrington et al. 2006, Kaper et al. 2004, Venter et al. 2006).

The group of *Enterobacteriaceae* also includes many other potential opportunistic pathogens. *Enterobacter sakazakii* is an opportunistic pathogen which causes meningitis and enteritidis, mostly in infants. Many of the cases have been associated with contaminated reconstituted infant formula (Drudy et al. 2006, Edelson-Mammel et al. 2006). Along *Enterobacteriaceae* many other Gram-negative bacteria, like *Aeromonas* spp. are opportunistic pathogens (Isonhood and Drake, 2002). *Helicobacter pylori* has been reported to colonize the gastric epithelium of at least 50% of the world`s human population, playing a causative role in the development of chronic gastritis and peptic ulcers (Terebiznik et al. 2006).

Role of Other Gram-Negative Bacteria as Food Contaminants

Spoilage bacteria produce as by-products of their growth and metabolism compounds (e.g. secondary metabolites) which can affect the quality of foods, causing either organoleptical, structural, chemical or microbiological changes in the products (Huis in`t Veld, 1996, Gram et al. 2002). These compounds include off-flavours (e.g. volatile compounds), proteolytic and lipolytic enzymes, toxins and biogenic amines (Silla Santos 1996, Rokka et al. 2004). Besides enteric pathogens the group of *Enterobacteriaceae* includes species that cause food spoilage (for reviews see Baylis, 2006). Many of the spoilage bacteria are also opportunistic pathogens able to cause disease in favourable conditions. Certain members of *Hafnia*, *Serratia* and *Enterobacter* species are associated with the production of biogenic amines and off-flavours (Gram et al. 2002, Smolander et al. 2004) as well as cases of human diseases (Drudy et al. 2006).

In aerobic conditions *Pseudomonas* spp. have been reported to be the most common spoilage microbes in foods with high protein content, such as meat, poultry, fish, milk and some dairy products (Huis in`t Veld, 1996). Several members of *Pseudomonas* and *Stenotrophomonas* species produce proteolytic and lipolytic enzymes which cause spoilage of e.g. milk (Munsch-Alatossava and Alatossava, 2006). Smolander et al. (2004) reported that in modified atmosphere packed (MAP) unmarinated broiler chicken cuts stored under unbroken cold-chain the spoilage was mainly caused by lactic acid bacteria, whereas in interrupted cold-chain enterobacteria and hydrogen sulphide producing bacteria were able to rapidly multiply and produce volatile compounds and biogenic amines causing spoilage of the food (Smolander et al. 2004, Rokka et al. 2004). In addition, spoilage detected by electronic nose (volatile compounds) correlated with the amount of *Enterobacteriaceae* in the MAP products (Rajamäki et al. 2006).

In food matrix microbes grow in a compact ecosystem in which they can exploit metabolites produced by other microbes (metabiosis) or there can be antagonistic interactions between microbes (Fleet, 1999, Gram et al. 2002). The ability of *Salmonella* and other bacteria to survive in the food chain is due, in part, to their ability to respond effectively to environmental changes (Anriany, 2001, 2006; McMahon et al. 2007). In several cases a correlation between specific stress survival and virulence-associated phenotypes has been found (Gaynor et al. 2005). Stress adapted cells are also more tolerant towards desinfection treatments (Loughlin et al. 2002).

In nature and in process environments microbes grow as multi-species communities (biofilms) attached to surfaces where they are protected from the action of biocides and disinfectant agents (Davey, 2000; Guerrero et al. 2002). Cells of the community have close interactions with each other (multiple attachments, cell-cell signalling) and cooperation for nutrients and metabolic compounds (Davey, 2000; Young, 2006). In addition, cells in biofilm exhibit an altered phenotype (e.g. slower growth rate and gene expression) compared to free swimming (planktonic) cells (Young, 2006). Biofilm structure provides cells protection from an exposure to external agents (e.g. biocides, antibiotics) (Campanag et al. 2002, Donlan and Costerton, 2002, Snyder et al. 2006). Especially extracellular polysaccharides (EPS) play various roles in the structure formation and function of different biofilm communities: e.g. they exclude and/or influence the penetration of antimicrobial agents, and provide protection against a variety of environmental stresses, such as UV radiation, pH shifts, osmotic shock, and desiccation (Champanag et al. 2002, Davey and O`Toole, 2000, Robertson and Firestone, 1992, Young 2006). In addition, polysaccharide capsules present on bacterial surfaces strengthen interactions between microbes and biofilm matrix (Sutherland, 1999, Snyder et al. 2006). Since Gram-negative bacteria are efficient producers of extracellular matrix and polysaccharide capsules they have an important role in the formation and stabilization of biofilm structures on surfaces (Zogaj et al. 2001, White et al. 2006). Schooling and Beveridge (2006) recently reported that membrane vesicles produced by *Pseudomonas* sp. are an important part of biofilm structures. Contamination from process surface biofilms is one effective route for food contamination by pathogens or spoilage microbes (Ganesh Kumar and Anand, 1998, Reij et al. 2004).

Gram-Negative Cell Structures

The Outer Membrane

Control of Gram-negative bacteria is hampered by their relative resistance to many external agents compared to Gram-positive bacteria (Nikaido, 1999). Cells of Gram-negative bacteria are surrounded by an additional membrane (outer membrane, OM), which provides the bacterium with a hydrophilic surface and functions as a permeability barrier for many external agents (Helander et al. 1997c, Vaara 1992, 1999, 2003, Nikaido, 2003). The effect is mainly caused by the presence and features of lipopolysaccharide (LPS) molecules in the outer leaflet of the membrane, resulting in an inherent resistance to hydrophobic antibiotics, detergents and hydrophobic dyes in many Gram-negative bacteria (Nikaido and Vaara, 1985, Nikaido, 2003). The OM of common Gram-negative bacteria, like *E. coli*, is constructed asymmetrically (Beveridge, 2001). The lipid structure of the outward directed membrane leaflet is formed dominantly by lipid A regions of the LPS molecules which make up 75 % of the total membrane surface and form specific contacts to integral outer membrane proteins (Omp) like porins (Alexander and Rietschel, 2001, Bos and Tommassen, 2004). Bacterial lipoproteins anchor the OM to the periplasmic peptidoglycan layer (Bos and Tommassen, 2004). Divalent cations are tightly associated to the anionic membrane proximal regions of LPS molecules strengthening the structures (Vaara, 1992). Some Gram-negative bacteria are known to contain glycosphingolipids instead of LPS in their OM (Kawahara et al. 1991). In addition to LPS, some enterobacteria have present on their OM surface an additional glycolipic, the enterobacterial common antigen (Erbel et al. 2003). Some species and strains have also capsular polysaccharides on their cell surface (Snyder et al. 2006).

Gram-negative bacteria regulate OM permeability characteristics with hydrophilic channels known as porins, which allow nutrients with relatively small molecular weight (< 600 Daltons) to enter the inner parts of the cell (Nikaido, 2003). These water-filled pores generally exclude the entry of hydrophobic substances (Nikaido, 2003). The influx of lipophilic compounds into gram-negative cells is limited not only by the OM permeability barrier but also by their active efflux by "multidrug" efflux pumps, which are usually energized by the proton motive force (Poole, 2002, Piddock, 2006). *Pseudomonas* species have been reported to be resistant to many biocides and antimicrobial agents (Poole, 2002, Walsh et al. 2000).

The components of the OM are synthesized inside the cell or at the inner leaflet of CM (Wu et al. 2006, Ruiz et al. 2006). During the last years there has been a marked advance in the identification of the steps of bacterial OM biosynthesis (Bos and Tommassen, 2004, Ruiz et al. 2006, Trent et al. 2006). LPS molecules localized on the outer leaflet of the outer membrane serve as the major surface component of the Gram-negative bacterial cell envelope (Nikaido, 1999, Trent et al. 2006). This glycolipid is essential for virtually all Gram-negative bacteria and represents one of the conserved microbial structures responsible for activation of the innate immune system (Reynolds et al. 2006). Therefore the structure, function, and biosynthesis of LPS have been an area of intense research (recently reviewed by Alexander and Rietschel, 2001, Trent et al. 2006). LPS (the main virulence factor in Gram-negative bacteria) of a number of bacteria is composed of three distinct regions -lipid A (endotoxic principle and anchoring molecule in the OM), a short core oligosaccharide, and O-antigen polysaccharide (highly variable O-antigen oligosaccharide consisting of repeating units) (Alexander and Rietschel, 2001). The basic structure of LPS, a covalently linked lipid and polysaccharide, is common in all studied LPS molecules, but otherwise there are large heterogeneities in the chemical structures of LPS depending on bacterial genera, species and strains (for review see Trent et al., 2006, Raetz and Whitfield, 2002, Yethon and Whitfield, 2001).

The lipid A domain, also known as endotoxin, anchors the LPS in the outer membrane and is the bioactive component recognized by Toll Like Receptor (TLR) 4 during human infection (primary immunostimulatory centre of LPS) (Trent, 2006). In severe cases endotoxins can cause lethal sepsis or septic shock (Alexander and Rietschel, 2001). Common type of lipid A is presented in *E. coli* where the hydrophilic backbone consists of a β-(1$\rightarrow$6)-linked 2-amino-2-deoxyglycopyranose (GlcN) disaccharide, carrying two phosphate groups at positions 1 and 4′ and four residues of (R)-3-hydroxy-myristic acid in ester- or amide linkage (Gronow and Brade, 2001). In addition, the disaccharide of the lipid A backbone carries 4-7 fatty acids with hydroxyl group, these fatty acids are essential structures of lipid A. The hydroxyl groups of some of the fatty acids are further esterified by nonhydroxyl fatty acids, creating the unique 3-acyloxylacul structure (Helander, 1996). Negatively charged phosphate groups are important for strengthening the LPS monolayer by linking molecules via ionic bridges with divalent cations (Vaara, 1999). The phopshate groups of the lipid A backbone may carry additional substituents such as 2-aminoethanol (EtN), 2-aminoethyl phosphate (P-EtN) and 4-amino-4deoxy-L-arabinopyranose (L-Ara*p*4N) (Helander et al. 1996). The general structure of LPS is presented in Figure 1.

Although the lipid A structures of various organisms show an impressive level of diversity, the biochemical synthesis of lipid A is considered to be a highly conserved process (Raetz and Whitfield 2002, Reynolds et al. 2006). Recent studies have shown that *S.* Typhimurium cells can respond to certain environmental conditions by producing enzymes that alter the lipid A structure (Reynolds, 2006, Venter et al. 2006). Lipid A modifications are primarily involved in the survival in environments with low level of divalent cations, in providing resistance to cationic peptides, in growth at low temperatures and in virulence (Frirdich and Whitfield, 2005, Reynolds et al. 2006). Members of the genus *Yersinia* have also been reported to be able to modify their lipid A structure as a response to temperature changes (Bengoechea et al. 2003, Rebell et al. 2004).

The complex heteropolysaccharide (core oligosaccharide and O-spesific chain) is covalently linked to the 6′of lipid A (Gronow and Brade, 2001). The core region can be further subdivided on the basis of structural features into two regions, the inner and outer core. The outer core typically consists of common hexose sugars, such as glucose, galactose, *N*-acetyl galactosamine and *N*-acetyl glucosamine and is generally more variable than the inner core region (Erridge et al. 2002). The inner core region is composed of at least one molecule of 3-deoxy-α-D-*manno*-oct-2-ulopyranosonic acid (usually also called 2-keto-3-deoxyoctulosonic aicd, Kdo) and two or more residues of L-glycero-α-D-*manno*-heptopyranose (L,D-Hep) (Alexander and Rietschel, 2001, Frirdich and Whitfield, 2005). Often additional anionic substituents like phosphate, diphosphate or diphosphoethanolamine (PPEtn) groups are added to the core structure (Alexander and Rietschel, 2001). LPS consisting of Lipid A and Kdo with negatively charged substituent is the smallest known structure to sustain growth and viability of Gram-negative bacteria (Gronow and Brade, 2001).

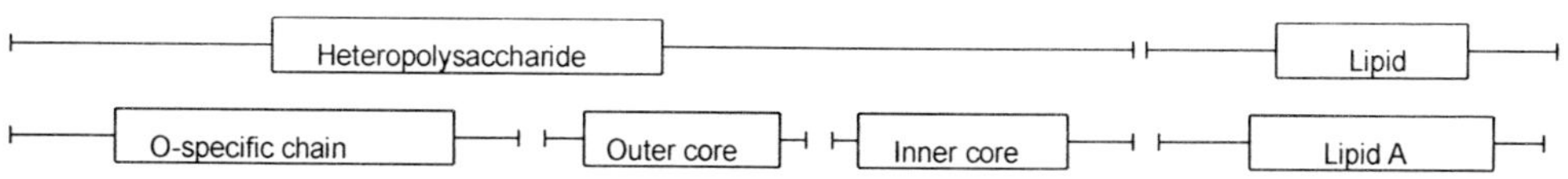

Figure 1. General structure of LPS.

The negative charges provided by the phosphate residues in the Hep region of *E. coli* and *Salmonella* are important in maintaining the barrier function of the OM by providing sites for cross-linking of adjacent LPS molecules with divalent cations or polyamines (Vaara, 1992, Frirdich and Whitfield, 2005). These negative charges, provided by residues of Kdo and phosphate, allow neighbouring LPS molecules to be crosslinked by divalent cations (Mg^{2+}, Ca^{2+}), structurally reinforcing the OM (Nikaido, 2001). Likewise, the negative charges play an important role in creating interactions between LPS and positive charges of OM proteins (OMPs). This part of the LPS has been found to be selectively targeted by several cationic antibiotics, positively charged host defence peptides and proteins such as the family of the mammalian defensins (Alexander and Rietschel, 2001). According to Walsh et al. (2000) the inner core phosphates of *P. aeruginosa* appear to play a key role in the intrinsic drug resistance of this bacterium. In addition, ethanolamine in the core of *P. aeruginosa* has been suggested to play a role in conferring resistance to cationic antimicrobial peptides (Knirel et al. 2006).

Many pathogenic bacteria have in their LPS an O-specific (O-antigen) chain which provides an additional shelter against the host´s defence mechanisms (Alexander and Rietschel, 2001). The O-antigen consists of up to 50 repeating units of mono- or oligosaccharides and is highly variable in structure (Gronow and Brade, 2001). Its presence results in a smooth colony appearance (S-form) in contrast to the rough morphology displayed by colonies of O-antigen-lacking bacteria (R-form) (Helander, 1996). As predominant surface structures O-antigens are highly immunogenic. O-spesific antibodies are thus versatile tools for serotyping (Popoff et al. 2004). The O-antigen is generally not essential for the survival of bacteria, but recent studies have shown that O-antigen plays an important role in the effective colonization of host tissues, resistance to complement-mediated killing and in the resistance to cationic antimicrobial peptides that are key elements of the immune system (Skurnik and Bengoechea, 2003). These phenomena have been reported e.g. for *Yersinia* (Skurnik and Bengoechea, 2003) as well as for *Salmonella* (Guard-Bouldin et al. 2004, Carroll et al. 2004). Certain pathogens (e.g. some *Campylobacter jejuni* strains) have been shown to produce low molecular weight LPS (or lipooligosaccharide) and are lacking the specific O-chain in their LPS (Preston et al. 1996).

PERMEABILIZERS

Although the OM of Gram-negative bacteria protects cells from many external agents, it is possible to specifically weaken it by various agents, collectively called permeabilizers, which disintegrate the LPS layer and increase the permeability of the OM to hydrophopic agents (Vaara, 1992, 1999). Permeabilizers themselves may not be bacteriocidal but they can potentiate activity of other compouds, thus acting synergistically (Vaara, 1999).

Examples of Permeabilizers

The classical example of permeabilizers is the chelator EDTA, which sequesters divalent cations that contribute to the stability of the OM by providing electrostatic interactions with proteins and LPS (Leive, 1965, Vaara, 1999). Treatment with EDTA releases a large proportion of LPS from the OM, exposing hydrophobic phospholipids, and creating a hydrophobic pathway for certain substances (Leive, 1965, Alakomi et al. 2003). This is noticeable as an increased susceptibility to hydrophobic agents (Hancock, 1998, Helander et al. 1997c, Nikaido, 1996, Vaara, 1999). Figure 2 shows a schematic view of the activity of EDTA. EDTA has been reported to potentiate the activity of cell wall degrading agents (e.g. lysozyme, nisin) and biocides against microbes (Helander et al. 1997, Shelef and Seiter, 1993, Walsh et al. 2003).

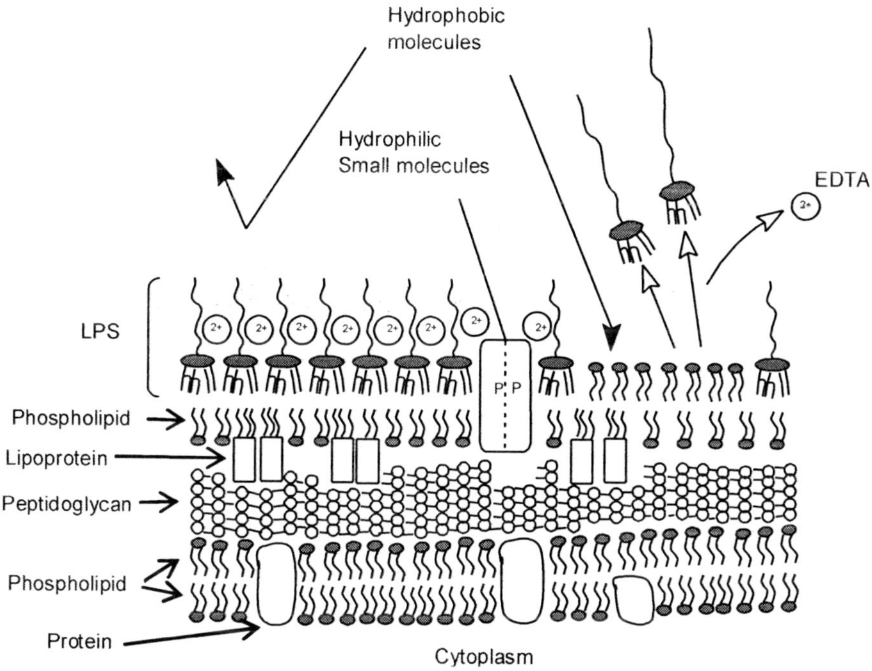

Figure 2. Cell wall architecture of Gram-negative bacteria. The intact outer membrane (left-hand side of the figure) functions as a permeability barrier against hydrophopic molecules, but allows the entry of small hydrophilic molecules through porin proteins (PP). Removal of stabilizing divalent cations (2+) from the lipopolysaccharide (LPS) layer by chelating agents such EDTA results in the release of LPS and appearance of phospholipids on the surface of the outer membrane (right-hand side of the figure), whereby hydrophobic compounds can penetrate the outer membrane and reach their site of action at the cytoplasmic membrane (modified from Helander et al. 1997c).

In addition, EDTA may act as a direct inhibitor of several species of bacteria or it may act synergistically with other antimicrobial agents to promote bacterial destruction (Shelef and Seiter 1993). Besides EDTA, a number of other permeabilizers are known, some of which act quite differently (Helander et al. 1997a, Alakomi et al. 2006). Table 1 summarises examples of known permabilizers and their known mechanisms of action against Gram-negative bacteria.

Microbicidal acids belong to the membrane-active substances (Paulus 1993). A major part of the inhibitory effects of weak acids used as food preservatives are related to the undissociated forms of acids which can penetrate cell membranes, causing perturbation of membrane function, acidification of the cytoplasm and inhibition of acid-sensitive enzymes (Hirshfield et al. 2003).

Table 1. Examples of permeabilizers and other substances possessing OM disrupting activity and their mode of action against Gram-negative bacteria

Agent	Mode of action	References
Chelators		
Ethylenediamine tetraacetic acid (EDTA)	Removes stabilizing cations from the OM, notably Ca^{2+} and Mg^{2+}. Releases LPS to the external medium and creates hydrophopic pathway.	Finch and Brown (1975); Leive (1965), Helander et al. (1997a); Alakomi et al. (2003); Shelef and Seiter (1993), Vaara, 1992, 1999, Walsh et al. (2003)
Na-hexametaphosphate	Removes stabilizing cations from the OM, notably Ca^{2+} and Mg^{2}.	Vaara (1992)
Na_2-pyrophosphate, Na-orthophosphate	Sensitizes cells to nisin.	Boziaris and Adams (1999)
Lactoferrin, transferrin (iron chelating compounds)	Releases LPS, increases sensitivity to rifampin.	Ellison et al. (1988)
Succimer, meso-2,3-dimercaptosuccinic (DMSA)	Destabilizes OM.	Alakomi et al. (2006)
Nitrilotriacetic acid (NTA)	Disintegrates of OM, Increases sensitivity to hydrophobic antibiotics	Vaara (1992); Ayres et al. (1998), Alakomi et al. (2006)
Polycationic substances		
Polymyxins Tris (monovalent)	Displaces cations from the OM, causes membrane damage.	Nikaido and Vaara (1985)
Polycationic peptides		
Polymyxin B nonapeptide (PMBN)	Permeabilizes the OM without significant release of LPS.	Vaara and Vaara (1983)
Polyethyleneimine (PEI)	Sensitizes target cells to hydrophobic antibiotics and to detergents; causes the formation of vesicular structures on the surface of OM, but no liberation of the membrane or its fragments.	Helander et al. (1997a), Alakomi et al. (2006)
Terpenoid and phenolic compounds found in herb plants ("essential oil components")		
Thymol, carvacrol	Destabilizes the OM.	Helander et al. (1998)
Gallic acid	Displaces cations from the OM, causes membrane damage.	Nohynek et al. (2006)
Phenol (also termed carbolic acid)	Weakens the OM.	Paulus (1993)
Phenolic berry extracts (cloudberry and raspberry)	Displaces cations from the OM, causes membrane damage.	Nohynek et al. (2006)
Membrane perturbing proteins and peptides		
Cationic ampliphilic peptides		Wiese et al. (2003)
Agent	Mode of action	References
Organic acids		
Benzoic acid Citric acid Malic acid Lactic acid Sorbic acid	Chelates cations from the OM, notably Ca^{2+} and Mg^{2+}, inducts release of LPS to the external medium.	Alakomi et al. (2000), Helander and Mattila-Sandholm (2000), Alakomi et al. (in press)

The undissociated forms of acids can interact with or pass through the membrane of the microbial cell which normally is negatively charged, thus being a barrier for the negatively charged forms of acids (Doores 1993). Weak organic acids like lactic acid and citric acid have been reported to permeabilize Gram-negative bacteria (Alakomi et al. 2000, Brul and Coote 1999). Table 2 summarizes chemical structure and pKa values of most common organic acids.

Lactic acid has been shown to be a powerful permeabilizer at concentrations of 5-10 mM, the effect being significantly stronger than that of a mineral acid (HCl) at similar pH values (Alakomi et al. 2000). The finding explains the great preservative potential of lactic acid and furthermore suggests that lactic acid present in naturally fermented products may potentiate the antimicrobial action of other compounds against gram-negative bacteria. Lactic acid has been identified as one of the major factors for antimicrobial activity of some probiotic strains, e.g. *Lactobacillus rhamnosus* GG (De Keersmaecker et al. 2006). In addition, Fayol-Messoudi et al. (2005) suggested that mechanism of the antimicrobial activity of probiotic *Lactobacillus* strains against *Salmonella* are multifunctional and may be due to the synergistic activity of lactic acid and secreted non-lactic molecules.

Organic acids, e.g. lactic acid and formic acid, have traditionally been used for the treatment of animal feeds and as food preservatives (Doores, 1993). Recently, Zhao et al. (2006) reported that various combinations of lactic acid, calcium sulphate and sodium benzoate/butyric acid effectively inactivated enterohemorrhagic *E. coli* in rumen content- or faeces-contaminated drinking water for cattle.

Table 2. Chemical structure and pKa values of most common organic acids. Data modified from (Doores, 1993)

Compound (systematic name)	Chemical structure	pKa at 25 °C
Benzoic acid (benzenecarboxylic acid)	C_6H_5COOH	4.19
Citric acid	$HOC(COOH)(CH_2COOH)_2$	3.13
Malic acid (DL-hydroxybutanedioic acid)	$HO_2CCH_2CH(OH)CO_2H$	3.40
Lactic acid (2-hydroxypropionic acid)	$CH_3CH(OH)COOH$	3.86
Sorbic acid (2,4-hexadienoic acid)	$CH_3CH{=}CHCH{=}CHCOOH$	4.76

However, there have been reports indicating that acid adaptation of pathogenic strains can lead to an increased resistance for antimicrobial agents and better survival in acidic conditions (Hirshfield et al. 2003, Bjornsdottir et al. 2006). Therefore, the effect of such adaptive responses on the permeability properties of Gram-negative bacteria should be studied.

Methods to Detect Permeabilization and Weakening of Gram-Negative Bacteria Cell Membranes

Several microbiological, chemical, chromatographic and fluorometric assays are used for the assessment of Gram-negative bacterial permeabilization. The permeabilizing activity of a certain compound against Gram-negative bacteria can be assessed by measuring the sensitization of bacteria to hydrophobic agents, e.g. antibiotics and other hydrophobic compounds by agar diffusion method or in liquid cultures by monitoring growth inhibition (Ayres et al. 1998, Boziaris and Adams, 1999, Alakomi et al. 2006). Measurement of an increase of the uptake of a hydrophopic probe (fluorochromes) is another widely applied method to monitor OM permeability. This is based on the use of fluorochomes, like 1-*N*-phenylnaphthylamine (NPN), which in aquatic environment are non-fluorescent, but once in contact with lipids their fluorescence increases (Loh et al. 1984, Helander and Mattila-Sandholm, 2000, Träuble and Overath, 1973). In addition, destabilization of the OM can be detected by measuring the release of OM specific compounds, e.g. LPS by monitoring LPS-specific fatty acids and phospholipids (Vaara, 1992, Helander et al. 1998) or radiolabelled LPS compounds from cell-free supernatants of treated cells (Hukari et al. 1986). A traditional way is to run SDS-PAGE gels to reveal LPS released in cell free culture supernatants of treated cells (Hitchcock and Brown, 1983). Chemical characterization of the released LPS fractions e.g. by ^{31}P-NMR (Helander et al. 1997b) reveal differences in the LPS heterogeneity of the cells. However, a number of permeabilizers are known not to release LPS, e.g. polyethylenimine (PEI) (Helander et al. 1997a).

Sensitization of Gram-negative cells to detergents (e.g. sodium dodecyl sulphate (SDS) and Triton X-100) and lysozyme induced cell lysis are also indications of the weakening of the OM (Ayres et al. 1993, Helander et al. 1997). According to Yasudu et al. (2002) an increase in the uptake of tetraphenylphosphonium ion (TPP+, a hydrophobic lipophilic ion) could be monitored from cells with distrupted OM permeability barrier. Detection of cytoplasmic membrane damage in Gram-negative cells indicates massive destabilization of OM. When studying combined and synergistic uses of permeabilizers and antimicrobial agents measurement of the CM permeabilization is useful. Leakage of low molecular weight cytoplasmic compounds from the cell interior can be used as an indicator of disorganization of the cytoplasmic membrane (Johnston et al. 2003). Potassium ion-sensitive electrodes, flame photometry and inductively coupled plasma spectrometry have been used to detect K+ leakage from cells after exposure to antimicrobials (Orlov et al. 2002, Yasudu et al. 2002, Johnston et al. 2003). Yasudu et al. (2002) combined a TPP+-selective electrode and K+ electrode to monitor changes simultaneously in the cytoplasmic and in the outer membrane permeability. In permeabilized cells the leakage of low molecular weight compounds is frequently followed by leakage of larger cellular constituents (e.g. ATP) or efflux of ABS_{260nm} absorbing material (Chen and Cooper, 2002). Changes in membrane depolarization can be measured by using fluorochromes, like 3,3′-dipropylthiocarocyanide ($diSC_3(5)$) which is taken up by bacterial cells according to their membrane potential and concentrates in the CM, where it quenches its own fluorescence. Agents that permeabilize the CM depolarize the membrane potential and cause release of the fluorochmore which results in an increase in the fluoresecence (Wu and Hancock, 1999). El-Kosasy (2006) recently introduced a potentiometric method for the assessment of Gram-negative bacteria permeability towards some antibiotics, like tobromycin. With Gram-negative bacteria the cytoplasmic β-galactosidase activity (monitored with a chromogenic substrate, O-nitrophenyl-β-D-galactopyranoside) can be used to monitor the level of the permeabilization (Silvestro et al. 2000).

Several researchers have also used in their studies constructed membrane vesicles to mimic bacterial and mammal membranes (Glukhov et al. 2005). These have been widely applied to study the efficacy of cationic antimicrobial peptides, e.g. defensins and polymyxin B (Wiese et al. 2003).

Microscopical methods, like Transmission Electron Microscopy (TEM) and Scanning Electrcon Microscopy (SEM), are useful tools in monitoring bacterial membrane structures and microbial samples. However, these methods need multistep pretreatments (e.g. dehydration) of the sample, which can alter fragile membrane structures. Recent developments in the Atomic Force Microscopy (AFM) techniques are providing new ways to examine bacteria without extensive pretreatments (recently reviewed Ubbink and Schär-Zammaretti, 2005). With AFM it is also possible to measure forces involved in bacterial interactions and attachment on surfaces (Gaboriaud et al. 2005). In AFM bacteria retain their cell shape and membranes intact (Bolshakova et al. 2004), allowing detailed inspection of various membrane structures.

AFM has been applied to study effects of antimicrobials, e.g. antibiotics on the surface of bacteria (Meincken et al. 2005). Alakomi et al. (2006) recently used AFM to monitor changes caused by permeabilizers, EDTA and PEI, on the OM of *Pseudomonas*. Visualization of *Pseudomonas* cells treated with EDTA or PEI by AFM revealed changes in the OM structure: PEI increased the surface area and bulges of the cells. Topographic images of EDTA-treated cells were compatible with events assigned for the effect of EDTA on outer membranes, i.e. release of LPS and disintegration of OM structure. In addition, the effect of EDTA treatment was visualized in phase contrast images as large areas with varying hydrophilicity on cell surfaces (Alakomi et al. 2006). Comparison of results obtained with various microscopic (e.g. TEM and AFM) and biophysical techniques give a wider view of the function and structure of the cell (Schär-Zammaretti and Ubbink, 2003, Ubbink and Schär-Zammaretti, 2005).

Factors Affecting the Activity of Antimicrobials and Permeabilizers

Efficacy of antimicrobial agents in various matrices is influenced by many factors. Among the most important effects are the physiological condition of the target microbe (hydrophobicity of the cell surface and cell membrane structures) as well as the previous adaptation of the target cells to stressful conditions. The composition of the matrix where the antimicrobial agent should be active, e.g. amount of divalent cations or lipids in food, is an additional factor as many antimicrobial compounds can react with them and thereby their activity can be significantly decreased.

The incidence of multi-drug resistant (MDR) bacteria and the emergence of new and opportunistic pathogens continuously sets new demands for more effective means for pathogen control (Levy, 2002). Reports about the reduced susceptibility of microbes to various biocides are also increasing (Walsh et al. 2003). The basic mechanisms of antibiotics are generally well documented (Russell, 2002). Biocides are likely to have multiple target sites within the cell and their mechanism of actions are often multifunctional (Russell, 2002, Walsh et al. 2003). Efflux pumps contribute in a major way to the microbial resistance to antibiotics and biocides (Gilbert and McBain 2003, Levy, 2002, Poole, 2002). Biofilm ecosystems are known to possess decreased sensitivity for antimicrobials (O'Toole 2000, Donlan and Costerton, 2002, Simoés et al. 2003, Gilbert and McBain 2003). MIC (Minimal Inhibitory Concentration) values for microbes in the stationary phase or in biofilms have been reported to be orders of magnitude higher than those of actively growing cells (Gilbert and McBain 2003).

It has been suggested that a general stress response to nutrient depletion and onset of stationary phase initiates the adaptation of resting or dormant phenotypes in bacteria. These dormant phenotypes are analogous to endospores, which are resistant to numerous physical and chemical agents, including biocides (Gilbert and McBain 2003). In addition, induction of starvation proteins and alterations in the membrane composition has been reported to increase resistance to environmental stresses (Walsh et al. 2003).

CONCLUSION

Control of harmful microbes in different environments and matrices is difficult. Although there on the other hand is a need to minimize the amount of antimicrobial agents used, e.g. by using hurdle concept, there also is a threat of increasing antimicrobial resistance while doing so. There are some reports indicating that increased use of bacteriostatic (sublethal), rather that bacteriocidal (lethal), food preservation systems may contribute to the development and dissemination of antibiotic resistance in food-related pathogens (McMahon et al. 2007). Hence, additional research on microbial adaptation, stress response as well as understanding the mechanisms of action of antimicrobial agents (and factors affecting their mode) both in laboratory and practical applications is needed. Microbial mutant strains isolated or produced under various stress conditions can be a useful tool in this research.

Mechanistic studies on the action of antimicrobial chemicals advance our understanding of their potential applications. Most biocides must traverse the outer cell layer(s) of microbes to reach their target sites (Russell, 2002). Therefore, permeabilizers can enhance the activity of biocides and other antimicrobial agents. Alternative and novel biocide formulations are needed to restrict the growth of harmful microbes in sites were traditional biocides are ineffective. The application of an effective biocide / permeabilizer combination could aid in the destruction of microbial biofilms while allowing the use of reduced concentrations of the biocide. In order to enable the enhancement of the activity of biocides knowledge of the mechanism of permeabilizers and factors influencing their activity is essential. Also; the efficacy of the formulated biocide products must be further evaluated on various matrices with complex microbial communities. Especially in food applications before antimicrobial/permeabilizers are applied, the compounds should be reviewed for the toxicological data and possible approval for the use in foods. In addition, effect of the compounds on the organoleptic quality of the foods needs to be assessed.

References

Alakomi, H-L., Skyttä, E., Saarela, M., Mattila-Sandholm, T., Latva-Kala, K. and Helander, I.M. (2000). Lactic acid permeabilizes gram-negative bacteria by disrupting the outer membrane. *Applied and Environmental Microbiology*. 66, 2001 - 2005.

Alakomi, H-L., Saarela, M. and Helander, I. (2003). Effect of EDTA on *Salmonella enterica* serovar Typhimurium involves a component not assignable to lipopolysaccharide release. *Microbiology*. 149, 2015-2021.

Alakomi, H-L., Puupponen-Pimiä, R., Aura, A-M., Helander, I.M, Nohynek, L., Oksman-Caldentey, K-M. and Saarela, M. Weakening of *Salmonella* with selected microbial metabolites of berry-derived phenolic compounds and organic acids. (In press).

Alakomi, H-L., Paananen, A., Suihko, M-L., Helander, I.M. and Saarela, M. (2006). Weakening effect of cell permeabilizers on Gram-negative bacteria causing biodeterioration. *Applied and Environmental Microbiology*. 72, 4695-4703.

Alexander, C. and Rietschel, E.T. (2001). Bacterial lipopolysaccharides and innate immunity. *Journal of Endotoxin Research*. 7, 167-202.

al-Saigh, H., Zweifel, C., Blanco, J., Blanco, J.e:, Blanco, M., Usera, M.A. and Stephan, R. (2004). Fecal shedding of *Escherichia coli* O157, *Salmonella*, and *Campylobacter* in Swiss cattle at slaughter. *Journal of Food Protection*. 4, 679-684.

Anonymous, 2006. *Infectious Diseases in Finland 2005*. Publications of the National Public Health Institute. B 17/2006. p. 60.

Anriany, Y.A., Weiner, R.W., Johnson, J.A., de Rezende, C. and Joseph, S.W. (2001). *Salmonella enterica* serovar Typhimurium DT104 displays a rugose phenotype. *Applied and Environmental Microbiology*. 9, 4048-4056.

Anriany, Y.A, Sahu, S.N., Wessels, K.R., McCann, L.M. and Joseph, S.W. (2006). Alteration of the rugose phenotype in *waa*G and *ddh*C mutants of *Salmonella enterica* serovar Typhimurium DT205 is associated with inverse production of curli and cellulose. *Applied and Environmental Microbiology*. 7, 5002-2012.

Antikainen, J., Anton, L., Sillanpää, J. and Korhonen, T. K. (2002). Domains in the S-layer protein CbsA of *Lactobacillus crispatus* involved in adherance to collagens, laminin and lipoteichoic acids and in self-assembly. *Molecular Microbiology*. 46, 381-394.

Ayres, H.M., Furr, J.R. and Russell, A.D. (1993). A rapid method of evaluating permeabilizing activity against *Pseudomonas aeruginosa. Letters in Applied Microbiology*. 17, 149-151.

Ayres, H.M., Payne, D.N., Furr, J.R. and Russell, A.D. (1998). Use of the Malthus-AT system to assess the efficacy of permeabilizing agents on the activity of antimicrobial agents against *Pseudomonas aeruginosa. Lettesr Applied Microbiology*. 6, 422-426.

Baylis, C.L. (2006). Enterobacteriacea. Chapter 22. In *Food Spoilage Microorganisms*. Ed. Blackburn, C. de W. Woodhead Publishing. CRC Press. pp. 579-623.

Bengoechea, J.A., Brandenburg, K., Arraiza, M.D., Seydel, U., Skurnik, M. and Moriyón, I. (2003). Pathogenic *Yersinia enterocolitica* strains increase the outer membrane permeability in response to environmental stimuli by modulating lipopolysaccharide fluidity and lipid A structure. *Infection and Immunity*. 71, 2014-2021.

Beveridge, T.J. (2001). Use of the Gram stain in microbiology. *Biotechnic and Histochemistry*. 76, 111-118.

Bjornsdottir, K., Breidt, F., Jr. and McFeeters, R.F. (2006) Protective effects of organic acids on survival of *Escherichia coli* O157:H7 in acidic environments. *Applied and Environmental Microbiology*. 72, 660-664.

Bolshakova, A. V., Kiselyova, O.I. and Yaminsky, I.V. (2004). Microbial surfaces investigated using Atomic Force Microscopy. *Biotechnology Progress*. 20, 1615-1622.

Bos, M. P. and Tommassen, J. (2004). Biogenesis of the Gram-negative bacterial outer membrane. *Current Opinions in Microbiology*. 7, 610-616.

Boziaris, I.S. and Adams, M.R. (1999). Effect of chelators and nisin produced in situ on inhibition and inactivation of Gram negatives. *International Journal of Food Microbiology*. 53, 105-113.

Brennan, P.J. and Nikaido, H. (1995). The envelope of mycobacteria. *Annual Review of Biochemistry*. 64, 29-63.

Brul, S. and Coote, P. (1999). Preservative agents in foods. Mode of action and microbial resistance mechanisms. *International Journal of Food Microbiology*. 50, 1-17.

Cabeen, M.T. and Jacobs-Wagner, C. (2005). Bacterial cell shape. *Nature Reviews. Microbiology*. 3, 601-610.

Carroll, P., La Ragione, R.M., Sayers, A.R. and Woodward, M.J. (2004). The O-antigen of *Salmonella enterica* serotype Enteritidis PT4: a significant factor in gastrointestinal colonisation of young but not newly hatched chicks. *Veterinary Microbiology*. 102, 73-85.

Campanac, C., Pineau, L. Payard, A., Baziard-Mouysset, G. and Roques, C. (2002). Interactions between biocide cationic agents and bacterial biofilms. *Antimicrobial Agents Chemotherapy*. 46, 1469-1474.

Chen, C.Z. and Cooper, S.L. (2002). Interactions between dendrimer biocides and bacterial membranes. Biomaterials. 23, 3359-3368.

Chiu, C-H., Su, L-H. and Chu, C. (2004). *Salmonella enterica* serotype Chloraesuis: Epidemiology, pathogenesis, clinical disease and treatment. *Clinical Microbiology Reviews*. 2, 311-322.

Cogan, T.A., Jorgensen, F., Lappin-Scott, H.M., Benson, C.E., Woodward, M.J. and Humprey, T.J. (2004). Flagella and curli fimbriae are important for the growth of *Salmonella enterica* serovars in hen eggs. *Microbiology*. 150, 1063-1071.

Daniel, R.A. and Errington, J. (2003). Control of cell morphogenesis in bacteria: two distinct ways to make a rod-shaped cell. *Cell*. 113, 767-776.

Davey, M. E. and G.A. O'Toole. (2000). Microbial biofilms: from ecology to molecular genetics. *Microbiology and Molecular Biology Reviews*. 4, 847-867.

Das, E., Gürakan, C. and Bayindirh, A. (2006). Effect of controlled atmosphere storage, modified atmosphere packaking and gaseous ozone treatment on the survival of *Salmonella* Enteritidis on cherry tomatoes. *Food Microbiology*. 23, 430 -438.

De Keersmaecker, S.C., Verhoeven, T.L., Desair, J., Marchal, K., Vanderleyden, J. and Nagy, I. (2006).. Strong antimicrobial activity of Lactobacillus rhamnosus GG against *Salmonella typhimurium* is due to accumulation of lactic acid. *FEMS Microbiological Letters*. 259, 89-96.

Donlan, R. M. and Costerton, J.W. (2002). Biofilms: Survival mechanisms of clinically relevant microorganisms. *Clinical Microbiology Reviews*. 2, 167-193.

Doores, S. (1993). Organic acids. In Davidson, P.M. and Branen, A.L. (Eds) Antimicrobials in Food (Chapter 4). Marcel Dekker Ltd., New York, 2 ed. pp. 95-136.

Drudy, D., O'Rourke, M., Murphy, M., Mullane, N.R., O'Mahony, R., Kelly, L., Fischer, M., Sanjag, S., Shannon, P., Wall, P., O'Mahony, M., Whyte, P. and Fanning, S. (2006). *International Journal of Food Microbiology*. 110, 127-134.

Duffy, E.A., Lucia, I.M., Kells, J.M., Castillo, A:, Pillai, S.D. and Acuff, G.R. (2005). Concentrations of *Escherichia coli* and genetic diversity and antibiotic resistance profiling of *Salmonella* isolated from irrigation water, packing shed equipment, and fresh produce in Texas. *Journal of Food Protection*. 68, 70-79.

Edelson-Mammel, S., Porteous, M.K. and Buchanan, R.L. (2006). Acid resistance of twelve strains of *Enterobacter sakazakii*, and the impact of habituating the cells to an acidic environment. *Journal of Food Science*. 71, M210 – M207.

EFSA, European Food Safety Authority, (2006). Trends and source of zoonoses, zoonotic agents and antimicrobial resistance in the European Union in 2004. *EFSA Journal*. 310, 10: 23.95.

El-Kosasy, Am.M. (2006). Potentiometric assessment of Gram-negative bacterial permeabilization of tobramycin. *Journal of Pharmaceutical and Biomedical Analysis*. 42, 389-394.

Ellison III, R.T., Giehl, T.J. and LaForce, F.M. 1988. Damage of the outer membrane of enteric Gram-negative bacteria by lactoferrin and transferrin. *Infection and Immunity*. 56, 2774-2781.

Erbel, P.J.A., Barr, K., Gao, N., Gerwig, G.J., Rick, P.D. and Gardner, K.H. (2003). Identification and biosynthesis of cyclic enterobacterial common antigen in *Escherichia coli*. *Journal of Bacteriology*. 185, 1995-2004.

Erridge, C., Bennett-Guerrero, E. and Poxton, I.R. (2002). Structure and function of lipopolysaccharides. *Microbes and Infection*. 4, 837-851.

Fayol-Messoudi, D., Berger, C.N., Coconnier-Polter, M.H., Lievin-Le Moal, V. and Servin, A.L. (2005). pH-, lactic acid-, and non-lactic acid-dependent activities of probiotic lactobacilli against *Salmonella enterica* serovar Typhimurium. 71, 6008-6013.

Finch, J.E. and Brown, M.R. (1975). The influence of nutrient limitation in a chemostat on the sensitivity of Pseudomonas aeruginosa to polymyxin and EDTA. *Journal of antimicrobial and Chemotherapy*. 1, 379-386.

Fleet, G. H. (1999). Microorganisms in food ecosystems. *International Journal of Food Microbiology*. 50, 101-117.

Foster, J.W. (2004). *Escherichia coli* acid resistance: Tales of an amateur acidophile. *Nature Reviews. Microbiology*. 2, 898-907.

Fredriksson-Ahomaa, M., Stolle, A. and Korkeala, H. (2006). Molecular epidemiology of *Yersinia enterocolitic*a infections. *FEMS Immunology and medical microbiology*. 47, 315-329.

Frirdich, E. and Whitfield, C. (2005). Lipopolysaccharide inner core oligosaccharide structure and outer membrane stability in human pathogens belonging to the Enterobacteriaceae. *Journal of Endotoxin Research*. 11, 133-144.

Gaboriaud, F., Bailet, S., Dague, E. and Jorand, F. (2005). Surface structure and nanomechanical properties of *Shewanella putrefaciens* bacteria at two pH values (4 and 10) determined by Atomic Force Microscopy. *Journal of Bacteriology*. 11, 3864-3868.

Ganesh Kumar, C. and Anand, S.K. (1998). Significance of microbial biofilms in food industry: a review. *International Journal of Food Microbiology*. 42, 9-27.

Gasem, M.H., Dolmans, W.M.V., Keuter, M. and Djokomoeljanto, R. (2002). Poor food hygiene and housing as risk factors for typhoid fever in Semerang, Indonesia. *Tropical Medicine and International Health*. 6, 484-490.

Gaynor, E.C., Wells, D.H., MacKichan, J.K. and Falkow, S. (2005). The *Campylobacter jejuni* stringent response controls specific stress survival and virulence-associated phenotypes. *Molecular Microbiology*. 56(1), 8-27.

Gibson, D.L., White, A.P., Snyder, S.D., Martin, S., Heiss, C., Azadi, P., Surette, M. and Kay, W.W. (2006). *Salmonella* produces an O-antigen capsule regulated by AgfD and important for environmental persistence. *Journal of Bacteriology*. 188, 7722-7730.

Gilbert, P. and McBain, A.J. (2003). Potential impact of increased use of biocides in consumer products on prevalence of antibiotic resistance. *Clinical Microbiology Reviews*. 16, 189-208.

Glukhov, E., Stark, M., Burrows, L.L. and Deber, C.M. (2005). Basis for selectivity of cationic antimicrobial peptides for bacterial versus mammalian membranes. *Journal of Biological Chemistry*. 280, 33960-33967.

Gram, L., Ravn, L., Rasch, M., Bartholin Bruhn, J., Christensen, A.B. and Givskov, M. (2002). Food spoilage- interactions between food spoilage bacteria. *International Journal of Food Microbiology*. 78, 79-97.

Gronow, S. and Brade, H. (2001). Lipopolysaccharide biosynthesis: which steps do bacteria need to survive? *Journal of Endotoxin Research*. 7, 3-23.

Guard-Bouldin, J., Gast, R.K., Humprey, T.J., Henzler, D.J., Morales, C. and Coles, K. (2004). Subpopulation characterictis of egg-contaminating *Salmonella enterica* serovar Enteritidis as defined by the lipopolysaccharide O chain. *Applied and Environmental Microbiology*. 70, 2756-2763.

Guard-Petter, J. (2001). The chicken, the egg and *Salmonella enteritidis*. *Environmental Microbiology*. 3, 421-430.

Guerrero, R., Piqueras, M. and Berlanga, M. (2002). Microbial mats and the search for minimal ecosystems. *International Microbiology*. 5, 177-188.

Gutsmann, T., Hagge, S.O., David, A., Roes, S., Böhling, A., Hammer, M.U. and Seydel, U. (2005). Lipid-mediated resistance of Gram-negative bacteria against various pore-forming antimicrobial peptides. *Journal of Endotoxin Research*. 11, 167-173.

Hancock, R.E. (1998). Resistance mechanisms in *Pseudomonas aeruginosa* and other non-fermentative gram-negative bacteria. *Clinical Infectious Diseases*. 27(Suppl.I), S93-99.

Harrington, S.M., Dudley, E.G. and Nataro, J.P. (2006). Pathogenesis of enteroaggregative *Escherichia coli* infection. *FEMS Microbiology Letters*. 254, 12-18.

Helander, I.M., Mäkelä, P.H., Westphal, O. and Rietschel, E.T. (1996) Lipopolysaccharides. In R.A. Meyers (Ed), Encyclopedia of Molecular Biology and Molecular Medicine, vol. 3, p. 462-471. VCH, Weinheim, Germany.

Helander, I.M., Alakomi, H-L., Latva-Kala, K. and Koski, P. (1997a) Polyethyleneimine is an effective permeabilizer of gram-negative bacteria. *Microbiology*. 143, 3193—99.

Helander, I.M., Kilpeläinen, I. and Vaara, M. (1997b) Phosphate groups in lipopolysaccharides of *Salmonella typhimurium rfaP* mutants. *FEBS Letters*. 409, 457—460.

Helander, I.M., von Wright, A. and Mattila-Sandholm, T. (1997c). Potential of lactic acid bacteria and novel antimicrobials against gram-negative bacteria. *Trends in Foods Science and Technology*. 8, 146-150.

Helander, I.M, Alakomi, H.-L., Latva-Kala, K., Mattila-Sandholm, T., Pol, I., Gorris, L.G.M. and von Wright, A. (1998) Characterization of the action of selected essential oil components on gram-negative bacteria. *Journal of Agricultural and Food Chemistry.* 46, 3590—3595.

Helander, IM and Mattila-Sandholm, T. (2000) Fluorometric assessment of Gram-negative bacterial permeabilization. *Journal of Applied Microbiology.* 88, 213—9.

Heumann, D. and Roger T. (2002). Initial responses to endotoxins and Gram-negative bacteria. *Clinica Chimica Acta.* 323, 59-72.

Hirshfield, I.N., Terzulli, S. and O`Byrne, C. (2003). Weak organic acids: a panoply of effects on bacteria. *Science Progress.* 86, 245-269.

Hitchcock, P.J. and Brown, T.M. (1983). Morphological heterogeneity among *Salmonella* lipopolysaccharide chemotypes in silver-stained polyacrylamide gels. *Journal of Bacteriology.* 154, 269-277.

Holland, I.B. (2004). Translocation of bacterial proteins – an overview. *Biochemica et Biophysica Acta.* 1694, 5-16.

Holt, J.G., Krieg, N.R., Sneath, P.H.A., Staley, J.T. and Williams, S.T. (1994). Bergey`s Manual of Determinative Bacteriology. Ninth edition. Williams and Wilkins. Baltimore, Maryland. p. 787.

Huis in't Veld, J.H.J. (1996). Microbiological and biochemical spoilage of foods: an overwiev. *International Journal of Food Microbiology.* 33, 1-18.

Hukari, R, Helander, I.M and Vaara, M. (1986) Chain length heterogeneity of lipopolysaccharide released from *Salmonella typhimurium* by ethylenediaminetetraacetic acid or polycations. *European Journal Biochemistry.* 154, 673—676.

Humphrey, T. (2004). *Salmonella,* stress responses and food safety. *Nature Reviews. Microbiology.* 2, 504-509.

Isonhood, J.H. and Drake, M. (2002). *Aeromonas* species in foods. *Journal of Food Protection.* 65, 575-582.

Johnston, M.D., Hanlon, G.W., Denyer, S.P. and Lambert, R.J.W. (2003). Membrane damage to bacteria caused by single and combined biocides. *Journal of Applied Microbiology.* 94, 1015-1023.

Jorgensen, F., Leach, S., Wilde, S.J., Davies, A., Steward, G.S.A.B. and Humphrey, T. (2000). Invasiveness in chickens, stress resistance and RpoS satus of wild-type *Salmonella enterica* subsp. *enterica* serovar Typhimurium definitive type 104 and serovar Enteritidis phage type 4 strains. *Microbiology.* 146, 3227-3235.

Kaper, J.B., Nataro, J.P. and Mobley, L.T. (2004). Pathogenic *Esherichia coli. Nature reviews. Microbiology.* 2, 123-140.

Kaufmann, S.H., Raupach, B. and Finlay, B.B. (2001). Introduction: microbiology and immunology: lessons learned from *Salmonella. Microbes and Infection.* 3, 1177-81.

Kawahara, K. Seydel, U., Matsuura, M., Danbara, H., Rietschel, E. Th, and Zähringer, U. (1991). Chemical structure of glycosphingolipids isolated from *Sphingomonas paucimobilis. FEBS Letters.* 292, 107-110.

Knirel, Y.A, Bystrova, O.V., Kocharova, N.A., Zähringer, U. and Pier, G.B. (2006). Conserved and variable structural features in the lipopolysaccharide of *Pseudomonas aeruginosa. Journal of Endotoxin Research.* 12, 324-336.

Lazdunski, A.M., Ventre, I. and Sturgis, J.N. (2004). Regulatory circuits and communication in Gram-negative bacteria. *Nature Reviews.* 2, 581-591.

Liebana, E., Garcia-Migura, L., Clouting, C., Clifton-Hadley, F.A., Breslin, M. and Davies, R.H. (2003). Molecular fingerprinting evidence of the contribution of wildlife vectors in the maintenance of *Salmonella* Enteritidis infection in layer farms. *Journal of Applied Microbiology*. 94, 1024-1029.

Little, C.L. and Knochel, S. (1994). Growth and survival of *Yersinia enterocolitica*, *Salmonella* and *Bacillus cereus* in brie stored at 4, 8 and 20 degrees Celsius. *International Journal of Food Microbiology*. 24, 137-145.

Leive, L. (1965). Release of lipopolysaccharide by EDTA treatment of *E. coli*. Biochem Biophys Res Commun 21, 290—6.

Le Minor, L. (1999). The genus *Salmonella*. Chapter in Prokaryotes. Release 3.0. Springer-Verlag, New-York. LCC.

Levy, S.B. (2002). Active efflux, a common mechanisms for biocide and antibiotic resistance. *Journal of Applied Microbiology Symposium Supplement*. 92, 65S-71S.

Loh, B, Grant, C and Hancock, REW. (1984) Use of the fluorescent probe 1-*N*-phenylnaphthylamine to study the interactions of aminoglycoside antibiotics with the outer memebrane of *Pseudomonas aeruginosa*. *Antimicrobial Agents and Chemotherapy*. 26, 546—51.

Loughlin, M.F., M. V. Jones, and P.A. Lambert. (2002). *Pseudomonas aeruginosa* cells adapted to benzalkonium chloride show resistance to other membrane-active agents but not to clinically relevant antibiotics. *Journal of Antimicrobial and Chemotherapy*. 49, 631-639.

Maillard, J. Y. (2002). Bacterial target sites for biocide action. *Journal of Applied Microbiology. Symposium Supplement*. 90, 16S-27S.

Marcus, S.L., Brumell, J.H., Pfeifer, C.G. and Finlay, B.B. (2000). *Salmonella* pathogenicity islands: big virulence in small packages. *Microbes and Infection*. 2, 145-156.

Martinez-Urtaza, J., Saco, M., de Novoa, J., Perez-Pineiro, P., Peiteado, J., Lozano-Leon, A. and Garcia-Martin, O. (2004). Influence of environmental factors and human activity on the presence of *Salmonella* serovars in a marine environment. *Applied and Environmental Microbiology*. 70, 2089-2097.

Matsumoto, K., Kusaka, J., Nishibori, A. and Hara, H. (2006). Lipid domains in bacterial membranes. *Molecular Microbiology*. 61, 1110-1117.

McMahon, M.A.S., Xu, J., Moore, J.E., Blair, I.S. and McDowell, D.A. (2007). Environmental stress and antibiotic resitance in food-related pathogens. *Applied and Environmental Microbiology*. 73, 211-217.

Meincken, M., Holroy, D.L. and Rautenbach, M. (2005). Atomic force microscopy study of the effect of antimicrobial peptides on the cell envelope of *Escherichia coli*. *Antimicrobial Agents and Chemotherapy*. 49, 4085-4092.

Meroueh, S.O., Bencze, K.Z., Hesek, D., Lee, M., Fisher, J.F., Stemmler, T.L., and Mobashery, S. (2006). Three-dimensional structure of the bacterial cell wall peptidoglycan. *PNAS*. 12, 4404-4409.

Merrell, D.S. and Falkow, S. (2004). Frontal and stealth attack strategies in microbial pathogenesis. *Nature*. 430, 250-256.

Monack, D.M., Mueller, A. and Falkow, S. (2004). Persistent bacterial infections: the interface of the pathogen and the host immune system. *Natures Review in Microbiology*. 2, 747-65.

Munsch-Alatossava, P. and Alatossava, T. (2006). Phenotypic characterization of raw milk-associated psychrotrophic bacteria. *Microbiological Research*. 161, 334-346.

Nikaido, H. (1996). Multidrug efflux pumps of gram-negative bacteria. *Journal of Bacteriology*. 178, 5853-5859.

Nikaido, H. (1999). Microdermatology: cell surface in the interaction of microbes with the external world. *Journal of Bacteriology*. 181, 4-8.

Nikaido, H. (2001). Preventing drug access to targets; cell surface permeability barriers and active efflux in bacteria. *Seminars in Cell and Development Biology*. 12, 215-223.

Nikaido, H. (2003). Molecular basics of bacterial outer membrane permeability revisited. Microbiol. Molecular Biology Reviews. 4, 593-656.

Nikaido, H. and Vaara, M. (1985). Molecular basis of bacterial outer membrane permeability. *Microbiological Reviews*. 49, 1-32.

Niskanen, T., Johansson, T., Kuusi, M., Raahenmaa, M. Siitonen, A. and Tuominen, P. (2006). Foodborne and waterborne outbreaks in Finland 2005. Evira Publications 2/2006. (In Finnish). p. 52.

Nohynek, L., Alakomi, H-L., Kähkönen, M.P., Heinonen, M., Helander, I.M., Oksman-Caldentey, K.M. and Puupponen-Pimiä, R.H. (2006). Berry phenolics: antimicrobial properties and mechanisms of action against severe human pathogens. *Nutrition and Cancer*. *54*, 18 32.

Nuotio, L., Schneitz, C., Halonen, U. and Nurmi, E. (1992). Use of competative exclusion to protect newly-hatched chicks against intestinal colonisation and invasion by *Salmonella enteritidis* PT4. *British Poultry Science*. 33, 775-779.

Ohmizo, C., Yata, M. and Katsu, T. (2004). Bacterial cytoplasmic membrane assay using ion-selective electrodes. *Journal of Microbiological Methods*. 59, 173-179.

Olsen, S.J., Bishop, R., Brenner, F.W., Roels, T.H., Bean, N., Tauxe, R.V. and Slutsker, L. (2002). The changing epidemiology of Salmonella: Trends in serotypes isolated from humans in the United States, 1987-1997. *The Journal of Infectious Diseases*. 183, 73-761.

Orlov, D.S., Nguyen, T. and Lehrer, R.I. (2002). Potassium release, a useful tool for studying antimicrobial peptides. *Journal of Microbiological Methods*. 49, 325-328.

Pang, T., Levine, M.M., Ivanoff, B., Waln, J. and Finlay, B.B. (1998). Typhoid fever-important issues still remain. *Trends in Microbiology*. 4, 131-133.

Paulus, W. (1993). Microbicides for the protection of materials – A handbook. Chapman and Hall. London. UK. 496 pp.

Piddock, L.J.V. (2006). Multidrug-resistance efflux pumps – not just for resistance. *Nature Reviews. Microbiology*. 4, 629-636.

Plym Forshell, L. and Wierup, M. (2006). Salmonella contamination: a significant challenge to the global marketing of animal food products. *Scientific and Technical Review, OIE. World Organization for animal health*. 25, 541-554.

Preston, A., Mandrell, R.E., Gibson, B.W. and Apicella, M.A. (1996). The lipooligosaccharides of pathogenic gram-negative bacteria. *Critical Reviews Microbiology*. 22, 139-180.

Prigent-Combaret, C., Prensier, G., Le Thi, T.T., Vidal, O., Lejeune, P. and Dorel, C. (2000). Development of pathway for biofilm formation in curli-producing *Escherichia coli* strains: role of flagella, curli and colonic acid. *Environmental Microbiology*. 2, 450-464.

Poole, K. (2002). Outer membranes and efflux: The path to multidrug resistance in Gram-negative bacteria. *Current Pharmaceutical Biotechnology*. 3, 77-98.

Popoff, M.Y., Bockemühl, J. and Gheesling, L.L. (2004). Supplement 2002 (no. 46) to the Kauffman-White scheme. *Research in Microbiology*. 155, 568-570.

Raetz, CRH. and Whitfield, C. (2002). Lipopolysaccharide endotoxins. *Annual Review Biochemistry*. 71, 635-700.

Raivio, T.L. (2005). Envelope tress responses and Gram-negative bacterial pathogenesis. *Molecular Microbiology*. 56(5), 1119-1128.

Rajamäki, T., Alakomi, H-L., Ritvanen, T., Skyttä, E., Smolander, M. and Ahvenainen, R. (2006). Application of electronic nose for quality assessment of modified atmosphere packaged poultry meat. *Food Control*. 17, 5 - 13.

Raupach, B. and Kauffman, S.H.E. (2001). Immune responses to intracellular bacteria. Current Opinion in Immunology. 1, 417-428.

Rebell, R., Ernst, R.K., Gowen, B.B., Miller, S.I. and Hinnebush, B.J. (2004). Variation in lipid A structure in the pathogenic yersinia. *Molecular Microbiology*. 52(5), 1363-1373.

Reij, M.W., Den Aantrekker, E.D., ILSI Europe Risk Analysis in Microbiology Task Force. (2004). Recontamination as a source of pathogens in processed foods. *International Journal of Food Microbiology*. 91, 1-11.

Reynolds, C.M., Ribeiro, A.A., McGrath, S.C., Cotter, R.:J., Raetz, C.R.H: and Trent, M.S. (2006). An outer membrane enzyme encoded by Salmonella typhimurium *lpxR* that removes the 3′-acyloxyacyl moiety of lipid A. *The Journal of Biological Chemistry*. 31, 21974-21987.

Rezende de, C.E., Anriany, Y., Carr, L.E., Jospeh, S.W. and Weiner, R.M. (2005). Capsular polysaccharide surrounds smooth and rugose types of *Salmonella enterica* serovar Typhimurium DT104. *Applied and Environmental Microbiology*. 71, 7345-7351.

Robertson, E.B. and Firestone, M.K. (1992). Relationship between desiccation and exopolysaccharide production in a soil *Pseudomonas* sp. *Applied and Environmental Microbiology*. 58, 1284-1291.

Rokka, M., Eerola, S., Smolander, M., Alakomi, H-L. and Ahvenainen, R. (2004). Monitoring of the quality of modified atmosphere packaged broiler chicken cuts stored in different temperature conditions. B. Biogenic amines as quality-indicating metabolites. *Food Control*. 15, 601-607.

Rowley, G., Spector, M., Kormanec, J. and Roberts, M. (2006). Pushing the envelope: extracytoplasmic stress responses in bacterial pathogenesis. *Nature Reviews. Microbiology*. 4, 383-394.

Ruiz, N., Kahne, D. and Silhavy, T.J. (2006). Advances in understanding bacterial outer-membrane biogenesis. *Nature Reviews*. 4, 57-65.

Russell, A. D. (2002). Antibiotic and biocide resistance in bacteria: Introduction. *Journal of Applied Microbiology*. 9 Suppl, 1S-3S.

Sauli, I., Danuser, J., Geeraerd, A.H., Van Impe, J.F., Rüfenacht, J., Bissig-Choisat, B., Wenk, C. and Stärk, K.D.C. (2005). Estimating the probability and level of contamination with *Sajmonella* of feed for finishing pigs produced in Switzerland- the impact of the production pathway. *International Journal of Food Microbiology*. 100, 289-310.

Schär-Zammaretti, P. and Ubbink, J. (2003). Imaging of lactic acid bacteria with AFM – elesticity and adhesion maps and their relationship to biological and structural data. *Ultramicroscopy*. 97, 199-208.

Scheffers, D-J. and Pinho, M.G. (2005). Bacterial cell wall synthesis: New insights from localization studies. *Microbiology and Molecular Biology Reviews*. 4, 585-607.

Schleifer, K.H., Leuteritz, M., Weiss, N., Ludwig, W., Kirchhof, G. and Seidel-Rufer, H. (1990). Taxonomic study of anaerobic, gram-negative, rod-shaped bacteria from breweries: emended description of *Pectinatus cerevisiiphilus* and description of *Pectinatus frisingensis* sp. nov., *Selenomonas lacticifex* sp. nov., *Zymophilus raffinosivorans* gen. nov., sp. nov., and *Zymophilus paucivorans* sp. nov. *International Journal of Systematic Bacteriology*. 40,19-27.

Schneitz, C., Nuotio, L., Mead, G. & Nurmi, E. (1992). Competitive exclusion in the young bird: challenge models, administration and reciprocal protection. *International Journal of Food Microbiology*. 15, 241-244.

Schneitz, C. & Renney, D.J. (2003). Effect of a commerical competitive exclusion product on the colonization of *Salmonella infantis* in day-old pheasant chicks. *Avian Diseases*. 47, 1448-1451.

Schooling, S.R. and Beveridge, T.J. (2006). Membrane vesicles: an overlooked component of the matrices of biofilms. *Journal of Bacteriology*. 188, 5945-5957.

Schutze, G.E:, Sikes, J.D., Stefanova, R. and Cave, M.D. (1999). The home environment and salmonellosis in children. *Pediatrics*. 103, 1-5.

Schäffer, C. and Messner, P. (2005). The structure of secondary cell wall polymers: how Gram-positive bacteria stick their cell walls together. *Microbiology*. 151, 643-651.

Scott, J.R. and Zahner, D. (2006). Pili with strong attachments: Gram-positive bacteria do it differently. *Molecular Microbiology*. 62, 320-330.

Shelef, L.A. and Seiter, J.A. (1993). Indirect Antimicrobials. In Davidson, P. M. and Branen, A.L. (ed.). Antimicrobials in Food (Chapter 15), Marcel Dekker Ltd., New York. 2 ed. pp. 539-569.

Silla Santos, M.H. (1996). Biogenic amines: their importance in foods. *International Journal of Food Microbiology*. 29, 213-231.

Silvestro, L., Weiser, J.N. and Axelsen, P.H. 2000. Antibacterial and antimembrane activities of cecropin A in *Escherichia coli*. *Antimicrobial Agents and Chemotherapy*. 44, 602-607.

Skurnik, M. and Bengoechea, J.A. (2003). The biosynthesis and biological role of lipopolysaccharide O-antigens of pathogenic *Yersiniae*. *Carbohydrate Research*. 338, 2521-2529.

Smolander, M., Alakomi, H-L., Ritvanen, T., Vainionpää, J. and Ahvenainen, R. (2004). Monitoring of the quality of modified atmosphere packaged broiler chicken cuts stored in different temperature conditions A. Time-temperature indicators as quality-indicating tools. *Food Control*. 15, 217 - 229.

Snyder, D. S., Gibson, D., Heiss, C., Kay, W. and Azadi, P. (2006). Structure of a capsular polysaccharide from *Salmonella enteritidis*. *Carbohydrate Research*. 341, 2388-2397.

Solano, C. B., Garicia, J., Valle, C., Berasain, J.M., Ghigo, J.M., Gamazo, C. and Lasa, I. (2002). Genetic analysis of *Salmonella enterica* biofilm formation: critical role of cellulose. *Molecular Microbiology*. 43, 793-808.

Sørum, H. and L'Abée-Lund, T.M. (2002). Antibiotic resistance in food-related bacteria – a result of interfering with the global web of bacterial genetics. *International Journal of Food Microbiology*. 78, 43-56.

Stenberg, F., Chovanec, P., Maslen, S.L., Robinson, C.V., Ilag, L.L., von Heijne, G. and Daley, D.O. (2005). Protein complexes of the *Escherichia coli* envelopes. *Journal of Biological Chemistry*. 41, 34409-34419.

Sutherland, I.W. (1999). Microbial polysaccharide products. *Biotechnology and Genetic Engineering Reviews*. 16, 217-229.

Su, L-H., Chiu, C-H., Chu, C. and Ou, J.T. (2004). Antimicrobial resistance in nontyphoid *Salmonella* serotypes: A global challenge. *Clinical Infectious Diseases*. 39, 546-551.

Tauxe, R.V. (2002). Emerging foodborne pathogens. *International Journal of Food Microbiology*. 78, 31-41.

Terebiznik, M.R., Vazquez, C.L., Torbicki, K., Banks, D., Wang, T., Hong, W., Blanke, S.R., Colombo, M.I. and Jones, N.L. (2006). *Helicobacter pylori* VacA toxin promotes bacterial intracellular survival in gastric epithelial cells. *Infection and Immunity*. 74, 6599-6614.

Tindall, B.J., Grimont, PA.D., Garrity, G.M. and Euzéby, J.P. (2005). Nomenclature and taxonomy of the genus *Salmonella*. *International Journal of Systematic and Evolutionary Microbiology*. 55, 521-524.

Trent, M.S., Stead, C.M., Tran, A.X. and Hankins, J.V. (2006) Diversity of endotoxin and its impact on pathogenesis. *Journal of Endotoxin Research*. 12, 205-223.

Träuble, H and Overath, P. (1973) The structure of *Escherichia coli* membranes studied by fluorescence measurements of lipid phase transition. *Biochimica Biophysica Acta* 307, 491-512.

Ubbink, J. and Schär-Zammaretti, P. (2005). Probing bacterial interactions: integrated approaches combining atomic force microscopy and biophysical techniques. *Micron*. 36, 293-320.

Vaara, M. (1992). Agents that increase the permeability of the outer membrane. *Microbiological Reviews*. 56, 395-411.

Vaara, M. (1999). Lipopolysaccharide and the permeability of the bacterial outer membrane. *In* H. Brade, S.M. Opal, S.N. Vogel, D.C. Morrison (ed.), Endotoxin in Health and Disease, Marcel Dekker, Inc., New York and Basel, p. 31—38.

Vaara, M. and Vaara, T. (1983). Polycationic sensitize enteric bacteria to antibiotics. *Antimicrobial Agents and Chemotherapy*. 24, 107-113.

van Asselt, E.D. and Zwietering, M.H. (2006). A systematic approach to determine global thermal inactivation parameters for various food pathogens. *International Journal of Food Microbiology*. 107, 73-82.

Venter, P., Abraham, M.., Lues, J.F.R. and Ivanov, I. (2006). The influenze of sanitizers on the lipopolysaccharide composition of *Escherichia coli*. *International Journal of Food Microbiology*. 111, 221-227.

Voetsch, A.C., Van Gilder, T.J., Angulo, F.J., Farley, M.M., Shallow, S., Marcus, R., Cieslak, P.R., Deneen, V.C. and Tauxe, R.V. (2004). FoodNet estimate of the burden of illness caused by nontyphoidal *Salmonella* infections in the United States. *Clinical Infectious Diseases*. 38 Suppl 3, S127-S134.

Vollaard, A.M., Ali, S., van Asten, H.A., Widjaja, S., Visser, L.G., Surjadi, C. and van Dissel, J.T. (2004). Risk factors for typhoid and paratyphoid fever in Jakarta, Indonesia. JAMA, 291, 2607-2615.

Vollmer, W. and Höltje, J-V. (2001). Morphogenesis of *Escherichia coli. Current Opinion in Microbiology*. 4: 625-633.

Wagenaar, J.A., Mevius, D.J. and Havelaar, A.H. (2006). *Campylobacter* in primary animal production and control strategies to reduce the burden of human campylobacteriosis. *Scientific and Technical Review, OIE. World Organization for animal health*. 25, 581-594.

Walsh, A. G., Matewish, M.J., Burrows, L.L., Monteiro, M.A., Perry, M.B. and Lam, J.S. (2000). Lipopolysaccharide core phosphates are required for viability and intrinsic drug resistance in *Pseudomonas aeruginosa. Molecular Microbiology*. 35, 712-727.

Walsh, S.E., Maillard, J-Y., Russell, A.D., Catrenich, C.E., Charbonneau, D.L.and Bartolo, R.G. (2003). Development of bacterial resistance to several biocides and effects on antibiotic susceptibility. *Journal of Hospital Infection*. 55, 98-107.

White, A.P., Collins, S.K., banser, P.A., Gibson, D.L., Paetzel, M., Strynadka, N.C.J. and Kay, W.W. (2001). Structure and characterization of AgfB from *Salmonella enteritidis* thin aggregative fimbria. *Journal of Molecular Biology*. 311, 735-749.

White, A.P., Gibson, D.L., Collinson, S.K., Banser, P.A. and Kay, W.W. (2003). Extracellular polysaccharide associated with thin aggregative fimbria of *Salmonella enterica* serovar Enteritidis. *Journal of Bacteriology*. 18, 5398-5407.

White, P.P., Gibson, D.L., Kim, W., Kay, W.W. and Surette, M.G. (2006). Thin aggregative fimbria and cellulose enhance long term survival and persistence of Salmonella. *Journal of Bacteriology*. 9, 3219-3227.

Wiese, A., Gutsmann, T. and Seydel, U. (2003). Towards antibacterial strategies: studies on the mechanisms of interactions between antimicrobial peptides and model membranes. *Journal of Endotoxin Research*. 9, 67-84.

Wu, M. and Hancock, R.E.W. (1999). Interaction of the cyclic antimicrobial cationic peptide bactenecin with the outer and cytoplasmic membranes. *The Journal of Biological Chemistry*. 274, 29-35.

Wu, T., McCandlish, A.C., Gronenberg, L.S:, Chng, S-S. Silhavy, T.J. and Kahne, D. (2006). Identification of a protein complex that assembles lipopolysaccharide in the outer membrane of *Escherichia coli. PNAS*. 31, 11754-11759.

Yasudu, K., Ohmizo, C. and Katsu, T.(2003). Potassium and tetraphenylphosphonium ion-selective electrodes of monitoring changes in the permeability of bacterial outer and cytoplasmic membranes. *Journal of Microbiological Methods*. 54, 111-115.

Yethon, J.A., and Whitfield, C. (2001). Lipopolysaccharide as a target for the development of novel therapeutics in gram-negative bacteria. *Current Drug Targets- Infectious Disorders*. 1, 91-106.

Young, K.D. (2006). The selective value of bacterial shape. *Microbiology and Molecular Biology Reviews*. 70, 660-703.

Zhao, T., Zhao, P., West, J.W., Bernard, J.K., Cross, H.G. and Doyle, M.P. (2006). Inactivation of enterohemorrhagic Escherichia coli in rumen content- or feces-contaminated drinking water for cattle. *Applied and Environmental Microbiology*. 72, 3268-3273.

Zogaj, X., Nimtz, M., Rohde, M., Bokranz, W. and Römling, U. (2001). The multicellular morphotypes of *Salmonella typhimurium* and *Escherichia coli* produce cellulose as the second component of the extracellular matrix. *Molecular Microbiology*. 39, 1452-1463.

In: Food Microbiology Research Trends
Editor: Marta V. Palino, pp. 217-244

ISBN 978-1-60021-723-4

Chapter 7

Real-Time Detection of Foodborne Pathogen Contamination in Food Using GC/MS and Electronic Sensors Integrated with Pattern Recognition Techniques

Ubonrat Siripatrawan*

Department of Food Technology, Faculty of Science,
Chulalongkorn University, Bangkok, Thailand

Abstract

The Conventional techniques to determine the presence of foodborne pathogens in food products are laborious and time-consuming. Recently, research has been focused on the development of real-time and accurate techniques to identify pathogens in food products. The real-time detection of foodborne pathogen contamination in food product using electronic sensors coupled with pattern recognition techniques has been developed. The metabolic volatile compounds used as indicators for the target microorganism were collected using a solid phase microextraction (SPME) device and identified using gas chromatograph/ mass spectrometry (GC/MS). The electronic nose with an array of metal oxide sensors was used to monitor changes in the composition of the headspace gas phase relating to the biochemical products of the target microorganism. Nevertheless, the data obtained from SPME/GC/MS or electronic sensor array responses alone were not sufficient to distinguish samples with and without contamination. Pattern recognition techniques were used to improve classification of the data output from the electronic sensor array. The potential to predict the number of the target microorganism in the samples was also investigated. In connect with pattern recognition analyses, the electronic sensors and SPME/GC/MS proved to have the ability to detect the target microorganism in food samples.

Keywords: detection; foodborne pathogen; electronic sensors; pattern recognition.

---

* Corresponding author: Email: ubonratana.s@chula.ac.th

INTRODUCTION

Foodborne pathogens in food cause approximately 6.5-33 million cases of human illnesses and up to 5000 deaths [The Center for Disease Control and Prevention (CDC), 1999]. The Economic Research Service (ERS) [2000] estimated the cost from bacterial foodborne illnesses as $ 6.9 billion in the United States each year. Fruits and vegetables can become contaminated with pathogenic microorganisms during growing, harvesting, post-harvest, handling, processing and distribution [The Council for Agriculture and Science Technology, 1999; Beuchat, 2002]. Bacteria normally reside in the intestinal tracts of animals, including humans, and may contaminate raw fruits and vegetables through contact with feces, sewage, untreated irrigation water or surface water. Many microbial contaminants are part of the environment and vegetables may be inadvertently contaminated. Washing fruits and vegetables with contaminated water and handling of produce by infected workers, vendors and consumers help spread pathogenic microorganisms [Nguyen-the and Carlin, 1994]. Increased global food trade in raw fruits and vegetables, as well as increased international travel, could also increase the risk of foodborne illnesses linked to microbial contamination of imported fresh produce [Calvin, 2003].

Ready-to-eat fresh vegetables such as alfalfa and other seed sprouts are under increased demand by consumers due to their popular perception as a healthy and convenient food [Bari et al., 1999; Backer et al., 2000; Breuer et al., 2001; CDC, 2006]. However, raw sprouts are common vehicles for fresh produces associated bacterial foodborne illness. In recent years, packaged raw sprouts have been identified as a vehicle associated with outbreaks of *Escherichia coli* 0157:H7 [Bari et al., 1999; Breuer et al., 2001], *Listeria monocytogenes* [Palmai and Buchanan, 2002] and *Salmonella* [U.S. Food and Drug Administration, 1999; Backer et al., 2000; Gill et al., 2003; Kim et al., 2003; Van Beneden et al., 2001; CDC, 2006] infections as well as other communicable diseases in recent years.

Conventional techniques, such as the standard plate count, to determine the presence of these pathogens in food products are laborious and time-consuming. Consequently, research has been focused on the development of rapid and accurate techniques to identify pathogens in food products [Alugupalli, Larsson and Slosarek, 1992]. Microorganisms can be characterized by identification of specific metabolites generated by specific biochemical pathways [Canhoto et al., 2004; Marilley et al., 2004]. The selection of volatiles for use as incipient disease indicators has been reviewed in terms of the composite rate of pathogenic destruction within food products [Schweizer-Berberich, Vaihinger and Gopel, 1994; Borjesson, 1996; Jonsson, 1997].

Identification of microorganisms based on their characteristics has been determined using Gas Chromatography (GC) [Decallonne et al., 1991; Wauthoz, Lioui and Decallonne, 1995; Mosca et al., 2006] and Gas Chromatography/Mass Spectrometry (GC/MS) [Decallonne et al., 1991; Basile et al., 1998; Haugen et al., 2006]. Gas chromatographic methods have been qualitatively and quantitatively employed to determine the specific catabolic products of various specific compounds such as fatty acids, glucose, amino acids, or other metabolic compounds produced by microorganisms. Gas chromatographic analysis of microorganisms or their products is useful for taxonomic differentiation of the organisms and for diagnostic aims. Gharaibeh and Voorhees [1996] classified 10 bacterial species by extracting and derivatizing fatty acids from whole bacterial cells, followed by GC/MS analysis. GC analysis of volatile metabolic compounds produced by microorganisms in growth media is also useful for the identification of microbial species [Alugupalli et al., 1992; Forsyth and Dusseault, 1997; Siripatrawan, Linz and Harte, 2004 b]. Analyses of volatile organic compounds with GC techniques usually begin with concentration of the analytes of interest though liquid-liquid extraction, purge-and-trap, headspace, or various other techniques. These procedures typically require a long time, complicated equipment and/or large amounts of organic solvents. The development of alternative analytical techniques that are rapid and simple has become increasingly important to reduce per-sample-time investment and to conduct real time analyses.

Solid Phase Microextraction

Solid phase microextraction (SPME) is a rapid sampling technique that is well adapted to GC analysis. SPME has been applied to the analysis of volatile and nonvolatile compounds in gaseous and liquid samples. SPME can eliminate the need for solvents or a complicated apparatus for concentrating volatile compounds in liquid samples or in gas samples [Bao et al., 1999; Song, Fan and Beaudry, 1998; Zhang and Pawliszyn, 1993]. SPME can extract analytes from a variety of matrices by partitioning from a liquid or gaseous sample into an immobilized stationary phase. SPME is compatible with analyte separation/detection by gas chromatography and provides linear results for a wide concentration of analytes. Headspace SPME is based on the equilibrium of analytes among three phases of the system including the polymeric liquid coating, the headspace and the aqueous solution [Song, Fan and Beaudry, 1998]. Theoretically, SPME extraction is considered to be completed when the analyte concentration has reached distribution equilibrium between the sample matrix and the fiber coating. When equilibrium is reached, the extracted amount is constant within the limits of experimental error and is independent on additional extraction time [Gorecki and Pawliszyn, 1997; Gorecki, Kahled and Pawliszyn, 1998; Zhang and Pawliszyn, 1993].

In SPME, equilibrium is established among the concentrations of an analyte in the sample, in the headspace above the sample, and in the polymer coating on the silica fiber. The amount of analyte adsorbed by the fiber is dependent on the thickness of the polymer coating and on the distribution constant for the analyte. Selectivity can be altered by changing the type of polymer coating on the fiber or the coating thickness to match the characteristics of the analytes of interest. Full equilibration is not necessary for high accuracy and precision from SPME, but consistent sampling time and other sampling parameters are essential. Desorption of an analyte from the SPME fiber depends on the boiling point of the analyte, the thickness of the coating on the fiber and the desorption temperature [Gorecki and Pawliszyn, 1997; Gorecki, Kahled and Pawliszyn, 1998; Bao et al., 1999]. The advantage of using SPME are its simplicity, absence of solvent and rapid response. SPME minimizes sample preparation and concentrates volatile analytes in a solvent-free manner. Headspace SPME has been shown to be more sensitive than conventional headspace analysis of similar samples performed with an airtight syringe, and favorable to the commonly used purge and trap type analysis [Gorecki and Pawliszyn, 1997].

The SPME/GC technique has been utilized in the areas of food/flavor chemistry as pesticide determination in wine [Nogueira and Nascimento, 1999], examination of volatile compounds produced by tomato and strawberry fruit [Song, et al., 1998], analysis of alcohols and esters in beer [Jelen et al., 1998], monitoring of volatiles of apple (Song et al., 1997), determination of methylcyclopentadienly manganese tricarbonyl, a gasoline antiknock additive in beverages (Forsyth and Dusseault, 1997), study of off-flavors in milk [Marsili, 1999], analysis of rancidity in dry-cured ham (Ruiz et al., 1998), detection of trace levels of taste and odor compounds in water (Bao et al., 1999) and characterization of volatile metabolites emitted by microorganisms [Borjesson et al., 1993; Vergnais et al., 1998; Siripatrawan, Linz and Harte, 2004b].

Recently, research has focused on development of rapid and accurate techniques to identify pathogens in food products [Alugupalli, Larsson and Slosarek, 1992; Council for Agricultural Science and Technology, 1999]. Microorganisms can be characterized by identification of specific metabolites generated by specific biochemical pathways [Canhoto et al., 2004; Marilley et al., 2004]. The selection of volatiles for use as incipient disease indicators has been reviewed in terms of the composite rate of pathogenic destruction within food products [Schweizer-Berberich, Vaihinger and Goper, 1994; Borjesson, 1996; Jonsson, 1997]. This concept has been actualized in electronic sensors or "electronic nose".

Electronic Nose

The electronic nose is an analytical instrument having an array of sensors, used for the specific analysis, identification, and recognition of complex odors and volatile organic compounds [Maul et al., 1998; Stetter et al., 2000; Brudzewski, Osowski and markiewicz, 2004]. There are three main types of sensors including metal oxide, conducting polymer and quartz crystal micro-balance sensors, which can be used to detect specific gases and vapors. A basic requirement is that the sensors in an electronic sensor array must show partial sensitivity, i.e., that they can respond broadly to a range or class of gases rather than to a specific one. [Lucas, Poling and Bennincasa, 1998].

Metal oxides are semiconducting materials which are gas-sensitive. These sensors comprised a thin layer (50 μm) of an oxide film deposited on a ceramic tube or plate. The selectivity of the sensors is related to different catalytic amounts of a doping metal introduced as a trace impurity on the sensor surface. Reaction of an odorant with a sensor changes its conductivity. There are two main types of semiconductors: n-type semiconductors (mainly composed of zinc, tin or iron oxides) that respond primarily to reducing compounds, whereas p-type semiconductors (mainly nickel or cobalt oxide) respond to oxidizing compounds. Other metal oxides (e.g. zinc oxide and tungsten oxide) are also available for this instrument [Lee et al., 2001; Mielle, 1996]. Metal oxide sensors are relatively resistant to humidity and to aging. Drawbacks include a slow baseline recovery for some high molecular weight compounds, lower specificity, potential of poisoning of the sensor, partial and irreversible destruction of the material owing to irreversible binding caused by the presence of sulfur compounds or weak acids within the sample [Gardner et al., 1992; Persaud and Pelosi, 1992; Hodgins and Simmonds, 1996; Kaipainen et al., 1997; Getino et al., 1998].

Electronic sensor technology has been shown to provide rapid, continuous monitoring of a wide array of different volatile chemicals [Stetter et al., 2000; Mart et al., 2005]. A considerable number of electronic sensor applications have been reported, including characterization of different types of vegetable oils [Gan et al., 2003], classification of changes in milk resulting from a variety of heat treatments [Brudzewski, Osowski and markiewicz, 2004], evaluation of the off-odor in wine [Ragazzo-Sanchez, Chalier and Ghommidh, 2005], detection of *E. coli* in nutrient media [Siripatrawan, Linz and Harte, 2004b], quality assessment of packaged broiler chicken [Rajamaki et al., 2005], detection of lard adulterant in refined, bleached, deodorized palm olein [Che Man et al., 2005] and quality measurement of smoked salmon [Haugen et al., 2006].

Electronic sensor technology is usually based on a hybrid sensor array system with different selectivity and sensitivity, with the result being a powerful analytical instrument especially for complex food analyses. However, analysis of volatile compounds using electronic sensor array or GC/MS often generates large and complicated data set and difficult to interpret if used directly. Data interpretation of multiple variables from several sensor responses usually requires the use of statistical interpretation for rapid verification of results. Useful information can usually be extracted using multivariate data analysis techniques such as principle component analysis [Gan et al. 2005; Thomas and Bro, 2005], discriminant factor analysis [Siripatrawan, Linz and Harte, 2004a] and artificial neural network [Siripatrawan, Linz and Harte, 2004b].

Pattern Recognition Techniques

Complicated data can be subjected to a variety of sophisticated pattern recognition techniques. A mathematical resolution of complex data is usually performed in far less time than it takes to conduct physical or chemical experiments [Domingo et al., 2000; Burden, Brereton and Walsh, 1997; Devabhaktuni, et al., 2001; Terra and Tinos, 2001; Siripatrawan, Linz and Harte, 2004a; Siripatrawan, Linz and Harte, 2006]. The direction of pattern recognition research is toward more sophisticated methodology, which is a direct result of tackling more difficult data analysis problems. A variety of powerful pattern recognition techniques such as Principal component analysis (PCA), Linear discriminant analysis (LDA), Multiple linear regression (MLR), Partial least squares regression (PLSR), Principal components regression (PCR) and Artificial neural network (ANN) have been successfully applied to generate qualitative and quantitative solutions for problems in pattern recognition and multivariate calibration [Wentzell, Andrews and Kowalski, 1997; Mawatari et al., 1991; Luco, 1999; Devabhaktuni et al., 2001; Watanabe et al., 2001].

LDA is a useful pattern recognition technique for multivariate data analysis [Rizzi et al., 2005]. Pattern recognition interpretation techniques provide complementary information which was simply impossible by conventional data analysis [Kokot et al., 1998]. LDA is related to both multivariate analysis of variance and multiple regression. LDA can be used not only to test multivariate differences among groups, but also to explore which variables are most useful for discriminating among groups if one subset of variables performs equally well as another, and which groups are most alike and most different [Li et al., 1999; Tomasko, Helms and Snapinn, 1999; Faber, 2000].

ANN is a mathematical model whose structure and function are inspired by the organization and function of human brain [Haykin, 1999]. A neural network is a massively parallel distributed processor made up of simple processing units, which has a propensity for storing experiential knowledge and making it available for use. The primary element of a neural network is the neuron. These neurons are arranged in input and output layers of one or more hidden processing layers. Neurons can be thought of as weighted transfer functions. The most common neural network approach to regression-type problems is multilayer perceptrons (MLP) [Devabhaktuni et al., 2001; Li et al., 2000]. For MLP, the network includes an input, one or more hidden layers, and an output layer. The inputs to a neuron include its bias and the sum of its weighted input. The output of a neuron depends on the neuron's inputs and on its transfer function. The output is computed starting from the input and then continuing through the hidden layers [Burden, Brereton and Walsh, 1997; Haykin, 1999; Coulibaly, Bobee and Anctil, 2001]. MLP is usually trained with a back propagation algorithm to solve difficult problems. Basically, back propagation learning comprises forward computation to calculate the output and backward computation to adjust the weight. The calculated output of the network is subtracted from the true output and the error is derived. The error is then propagated backward to correct the weight in order to make the output move closer to the true output [Bila et al., 1999; Devabhatuni et al., 2001; Haykin, 1999; Terra and Tinos, 2001].

This paper highlights some recent methods developed to identify and quantify a target microorganism in foods using GC/MS and electronic sensor analysis coupled with pattern recognition techniques. *E. coli* was used as the target microorganism because its presence in foods indicates fecal contamination, and the presence of pathogenic microorganisms. The electronic sensors were used to monitor the volatiles produced by *E. coli*. Microorganisms can be characterized by identification of specific metabolites associated with their activities. However, many metabolites may be common to several microorganism species. Therefore, the differences between samples often relate to a complex balance between pattern of volatiles rather than to a major change in one or two constituents. This underlines the important of using multivariate data analysis to extract the specific information necessary to target a specific microorganism. The ability of the electronic sensors and SPME/GC/MS to differentiate samples with and without *E. coli* was evaluated using LDA. ANN was then used as an experimental platform (in addition to the instrumental methods) to predict the number of *E. coli* by relating the electronic sensor array responses with the number of *E. coli* present in the samples. This work was carried out to help establish the feasibility of using the electronic sensors or SPME/GC/MS coupled with pattern recognition techniques including LDA and ANN to provide rapid and useful information related to food safety.

Methods

Preparation of Contaminated Vegetable

Alfalfa sprouts were chosen as the product component because the National Advisory Committee on Microbial Criteria for Foods (NACMCF) [1999] identified sprouts as a special problem due to the potential for pathogen growth during production, while there is increasing demand for sprouts due to their popularity as a healthy food [U.S. Food and Drug Administration, 1998; U.S. Food and Drug Administration, 1999; CDC, 2006].

E. coli ATCC 25922 was obtained from the American Type Culture Collection (ATCC, Rockville, MD). Preliminary experiments were conducted to determine the population of *E. coli* necessary in the dipping suspension to result in an initial population of ~10^5 CFU/g on sprouts. Previous studies also showed that the electronic sensors were able to detect volatiles produced by *E. coli* when the number of *E. coli* was higher than 10^5 CFU/g [Siripatrawan, Linz and Harte, 2004 a; Siripatrawan, Linz and Harte, 2006].

The sprouts were washed and drained several times before use. The sprouts were then placed in screened baskets, and submerged in the suspension containing *E. coli* for 3 min. The uninoculated control was similarly treated except sterile phosphate buffer was used in place of the inoculum. 50 g of sprouts were then packed into 500 ml glass jars and sealed. The samples were stored at 10 °C for 1, 2, and 3 days.

Microbiological Analysis

The microbial cell count was determined on the date of inoculation and periodically throughout storage at days 1, 2, and 3. Serial dilutions were prepared from the stock suspension, and Petri plates were inoculated with those dilutions expected to give countable colonies. Inocula consisting of each of a dilution serials were deposited on prepared plates in duplicate using 3M Petrifilm Aerobic Count Plates (3 M, St. Paul, MN) for determining aerobic bacteria and 3M Petrifilm *E. coli*/Coliform Count Plates (3 M, St. Paul, MN)containing Violet Red Bile nutrient agar as an indicator of glucuronidase activity for *E. coli*. All plates were incubated at 37 °C for 48 h. Plate counts are recorded as colony forming units (CFU/g).

SPME/GC/MS

The SPME fiber selected for collecting the volatile compounds was a 75 μm Carboxen/PDMS. Extraction temperature of 35 °C was used corresponding to the optimum temperature for analysis of *E. coli* volatile metabolites from similar samples using electronic sensor array as described in the recent paper [Siripatrawan, Linz and Harte, 2004a; Siripatrawan and Harte, 2007]. The extraction time at 35 °C was determined in the preliminary test by comparing the peak intensities of analytes with extraction times. Time at which the area under curve became constant was 10 min and was considered to be the optimum time. Once sampling was finished, the device was withdrawn and transferred to the injection port of the gas chromatograph (HP-6890, Hewlett-Packard Co., Wilmington, DEL). Absorbed volatiles were desorbed from the fiber coating by inserting the SPME fiber through a predrilled septum (Thermogreen LB-2, Supelco Co., Bellefonte, PA) and into a 0.75 mm ID inlet lined, splitless injection port of the GC. The GC column inlet was immersed in a liquid nitrogen bath during the desorption period, so that the volatiles were collected before entering the column for analysis. After 3 min of desorption, the fiber was retracted and removed as was the liquid nitrogen. The GC was manually started.

Volatiles were separated using a capillary column (SPB5, 30m x 0.1 mm id., 0.25 μm coating thickness, Supelco Co., Bellefonte, PA). The carrier gas used was ultrapurified Helium (99.99% purity) at a flow rate of 0.5 ml min^{-1}. The temperature program was isothermal for 2 min at 40 °C and raised to 240 °C at a rate of 50 °C min^{-1}. Time-of-flight (TOF) mass spectrometer (FCD-650, LECO Corp., St. Joseph, MI) was used for volatile detection. Mass spectra were collected at a rate of 40 spectra/s over a range of 30-400 m/z.

The ionization energy was 70 eV. Identification of volatile components was determined by comparison of collected mass spectra with those of authenticated standards and spectra in the NIST 98 mass spectral library (Search Version 1.6, The National Institute for Standards and Technology, Gaithersburg, MD).

Analysis of GC/MS Data

Data were made up of 120 samples with 8 subgroups (control (CTR), and sprouts inoculated with *E. coli* (EC) at different storage times: 0, 1, 2 and 3 days). LDA was carried out with SPSS Systat Software Version 10.0 (SPSS Inc, Chicago, IL). Classification was performed by assigning a pattern vector to the class with the closest Mahalanobis distance metric [Li et al., 1999; Tomasko, Helms and Snapinn, 1999], as given by

$$D_c^2 = (\mathbf{x}-\boldsymbol{\mu}_c)\, S\, (\mathbf{x}-\boldsymbol{\mu}_c)^T \tag{1}$$

where D_c^2 is a squared Mahalanobis distance between two data points, **x** is the pattern ***c*** vector being classified, $\boldsymbol{\mu}_c$ is the vector of means for subgroup ***c*** *(c = 1, ..., 8)*, T denotes the transformed vector, and *S* is the pooled variance-covariance matrix. Samples were attributed to the groups whose average Mahalanobis distance was similar to the average value of the data points of a certain group.

Electronic Sensor Analysis

A total of 240 samples, including non-inoculated alfalfa sprouts (control) and alfalfa sprouts inoculated with *E. coli* at time zero, were incubated for 1, 2, and 3 days prior to analysis. An electronic nose (Fox 3000, Alpha M.O.S., Hillsborough, NJ) was used for monitoring changes in volatiles produced by *E. coli* growing on the sprouts. The volatile analysis system, shown in Figure 1, combines a measurement chamber for generating the volatile compounds and a detection system made up of 12 metal oxide sensors (SYLG, SYG, SYAA, SYGH, SYGCTI, SYGCT, T301, P101, P102, P401, T702, and PA2). The descriptors associated with the sensors are shown in Table 1. This instrument was linked to an autosampler capable of analyzing a total of 64 samples. The optimum condition for operating electronic sensors is shown in Table 2. Samples were placed in the HS100 auto-sampler in arbitrary order. 5 ml was collected from the headspace of the samples and injected into the electronic sensor. The temperature of the injection syringe was 40°C. The delay time between two injections was 300 s. Each injection was repeated, with separate samples. The electronic signals from the sensors were digitized and then transferred to the control computer.

Table 1. Sensor types and volatile descriptors

	Sensors		Description
P- Type	**SY- Type**	**T- Type**	
P101 P102	SYGCT		non-polar volatiles: methane, propane
	SYAA		Organic solvents: Polar compounds, Ethanol
PA2			Organic solvents: Alcohol, solvents
		T301	Organic solvents: Polar compounds Ethanol
		T702	Alcohol and aromatic compounds
	SYG SYGH		Amines and amine containing compounds and ammonia derivatives
	SYGCTI		Ammonia and sulfur
P401,	SYLG		Fluoride and chloride: Fluorinated and chlorinated compounds, aldehydes

Table 2. Optimum condition for operating electronic sensors

HS 100	
Headspace generation time (s)	300
Headspace generation temperature (°C)	35
Syringe temperature (°C)	40
Syringe type (mL)	5
Vial type (mL)	20
SENSORS	
Acquisition time (s)	180
Acquisition period (s)	1
Delay (s)	180
Flow (mL/min)	250
Injection volume (μL)	5000
Sample volume (mL)	8

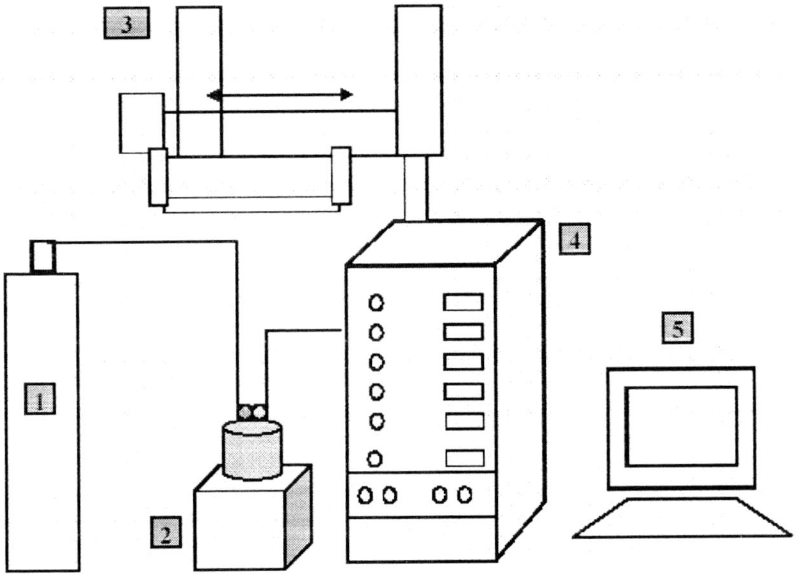

Figure 1. Electronic sensor array system.
(1) Synthetic air and a humidity regulator
(2) Chamber to control the temperature of the headspace
(3) An autosampler
(4) Channels of 12 electronic sensors
(5) Microcomputer for storing and processing sensor data

Analysis of Electronic Sensor Data

Data were made up of 240 samples from 8 subgroups (alfalfa sprouts (Control, CTR) and sprouts inoculated with *E. coli* (EC) on the first day of inoculation and incubated at 10 °C for 1, 2, and 3 days). Each subgroup had 30 replicate samples collected from several cultivations. Each sample was analyzed using 12 metal oxide sensors. The sensor responses of all 240 samples were arranged in a 240 x 12 matrix. Data classification was performed using LDA by assigning a pattern vector to the class with the closest Mahalanobis distance metric.

For ANN analysis, the data sets were partitioned into 2 subsets: a training set (70 % of the data), used to adjust the network weights, and a validation set (the remaining 30 % of the data), used to validate the model. The performance of the selected network was then assessed on a third independent set of data (unseen observations) called a test set. First, data patterns of known identity were presented to the ANN as exemplars. Once the ANN was trained, any data pattern could be presented to it and the output analyzed to find the most likely identity of the pattern. The network was trained by comparing predicted *E. coli* numbers with their known identify. When an optimized network is obtained, its ability to generalize must be tested by presenting it with an unseen independent set of data and comparing predicted data with known identity. All calculations were carried out using MATLAB 5.2 routines written by the authors, and making use of the toolbox provided by Mathworks (Mathworks, Inc., Natick, MA).

RESULTS AND DISCUSSION

Microbial Cell Counts

The cell counts of aerobic bacteria and *E. coli* on sprouts were determined. All samples had a high number of total aerobic bacteria. However, *E. coli* was not found in the control samples. After 3 days incubation, the numbers of *E. coli* increased from 3.5×10^5 to 2.1×10^7 CFU/g.

The alfalfa seeds were soaked in 20,000 ppm of calcium hypochlorite prior to germination as advised by U.S. Food and Drug Administration [1999] and NACMCF [1999]. This treatment has the potential to substantially reduce microbial contamination which can be passed on to the growing sprouts. Gill et al. [2003] suggested that chemical disinfection can reduce the human risk for disease posed by contaminated seed sprouts. However, the total number of aerobic bacteria was as high as 10^7 CFU/g. 2 % calcium hypochlorite showed the greatest reduction, but did not completely eliminate the natural microflora.

The number of aerobic bacteria and *E. coli* on the alfalfa seeds was determined. The number of aerobic bacteria was $\sim 10^1 - 10^2$ CFU/g, while no *E. coli* were found. The number of aerobic bacteria increased from $\sim 10^1 - 10^2$ CFU/g to $\sim 10^8$ CFU/g when the alfalfa sprouts were fully grown. This result is in good agreement with seed contamination experimental model by Jaquette, Beuchat and Mahon [1996] in which *Salmonella stanley* added to alfalfa seeds increased from approximately 2×10^3 bacteria/g of mature sprouts to 10^7 bacteria/g after 48–72 h incubation, without affecting the smell, taste or appearance of the sprouts.

The conditions during sprouting (e.g. time, temperature, water activity, pH, and nutrient level) may have promoted the growth of microflora [Jaquette, Beuchat and Mahon, 1996; Gill et al., 2003]. Thus, the risk of foodborne disease associated with sprouts increases during sprouting [Nguyen-the and Carlin, 1994; NACMCF, 1999].

SPME/GC/MS

E. coli volatile metabolites from the headspace of packaged alfalfa sprouts containing *E. coli* and stored at 10°C for 0, 1, 2 and 3 days were identified using SPME coupled to GC/MS. In headspace SPME, equilibrium was established among the volatiles in a sample, in the headspace above the sample, and in the polymer coated fiber. Library searches of the mass spectrum were used to identify individual compounds from the chromatograms, based on their mass spectra. The mass spectrum of each compound from the headspace of samples inoculated with *E. coli* was determined from the NIST 98 library reference spectra. From library matches, detected compounds in chromatograms included dimethyl sulfide, 4-penten-2-ol, dimethyl disulfide, 3-octanol, hepten-3-ol, isopropyl alcohol, O-acetimidoyl ethanol, 2-heptanol, propyl cyclopropane and indole. The alcohols detected in the samples may be from amino acid decarboxylation and deamination. Dimethyl sulfide and carbon disulfide may be present due to degradation of sulfur containing amino acids such as methionine and cysteine [Moat and Foster, 1998; Siripatrawan, Linz and Harte, 2004a].

Data were made up of 120 samples with 8 subgroups (control (CTR), and sprouts inoculated with *E. coli* (EC) and incubated at different storage times: 0, 1, 2 and 3 days). Each sample contained selected 10 volatile compound variables (dimethyl sulfide, 4penten-2-ol, dimethyl disulfide, 3-octanol, hepten-3-ol, isopropyl alcohol, O-acetimidoyl ethanol, 2-heptanol, propyl cyclopropane and indole). The data were arranged into a data matrix (120 x 8) for further data analysis. For GC/MS data analysis, each chromatogram (*i)* consisted of (*j*) selected identified peaks, having areas (x_{ij}) and assessed using TIC (total ion current). In this study, no internal standard was added, since the relative concentration of the specific compounds could be determined using selective normalization as follows:

$$^{norm}x_{ij} = \frac{x_{ij}}{x_i} \qquad (2)$$

where $^{norm}x_{ij}$ is the normalized selective peak area.

$$x_i = \sum_{j=1}^{m} x_{ij} \qquad (3)$$

where x_i is the total area of selected peaks.

Relative concentrations of the selected volatile compounds from all samples were randomly plotted on a three dimensional graph (Figure 2).

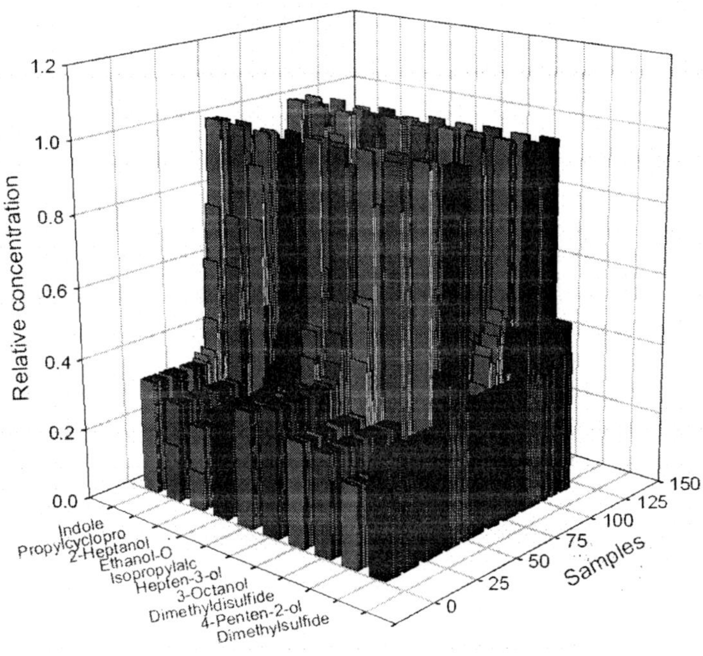

Figure 2. Relative concentrations of selected volatile compounds from the headspace of control and contaminated samples analyzed using SPME/GC/MS.

SPME/GC/MS Integrated with Pattern Recognition

Data obtained from GC/MS contained selected major 10 compounds. LDA is a method that differentiates between the within- and between-class scatters to derive class-specific feature spaces. For discriminant factors, the variables are chosen according to the characteristics that differing between the groups. These variables are then linearly combined and weighted so that the groups are forced to be as statistically distinct as possible by choosing the linear combination of variates that maximizes the one-way analysis of variance F-test, which tests the equality of the means for the linear combinations. LDA was used to visualize class separation between samples.

Using LDA for data classification, seven discriminant factors were extracted, which accounted for 100 % of the total variance in the data set. The results are given in Table 3. Eigenvalues are indicative of the relative importance of the discriminant factor in determining group separation. Although some information was available in later factors, the first two factors had the highest eigenvalue and account for as high as 94.30 % of the cumulative proportion of the total dispersion, thus, the data could be displayed using the first two factors.

Table 3. Eigenvalue and cumulative dispersion of discriminant factor of GC/MS data

Discriminant factor	Eigenvalue	Cumulative dispersion (%)
1	228.13	85.80
2	22.48	94.30
3	11.34	98.50
4	2.58	99.50
5	0.97	99.90
6	0.32	100.00
7	0.01	100.00

Table 4 lists the coefficients of each canonical discriminant factor. The first canonical variable (Factor 1) was the linear combination of the original variables (Equation 4) that best discriminate among the groups, the second canonical variable (Factor 2) was orthogonal to the first and was the next best combination of variables (Equation 5).

Factor 1 = 3.651 dimethyl sulfide + 0.474 4-penten-2-ol +
1.414 dimethyl disulfide + 1.544 3-octanol + 3.381 hepten-3-ol –
0.314 isopropyl alcohol – 2.549 O-acetimidoyl ethanol + (4)
0.103 2-heptanol + 0.161 propyl cyclopropane – 2.220 indole

Factor 2 = 0.861 dimethyl sulfide -0.202 4-penten-2-ol – (5)
0.781 dimethyl disulfide – 0.623 3-octanol + 0.782 hepten-3-ol -
0.071 isopropyl alcohol + 0.851 O-acetimidoyl ethanol –
0.172 2-heptanol – 0.082 propyl cyclopropane + 0.644 indole

Table 4. Coefficients of discriminant factor of GC/MS data

Volatile compound variables	Canonical discriminant factor						
	1	2	3	4	5	6	7
Dimethyl disulfide	3.651	0.861	2.192	0.131	0.553	0.637	2.018
4-penten-2-ol	0.474	-0.202	0.244	-0.274	0.203	0.155	1.476
Dimethyl disulfide	1.414	-0.781	1.174	0.580	0.161	0.273	1.000
3-octanol	1.544	-0.623	1.721	-0.449	0.447	0.313	0.886
Hepten-3-ol	3.381	0.782	2.143	0.134	0.579	-0.407	1.825
Isopropyl alcohol	-0.314	-0.071	0.311	-0.059	-0.913	-0.010	-0.283
O-acetimidoyl ethanol	-2.549	0.851	0.244	0.234	0.345	0.037	0.153
2-heptanol	0.103	-0.172	-0.700	-0.220	0.410	-0.003	-0.221
Propyl cyclopropane	0.161	-0.082	0.765	-0.057	0.066	0.148	0.771
Indole	-2.220	0.644	0.445	0.071	0.707	0.077	0.261

The first canonical variable was then plotted versus the second canonical variable for visual inspection of the data. Sample factor scores were plotted on a LDA graph, and the separation of class-labeled samples is shown in Figure 3. Classification was performed by assigning a pattern vector to the class with the closest Mahalanobis distance metric [Fisher and Belle, 1993; Fung, 1995]. Samples were attributed to the subgroups whose average Mahalanobis distance was similar to the average value of the data points of a certain subgroup.

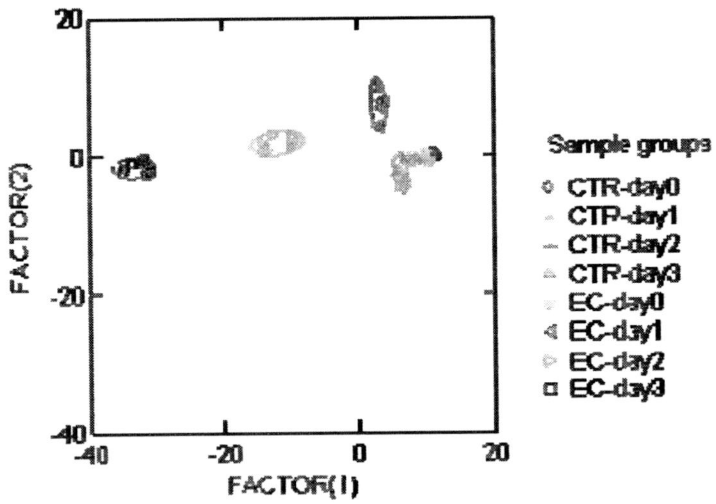

Figure 3. LDA of volatile compounds from the headspace of control and contaminated samples analyzed using SPME/GC/MS. The samples are labeled using the following scheme: CTR-day0, CRT-day1, CTR-day2 and CRT-day 3 are samples kept for 0, 1, 2, and 3 days, respectively; EC-day0, EC-day1, EC-day2 and EC-day3 are samples inoculated with *E. coli* on the first day of inoculation and after incubated for 1, 2, and 3 days, respectively.

LDA provided a graphic illustration in a two-dimensional plot which is more easily interpreted. In this study there were respectively eight subgroups (CTR-day0, CTR-day1, CTR-day2, CRT-day 3, EC-day0, EC-day1, EC-day2, and EC-day3). However, data were classified into 4 subgroups from the LDA pattern which accounted for the different numbers of *E. coli* in samples due to the varied storage periods.

The control samples were indistinguishable from samples on the date of inoculation, EC-day0, as all points were grouped under the same area in the LDA plot probably due to the low concentrations of volatile compounds produced by *E. coli*. Contaminated samples incubated for 1, 2 and 3 days occupied different areas, indicating that the SPME/GC/MS could differentiate between samples with different numbers of *E. coli* because the higher the number of *E. coli*, the higher the concentrations of volatile metabolites.

Electronic Sensors

The metal oxide sensors contain chemisorbed oxygen species which interact with odor molecules, thus, altering the conductivity of the oxide. These devices operate at elevated temperatures to avoid interference from water and to aid in rapid response and recovery times. The detection principle is based on measurement of the variation of the resistance of the metal oxide, thus, generating a measurable electronic signal. Metal oxide sensors have sensitivities in the ppm to ppb range (for special sensors) to a very broad range of chemical compounds. Typically, the electronic sensor array sensing system consists of three software modules; data acquisition and instrument control, data manipulation for extraction and preparation of signals and data processing.

Figure 4 shows a comparison between sensor responses of control sample and sprouts inoculated with *E. coli* and stored at 10 °C for 2 days. The figure also shows the differences between the pattern generated by a control sample (CTR) and the pattern generated by the various volatile metabolites produced by *E. coli* during the storage period. Each sensor changes in resistance when exposed to volatile compounds. In this study, twelve metal oxide sensors were used. The control samples had low responses from the sensor array. Different sensor signal intensities between samples are important to discriminate between samples.

Each sensor element changes in resistance (Γ_{max}) when exposed to volatile compounds. The information from electronic sensor analysis was extracted from the series of sensor resistances. In order to produce consistent data for the classification, the sensor response was presented with a volatile chemical relative to the base resistance in air, which is the maximum change in the sensor's electrical resistance divided by the initial resistance, as follows

$$\textit{relative resistance change} = \frac{\Delta\Gamma_{max}}{\Gamma_0} = \frac{\Gamma_{max} - \Gamma_0}{\Gamma_0} \qquad (6)$$

where $\Delta\Gamma_{max}$ = maximum change in the sensor's electrical resistance and Γ_0 = initial baseline resistance of the sensor [Persaud and Pelosi, 1992]. The baseline of the sensors was acquired in a synthetic air saturated steam water at fixed temperature. The relative resistance change was used for data evaluation because it gives the most stable result, and is more robust against sensor baseline variation.

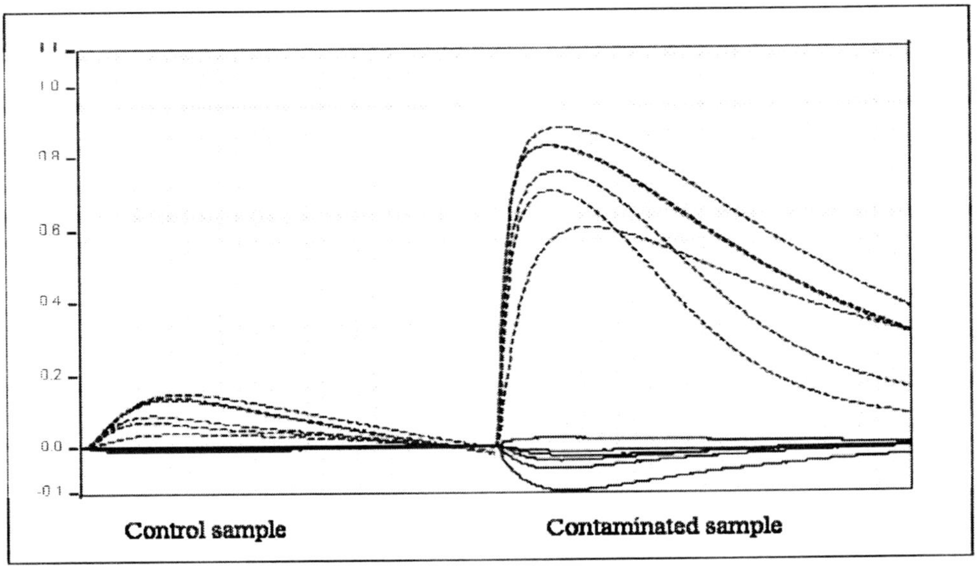

Figure 4. Sensor responses from the headspace of control and contaminated samples using 12 metal oxide sensors.

Electronic Sensors Integrated with Pattern Recognition

The data matrix comprised 240 samples from eight subgroups (CTR-day0, CTRday1, CTR-day2, CRT-day 3, EC-day0, EC-day1, EC-day2, and EC-day3) as analyzed using the 12 sensors (SYLG, SYG, SYAA, SYGH, SYGCTI, SYGCT, T301, P101, P102, P401, T702 and PA2). Figure 5 shows the responses of all samples in eight subgroups to the 12 metal oxide sensors. The sensor array has the potential to be a sensitive, fast, one-step method to characterize volatile compounds in a sample. However, the electronic sensors generated multi-dimensional data that was difficult to handle and visualize. Therefore, LDA was used as pattern recognition technique to discriminate between different sample groups based on voltametric responses. LDA was used to determine whether it is possible to separate two or more individual groups, given measurements for these individuals from several variables. LDA is a method that differentiates between the within- and between-class scatters to derive class-specific feature spaces. For discriminant factors, the variables are chosen according to the characteristics that differing between the groups. These variables are then linearly combined and weighted so that the groups are forced to be as statistically distinct as possible by choosing the linear combination of variates that maximizes the one-way analysis of variance F-test, which tests the equality of the means for the linear combinations [Tomasko, Helms and Snapinn, 1999; Lee et al., 2001].

Seven factors were extracted, which accounted for 100 % of the total variance in the data set. The results are given in Table 5. Although some information was available in later factors, the first two factors had the hightest eigenvalue and account for 89.70 % of the cumulative proportion of the total dispersion, thus, the data could be displayed using the first two factors.

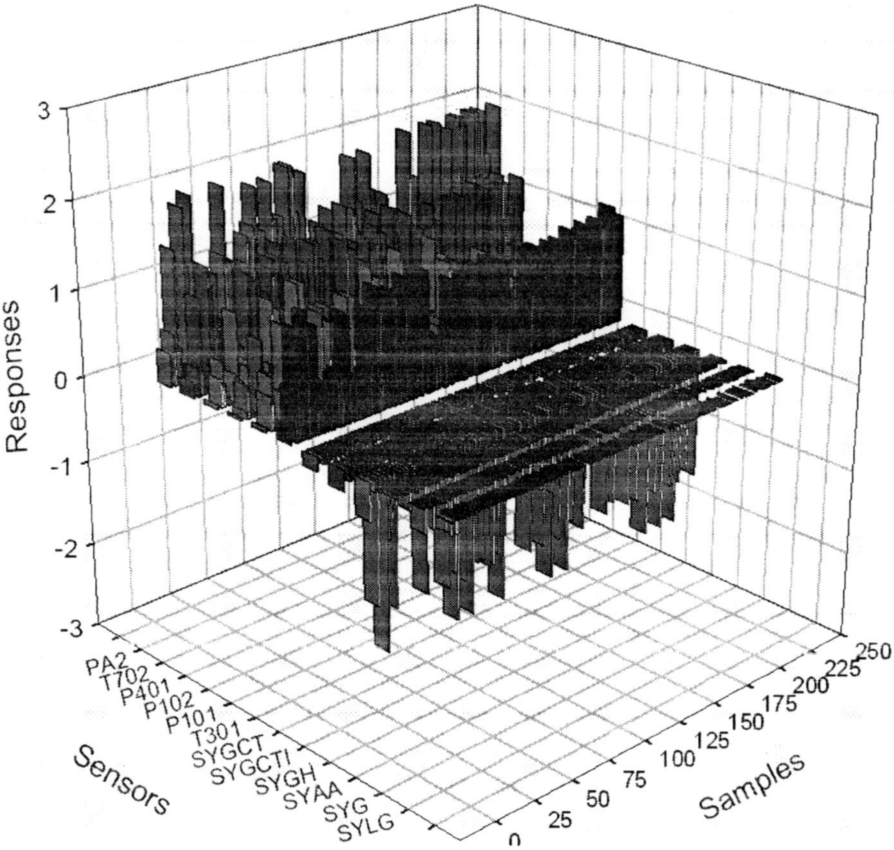

Figure 5. Sensor responses from the headspace of all control and contaminated samples using 12 metal oxide sensors.

Table 6 lists the coefficients of each canonical discriminant factor. The first canonical variable (Factor 1) was the linear combination of the original variables (Equation 7) that best discriminate among the groups; the second canonical variable (Factor 2) was orthogonal to the first and was the next best combination of variables (Equation 8).

$$\text{Factor 1} = 0.084\ \text{SYLG} + 0.540\ \text{SYG} + 0.200\ \text{SYAA} + 1.717\ \text{SYGH} + 0.306\ \text{SYGCTI} - 0.817\ \text{SYGCT} - 0.663\ \text{T301} - 14.933\ \text{P101} - 5.369\ \text{P102} + 20.314\ \text{P401} - 0.198\ \text{T702} - 0.198\ \text{PA2} \quad (7)$$

$$\text{Factor 2} = 2.732\ \text{SYLG} - 0.314\ \text{SYG} + 0.629\ \text{SYAA} + 2.157\ \text{SYGH} + 0.752\ \text{SYGCTI} - 0.547\ \text{SYGCT} + 0.640\ \text{T301} + 5.325\ \text{P101} - 2.239\ \text{P102} - 4.803\ \text{P401} + 0.378\ \text{T702} + 0.280\ \text{PA2} \quad (8)$$

The first canonical variable was then plotted versus the second canonical variable for visual inspection of the data. Sample factor scores were plotted on a LDA graph, and the separation of class-labeled samples is shown in Figure 6. Classification was performed by assigning a pattern vector to the class with the closest Mahalanobis distance. The samples (CTR-day0, CTR-day1, CTR-day2, CRT-day 3, EC-day0, ECday1, EC-day2, and EC-day3) were attributed to subgroups whose average Mahalanobis distance was similar to the average value of the data points of a certain subgroup.

Table 5. Eigenvalue and cumulative dispersion of discriminant factor of electronic sensor data

Discriminant factor	Eigenvalue	Cumulative dispersion (%)
1	370.79	65.70
2	135.33	89.70
3	30.97	95.20
4	22.36	99.20
5	4.18	99.90
6	0.55	100.00
7	0.01	100.00

Table 6. Coefficients of discriminant factor of electronic sensor data

Sensors	Canonical discriminant factor						
	1	2	3	4	5	6	7
SYLG	0.084	2.732	0.801	0.380	0.579	0.301	0.134
SYG	0.540	-0.314	-0.802	-0.216	-0.064	-1.279	0.116
SYAA	0.200	0.629	0.830	-0.265	0.666	0.564	-0.550
SYGH	1.717	2.157	-0.785	-0.495	-0.627	-0.337	0.490
SYGCTI	0.306	0.752	-0.896	-0.613	0.020	-0.182	-0.308
SYGCT	-0.817	-0.547	1.027	-0.078	0.324	0.168	0.601
T301	-0.663	0.640	0.096	0.231	-0.268	-0.296	-0.081
P101	-14.933	5.325	-8.712	4.773	13.325	1.483	4.853
P102	-5.369	-2.239	0.670	0.692	2.906	0.296	1.600
P401	20.314	-4.803	7.399	-4.737	-15.794	-2.014	-6.894
T702	-0.198	0.280	-1.423	-1.504	-0.579	-0.358	0.417

The data were classified into 5 subgroups from the LDA pattern which accounted for the different numbers of *E. coli* in samples due to the varied incubation times. The CTRday0, CRT-day1 and CRT-day2 were indistinguishable from samples on the date of *E. coli* inoculation, EC-day 0, as all points were grouped under the same area in the LDA plot due to the low concentrations of volatile compounds produced by *E. coli*. Contaminated samples incubated for 1, 2 and 3 days occupied different areas, indicating that the electronic sensors could differentiate between samples with different numbers of *E. coli* because the higher the number of *E. coli*, the higher the concentrations of volatile metabolites.

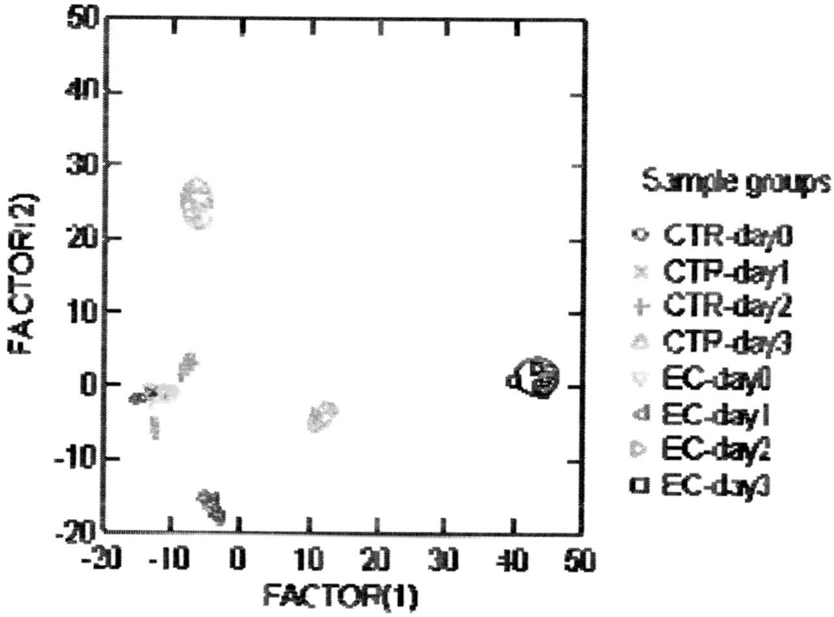

Figure 6. LDA of volatile compounds from the headspace of control and contaminated samples analyzed using 12 metal oxide sensors. The samples are labeled using the following scheme: CTR-day0, CRT-day1, CTR-day2 and CRT-day 3 are samples kept for 0, 1, 2, and 3 days, respectively; EC-day0, EC-day1, EC-day2 and EC-day3 are samples inoculated with *E. coli* on the first day of inoculation and after incubated for 1, 2, and 3 days, respectively.

Applying LDA, a good separation between all sample groups was observed. LDA could qualitatively classify the samples but did not provide quantitative information of the number of *E. coli* present in the samples. Therefore, ANN was used for the prediction of *E. coli* number.

Prediction of *E. coli* Number

A multilayer perceptron (MLP) neural network based on back propagation was used to predict number of *E. coli* from the volatile metabolites analyzed using the data from electronic sensor analysis. For the neural network, three steps were completed, including creating the network object, training the network and validating the network. Using this approach, data matrices were used as a learning set, and randomly selected data sets not used for calibration were instead used to validate the calibration technique. The performance of the selected network was then assessed by measuring its performance on a test set.

The network architecture created for the *E. coli* data matrix includes an input layer, one hidden layer of neurons, and an output layer (Figure 7).

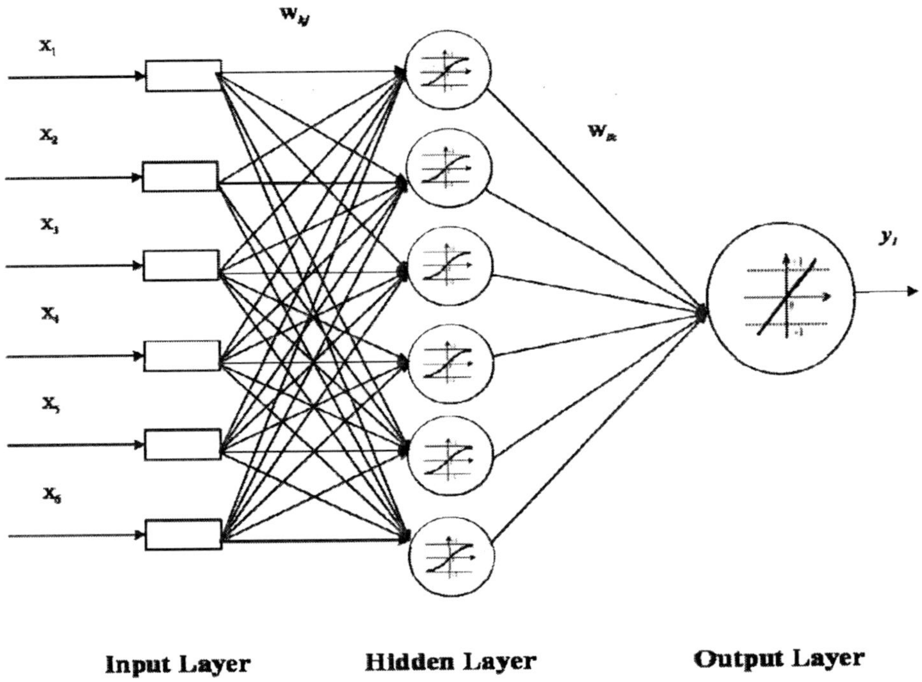

Figure 7. Architecture of ANN used for prediction the number of the target microorganism.

The indices j, k, and l refer to the input signals ($j = 1,..., m$) in the input layer, the neurons ($k = 1,..., p$) in the hidden layer, and the neuron ($l = 1,..., q$) in the output layer, respectively. There were 6 neurons ($m = 6$) in the input layer, 6 neurons ($p = 6$) in the hidden layer, and one neuron ($q = 1$) in the output layer. The transfer function, $\phi(v_k)$, in the hidden layer was a hyperbolic tangent (Equation 9) and a linear function was used in the output layer (Equation 11). The nonlinear hyperbolic tangent function can be calculated as follows.

$$y_k = \varphi(v_k) = \frac{1-\exp(v_k)}{1+\exp(v_k)} \tag{9}$$

with v_k being computed as m

$$v_k = b_k + \sum_{j=1}^{m} w_{kj} x_j \tag{10}$$

where y_k is the output of the hidden layer, $\phi(v_k)$ is the transfer function associated with the neuron k in the hidden layer, v_k is the sum of weighted input of neuron k, b_k is the bias, and x_j is the input signal. Use of bias b_k has the effect of applying an affine transformation in the model. A linear function can be calculated as follows.

$$y_l = \varphi(v_l) = b_l + \sum_{k=1}^{p} w_{lk} y_k \tag{11}$$

where y_l is the output of the output layer, $\phi(v_l)$ is the transfer function associated with neuron l in the output layer, y_k is the input to the neuron l, v_l is the sum of weighted input of neuron l, b_l is the bias, and w_{lk} is the weight connection of neuron k and neuron l.

The ANN was trained using selected parameters from data sets from several cultivations and was subsequently validated using independent data sets for estimating the concentration variables. The training started with different initial random weights, and was optimized during training. The performance function used during training of the feed-forward neural network was the mean sum of squares of the network errors (MSE).

$$MSE = \frac{1}{n}\sum_{i=1}^{n}(t_i - a_i)^2 \tag{12}$$

where a = network output, t = targets, and n = number of samples. The difference between target value and actual neural output was propagated back through the network to the input. The learning process described herein is referred to as error-correction learning. For error-correction learning, the error was minimized by adjusting the weight. Minimization of the error leads to a learning rule generally referred to as a gradient descent learning rule. In the learning process, the updated value of weight is computed by

$$w_{kj}(n+1) = w_{kj}(n) + \Delta w_{kj}(n) \tag{13}$$

$$\Delta w_{kj}(n) = \eta\, e_k(n) x_j(n) \tag{14}$$

where $\Delta w_{kj}(n)$ is the adjustment of weight applied to the synaptic weight at time step n, ek is the mean square error, xj is the element of the input vector, and η is the learning rate parameter. The learning rate parameter ($\eta = 0.05$) was selected to ensure that the convergence of the learning process was achieved. In order to ensure stability in the network and avoid convergence on a local minimum, the initial weights were randomly assigned to the network. One complete entire training process is called an epoch. The learning process continued epoch-by-epoch until the synaptic weights and bias level of the network stabilized and the average square error over the entire training set converged to the minimum value (MSE = 0.12). In this study, the network was trained for 50 epochs to obtain the acceptable output errors.

The performance of a trained network was assessed by measuring its performance on a third independent set of data (unseen observations) called a test set. Regression analysis between the network response and the corresponding targets was performed. Figure 8 shows the predictions versus true values of numbers of *E. coli* in alfalfa sprouts using electronic sensor array data and neural network algorithm. The correlation coefficient, R^2, between the outputs and targets was a measure of how well the variation in the output was explained by the targets and outputs. A regression coefficient ($R^2 = 0.98$) indicates a good fit between actual and predicted data. This study has shown that electronic sensor array analysis results and application of neural network can provide accurate information on the number of *E. coli*. These have potential in the real-time detection of *E. coli*.

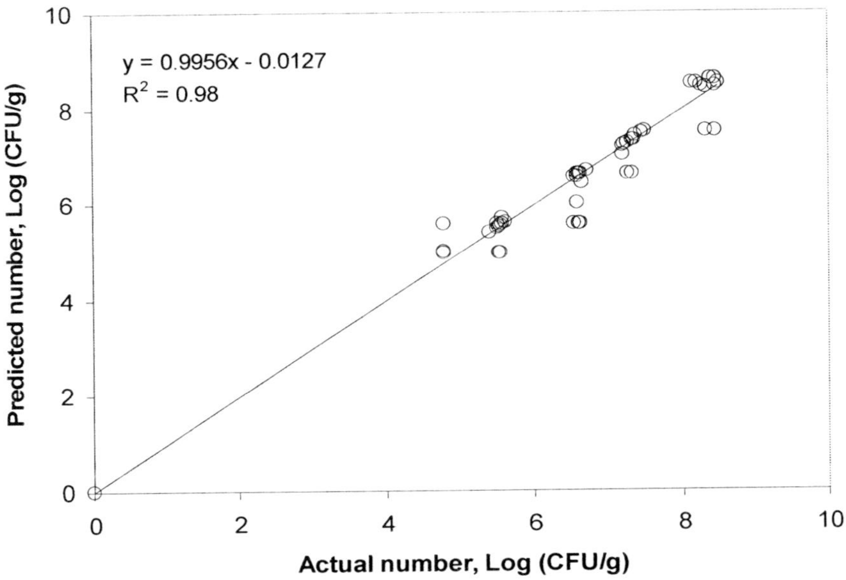

Figure 8. The predicted and actual numbers of the target microorganism using ANN algorithm.

Conclusion

The volatile metabolites from the headspace of samples inoculated with *E. coli* were analyzed using electronic sensor array and SPME/GC/MS. The differences between metabolic volatile profiles from the headspace of sprouts with and without *E. coli* inoculation can be used as possible indicators of *E. coli* contamination. SPME is a relatively simple technique that can be used to collect samples from the sample headspace. SPME minimizes sample preparation and eliminates the need for concentrating volatile compounds. GC/MS was combined with pattern recognition analysis to identify *E. coli* contamination in the sprouts. Using LDA for data classification provided complementary information that was unachievable by conventional data interpretation. LDA could visualize class separation between different sample subgroups.

The electronic sensor was used to monitor changes in the composition of the gas phase of biochemical products from *E. coli* volatile metabolites directly from the headspace above the samples (without culturing in standard media). Applying LDA, a good separation between all sample groups was observed. LDA could qualitatively classify the samples but did not provide quantitative information of the number of *E. coli* present in the samples. Therefore, ANN was used for the prediction of *E. coli* number. A MLP neural network approach was used as a measurement technique to relate the number of *E. coli* to the concentrations of volatile metabolic compounds. The MLP neural network with a back propagation algorithm has the potential of using volatile fingerprints to predict number of *E. coli* in unknown samples. Compared to colony counting methods, mathematical resolution can occur in less time. As a meaningful alternative method, it was determined that quantitation of *E. coli* can be directly accomplished from analysis of volatile compounds that demonstrate sameness or difference between samples.

In the current research, the electronic sensors have shown potential to detect specific *E. coli* volatile metabolites, even though the sprouts contained high aerobic counts. The primary advantage of the electronic sensors in a quality assurance method is in its speed of analysis, including data acquisition and interpretation. Rapid, significant data interpretation is possible using various multivariate data analyses.

ACKNOWLEDGMENTS

The author is highly grateful to Professor Bruce R Harte, School of Packging and Professor John E Linz, Department of Food Science and Human Nutritions, Michigan State University.

REFERENCES

Alugupalli, S., Larsson, L. and Slosarek, M. (1992). Application of gas chromatography-mass spectrometry for rapid detection of *mycobacterium xenopi* in drinking water. *Applied and Environmental Microbiology*, 11, 3538-3541.

Backer, H. D., Mohle-Boetani, J. C., Werner, S. B. , Abbott, S. L., Farrar, J. and Vugia, D. J. (2000). *High incidence of extra-intestinal infections in a Salmonella havana outbreak associated with alfalfa sprouts*. Public Health Report, 115, 339–345.

Bao, M., Griffini, O., Burrini, D., Santianni, D., Barbieri, K. and Maschini, M. (1999). Headspace solid-phase microextraction for the determinaiton of trace levels of taste and odor compounds in water samples. *The Analyst*. 124, 459-466.

Bari, M. L., Kusunoki, H., Furukawa, H., Ikeda, H., Isshiki, K., and Uemura, T. (1999). Inhibition of growth of *Escherichia coli* 0157:H7 in fresh radish sprout production by calcinated calcium. *Journal of Food Protection*, 62, 128-132.

Basile, F., Beverly, M. B., Abbas-Hawks, C., Mowry, C. D. and Voorhees, K. J. (1998). Direct mass spectrometric analysis of in situ thermally hydrolyzed and methylated lipids from whole bacterial cells. *Analytical Chemistry*, 70:1555-1562.

Bila, S., Harkouss, Y., Ibrahim, M., Rousset, J., N'Goya, E., Baillargeat, D., Verdeyme, S., Aubourg, M. and Guillon, P. (1999). An accurate wavelet neural-network-based model for electromagnetic optimization of microwave circuits. *International Journal of RF and Microwave Computer-Aided Engineering*, 93, 297-306.

Beuchat, L. R. (2002). Ecological factors influencing survival and growth of human pathogens on raw fruits and vegetables. *Microbes and Infection*, 4, 413-423.

Borjesson, T. (1996). Electronic nose for odor classification of grains. *Cereal Chemistry*, 73, 457-461.

Breuer,T., Benkel, D. H., Shapiro, R. L., Hall, W. N., Winnett, M. M., Linn, M. J., Neimann, J., Barrett, T. J., Dietrich, S., Downes, F. P., Toney, D. M., Pearson, J. L., Rolka, H.,

Slutsker, L. and Griffin, P. M. (2001). A multistate outbreak of *Escherichia coli* O157:H7 infections linked to alfalfa sprouts grown from contaminated seeds. *Emerging Infectious Diseases*, 7, 977-982.

Burden, F. R., Brereton, R. G. and Walsh, P. T. (1997). Cross-validation selection of test and validation sets in multivariate calibration and neural network as applied to spectroscopy. *The Analyst*. 122, 1015-1022.

Brudzewski, K., Osowski, S. and Markiewicz, T. (2004). Classification of milk by means of an electronic nose and SVM neural network. *Sensor Actuator* B, 98, 291-298.

Calvin, L. Produce, Food Safety, and International Trade: Response to U.S. Foodborne Illness Outbreaks Associated with Imported Produce. In: Buzby, J.C. *International Trade and Food Safety: Economic Theory and Case Studies: AER-828 Economic Research Service/USDA*; 2003; 74-96.

Canhoto, O., Pinzari, F., Fenelli, C. and Magan, N. (2004). Application of electronic nose technology for the detection of fungal contamination in library paper. *International Biodeterioration and Biodegradation*, 54: 303-309.

Centers for Disease Control and Prevention. (1999), *Food-Related Illness and Death in the United States*. Available from: http://www.cdc.gov/ecoli/2006/september/

Centers for Disease Control and Prevention. (2006). *Infections Associated with Eating Seed Sprouts*. Available from : http://www.cdc.gov/ecoli/2006/september/

Che Man, Y. B., Gan, H.L., NorAini, I., Nazimah, S.A.H. and Tan, C. P. (2005). Detection of lard adulteration in RBD palm olein using an electronic nose. *Food Chemistry*, 90, 829–835.

Coulibaly, P., Bobée, B., and Anctil, F. (2001). Improving extreme hydrologic events forecasting using a new criterion for artificial neural network selection. *Hydrological Processes,* 15, 1533-1536.

Council for Agricultural Science and Technology. Foodborne Pathogens: Review of Recommendations. October 1999. Available from : http://www.cdc.gov/ncidod/EID/vol9no4/99-0519.html.

Decallonne, J., Delmee, M., Wauthoz, P., Lioui, M E. and Lambert, R. (1991). A rapid procedure for the identification of Lactic acid bacteria based on the gas chromatographic analysis of the cellular fatty acids. *Journal of Food Protection*, 54, 217-224.

Devabhaktuni, V., Yagoub, M. C. E., Fang, Y., Xu, J. and Zhang, Q. (2001). Neural networks for microwave modeling: Model development issues and nonlinear modeling techniques. *International Journal of RF and Microwave Computer-Aided Engineering*, 11: 4-21.

Eley, A. R. (1996). *Microbial food poisoning: Infective bacterial food poisoning*. New York, Chapman and Hall.

Economic Research Service, U.S. Department of Agriculture. (2000). Economics of foodborne disease: Estimating the benefits of reducing foodborne disease. Available at : http://www.ers.usda.gov/Briefing/FoodborneDisease/ Accessed December 2006.

Faber, N. M. (2000). Improved computation of the standard error in the regression coefficient estimates of a multivariate calibration model. *Analytical Chemistry*, 72: 4675-4676.

Fisher, L. and Belle, G. V. (1993). *Discriminant factor analysis. p. 467-498. In A biostatics methodology for the health sciences*. John Wiley and Son, Inc., New York.

Forsyth, D. S. and Dusseault, L. (1997). Determination of methylcyclopentadienly manganese tricarbonyl in beverages by solid phase microextraction. *Food Additives and Contaminants,* 14: 301-307.

Fukushima H, Hashizume, T., Morita, Y., Tanaka, J., Azuma, K. and Mizumoto, Y. (1999). Clinical experiences in Sakai City Hospital during the massive outbreak of enterohemorrhagic *Escherichia coli* O157:H7 infections in Sakai City. *Pediatrics International.* 41, 213–217.

Fung, W. K. (1995). Diagnostics in linear discriminant analysis. *Journal of the American Statistical Association*, 90, 952-956.

Gan, H. L., Che Man, Y.B., Tan, C.P., NorAini, I. and Nazimah, S.A.H. (2005). Characterisation of vegetable oils by surface acoustic wave sensing electronic nose. *Food Chemistry,* 89, 507–518.

Getino, J., Ares, L., Robla, J. I., Carmen, M. D., Sayago, H. I., Fernandez, M. J., Rodrigo, J. and Gutierrez, F. J. Gas sensor arrays for contaminated soils, In Hurst, W. J., editor. Electronic Noses and Sensor Array Based Systems: design and applications. *Proceedings of the 5th International Symposium on Olfaction and the Electronic Nose*. PA. Technomic Publishing Company, Inc; 1998; 329-338.

Gharaibeh, A. A. and Voorhees, K. J. (1996). Characterization of lipid fatty acids in whole-cell microorganisms using in situ supercritical fluid derivatization/ extraction and gas chromatography/ mass spectrometry. *Analytical Chemistry*, 68, 2805-2810.

Gill, C. J., Keene, W. E., Mohle-Boetani, J. E., Farrar, J. A., Waller, P. L., Hahn, C. G. and Cieslak, R. P. (2003). Alfalfa seed decontamination in *Salmonella* outbreak. *Emerging Infectious Diseases*, 9, 474–479.

Gorecki, T. andPawliszyn, J. (1997). Effect of sample volume on quantitative analysis by solid-phase microextraction: Part 1. Theoretical Considerations. *The Analyst*, 122, 1079-1086.

Gorecki, T., Kahled, A. and Pawliszyn, J. (1998). The effect of sample volume on quantitative analysis by solid-phase microextraction: Part 2. Experimental verification. *The Analyst*, 123, 2819-2824.

Kokot, S., Grigg, M., Panayioton, H. and Phunong, T. D. (1998). Data interpretation by some common chemometrics methods. *Electronanlysis*, 10, 1081-1088.

Haugen, J. E., Rudi, K., Langsrud, S. and Bredholt, S. (2006). Application of gas-sensor array technology for detection and monitoring of growth of spoilage bacteria in milk: A model study. *Analytica Chimica Acta*, 565, 10-16.

Haykin, S. (1999). *Neural networks: A comprehensive foundation* (2nd ed). New Jersey, Prentice Hall.

Hodgins, D. and Simmonds, D. (1996). The Electronic nose and its application to the manufacture of food products. *Food Science and Technology Today*, 11, 179-185.

Jaquette C. B., Beuchat, L. R. and Mahon, B. E. (1996). Efficacy of chlorine and heat treatment in killing *Salmonella stanley* inoculated onto alfalfa seeds and growth and survival of the pathogen during sprouting and storage. *Applied and Environmental Microbiology*, 62, 2212–2215.

Jay, J. M. (2000). Modern food microbiology : Determining microorganisms and/or their products in foods. (6th ed., pp. 431-449). New York, Chapman and Hall.

Jonsson, A. (1997). Electronic nose for microbial quality classification of grains. *International Journal of Food Microbiology*. 35, 187-193.

Kaipainen, A., Lisuutari, Y., Lucus, Q. and Moy, L. (1997). A new approach to odor detection. *International Sugar Journal*, 99, 403-408.

Kim, C., Hung, Y. C., Brackett, R. E. and Lin, C.S. (2003). Efficacy of electrolyzed oxidizing water in inactivating *Salmonella* on alfalfa seeds and sprouts. *Journal of Food Protection*, 66, 208-214.

Lee, D., Jung, J., Lim, J., Hus, J. and Lee, D. (2001). Recognition of volatile organic compounds using sno2 sensor array and pattern recognition analysis. *Sensor Actuator* B, 77, 228-236.

Li, Y., Jiang, J. H., Chen, Z. P., Xu, C. J. and Yu, R. Q. (1999). Robust linear discriminant analysis for chemical pattern recognition. *Journal of Chemometrics*. 13: 3-13.

Li, Q., Yoa, X., Chen, X., Liu, M., Zhang, R., Zhang, X. and Hu, Z. (2000). Application of artificial neural networks for the simultaneous determination of a mixture of fluorescent dyes by synchronous fluorescence. *The Analyst*. 125, 2049-2053.

Lucas, Q., Poling, J. and Bennincasa, V. (1998). Analysis with hybrid system PCM/CP/MOS. *Seminars in Food Analysis*, 3:53-58.

Marilley, L., Amupuero, S., Zesiger, T. and Casey, M. G. (2004). Screening of aroma-producing lactic acid bacteria with electronic nose. International Dairy Journal, 14, 849-846.

Mart, M. P., Busto, O., Guasch, J. and Boqué, R. (2005). Electronic noses in the quality control of alcoholic beverages. *Trends in Analytical Chemistry*, 24, 57-66.

Martin, M.A., Santos, J. P. and Agapito, J. A. (2001). Application of artificial neural networks to calculate the partial gas concentration in a mixture. *Sensor Actuator* B, 77, 468-471.

Maul, F., Sargent, S. A., Balaban, M., Baldwin, E. A., Huber, D. J. and Sims, C. A. (1998).. Aroma volatile profiles from ripe tomatoes influenced by physiological maturity at harvest: an application for electronic nose technology. *Journal of the American Society for Horticultural Science*, 123, 1094-1101.

Moat, A. G. and Foster, J. W. *Microbial physiology* (2nd ed.) New York, Willey and Sons; 1988.

Mosca, A., Russo, F., Miragliotta, L., Iodice, M. A. and Miragliotta, G. (2006). Utility of gas chromatography for rapid identification of mycobacterial species frequently encountered in clinical laboratory, *Journal of Microbiological Methods*, In Press, Available online 13 November 2006.

National Advisory Committee on Microbial Criteria for Foods. Microbiological Safety Evaluations and Recommendations on Sprouted Seeds. 26 October 1999. Available from: http://www.cfsan.fda.gov/~mow/sprouts2.html

Nguyen-the, C. and Carlin, F. (1994). The microbiology of minimally processed fresh fruits and vegetables. *Critical Review in Food Science*. 34, 371-401.

Nogueira, J. M. and Nascimento, A. M. (1999). Analytical characterization of Madeira wine. *Journal of Agriculture and Food Chemistry*, 47, 566-575.

Palmai, M. and Buchanan, R. L. (2002). Growth of *Listeria monocytogenes* during germination of alfalfa sprouts. *Food Microbiology*, 19, 195-200.

Portnoy B. L, Goepfert, J. M. and Harmon, S. M. (1976). An outbreak of *Bacillus cereus* food poisoning resulting from contaminated vegetable sprouts. *American Journal of Epidemiology*, 103:589–94.

Persaud, K.C. and Pelosi, P. 1992. Sensor arrays using conducting polymers for an artificial nose. Pg.238-256. In *Sensors and Sensory Systems for an Electronic Nose*. Gardner, J.W., and Bartlett, P.N. 1992. Kluwer Academic Publisher, Netherlands.

Ragazzo-Sanchez, J. A., Chalier, P. and Ghommidh, C. (2005). Coupling gas chromatography and electronic nose for dehydration and desalcoholization of alcoholized beverages application to off-flavour detection in wine. *Sensor Actuator* B, 104, 301-308.

Rajamaki, T., Alakomi, H-L., Ritvanen, T., Skytta, E., Smolander, M. and Ahvenainen, R. (2005). Application of an electronic nose for quality asssesssment of modified atmospherer packaged poultry meat. *Food Control* (In Press).

Sales, F., Callao, M. P. and Rius, F. X. (2000). Multivariate standardization for correcting the ionic strength variation on potentiometric sensors. *The Analyst*. 125: 883-888.

Schweizer-Berberich, P., Vaihinger, S. and Gopel, W. (1994). Characterization of food freshness with sensor arrays. *Sensor Actuator* B, 18-19, 282-290.

Siripatrawan, U., Linz, J. E. and Harte, B. R. (2004 a). Rapid method for prediction of *Escherichia coli* numbers using an electronic sensor array and an artificial neural network. *Journal of Food Protection*, 67, 1604-1609.

Siripatrawan, U., Linz, J. E. and Harte, B. R. (2004 b). SPME/GC/MS coupled with DFA and ANN for prediction of *E. coli*. *Journal of Food Protection*, 67, 1597-1603.

Siripatrawan, U., Linz, J. E. and Harte, B. R. (2006). Electronic sensor array coupled with artificial neural network for detection of *Salmonella typhimurium*. *Sensor Actuator* B, 119, 64-69.

Siripatrawan, U.and Harte, B. R. (2007). Solid phase microextraction/Gas chromatograph/mass spectrometer integrated with chemometrics for detection of *Salmonella* Typhimurium contamination in a packaged fresh vegetable. *Analytica Chimica Acta*, 581, 63-70.

Song, J., Fan, L. and Beaudry, R. (1998). Application of solid phase microextraction and gas chromatography/time-of-flight mass spectrometry for rapid analysis of flavor volatiles in tomato and strawberry fruits. *Journal of Agriculture and Food Chemistry*. 46, 3721-26.

Stetter, J.P., Strathmann, S., McEntegart, C., Decastro, M. and Penrose, W. R. (2000). New sensor arrays and sampling systems for a modular electronic nose. *Sensor Actuator* B, 69: 410-419.

Terra, M. H. and Tinós, R. (2001). Fault detection and isolation in robotic manipulators via neural networks: A comparison among three architectures for residual analysis. *Journal of Robotic Systems*, 18: 357-374.

Thomas S. and Bro, R. (2005). A new approach for modeling sensor based data. *Sensor Actuator* B, 102, 719-729.

Tomasko, L. R., Helms, W. and Snapinn, S. M. (1999). A discriminant analysis extension to mixed models. *Statistics in Medicine*, 18, 1249-1260.

U.S. Food and Drug Administration. Press release. Consumers advised of risks associated with raw sprouts. P99-13. Guidance for industry guide to minimize microbial food safety hazards for sprouted seeds. 27 October 1999. Available from: http://www.fda.gov/bbs/topics/NEWS/NEW00684.html.

U.S. Food and Drug Administration. Interim Advisory on Alfalfa Sprouts. 31 August 1998. Available from : http://www.fda.gov/bbs/topics/ANSWERS/ANS00890.html

Van Beneden, C. A., Keene, W. E., Strang, R. A., Werker, D. H., King, A. S. and Mahon, B. (2001). Multinational outbreak of *Salmonella enterica* serotype Newport infections due to contaminated alfalfa sprouts. *Annals of International Medicine*, 135, 239–247.

Wauthoz, P., Lioui, M. E. and Decallonne, J. (1995). Gas chromatographic analysis of cellular fatty acids in the identification of foodborne bacteria. *Journal of Food Protection*, 58, 1234-1240.

Zhang, A., and Pawliszyn, J. (1993). Headspace solid-phase microextraction. *Analytical Chemistry*, 65, 1843-1852.

INDEX

A

access, 123, 211
accounting, 165
accuracy, 1, 19, 24, 27, 99, 108, 116, 123, 219
acetic acid, 40, 58, 59, 83, 159, 165, 167, 168, 170, 173, 185
acid, viii, ix, x, 15, 17, 28, 38, 39, 40, 41, 42, 43, 44, 45, 46, 49, 50, 51, 52, 53, 54, 55, 56, 57, 58, 59, 61, 62, 63, 65, 66, 67, 69, 72, 74, 75, 80, 82, 83, 87, 89, 115, 119, 131, 134, 136, 142, 144, 151, 155, 158, 159, 162, 164, 166, 168, 169, 170, 174, 175, 177, 179, 180, 181, 183, 184, 185, 186, 187, 189, 190, 197, 200, 201, 205, 207, 209, 211, 240
acidity, 40, 42, 51, 54, 56, 58, 62, 73, 162, 164, 173, 176, 178
acidosis, 45, 172
activation, 160, 193, 196
adaptation, 61, 201, 203, 204
additives, ix, 64, 155
adhesion, 64, 92, 119, 150, 151, 213
adjustment, 112, 237
adolescents, 67
adsorption, 26
adult population, 44
adults, 67, 69, 79, 138, 140, 192
aerobic bacteria, 52, 223, 227
aerosols, 117
AFM, 203, 213
Africa, 56, 59, 85, 156, 162, 174, 176
African continent, ix, 156
agar, 135, 202, 223
age, viii, 108, 131, 143, 145, 184
ageing, 60, 161
agent, 4, 5, 6, 12, 25, 26, 30, 42, 94, 95, 96, 159, 203
agglutination, 27
aging, 179, 220
agriculture, 19
alcohol(s), 17, 58, 159, 219, 227, 228, 229, 230
alcoholic liver disease, 160
aldehydes, 225
alfalfa, 141, 218, 224, 226, 227, 237, 239, 240, 241, 242, 243
algorithm, 222, 237, 238
allergy, 17, 180
alternative(s), 5, 9, 10, 11, 12, 14, 24, 25, 27, 67, 110, 115, 117, 119, 120, 167, 219, 239
amines, 64, 68, 71, 76, 78, 79, 80, 81, 83, 85, 86, 90, 164, 171, 172, 181, 182, 183, 194, 195, 212, 213
amino acid(s), 25, 40, 55, 58, 68, 87, 134, 159, 165, 171, 218, 227
ammonia, 225
amylase, 46, 58, 171, 186
anaerobe, 17
anaerobic bacteria, 191
anemia, 153
animals, 14, 15, 27, 32, 45, 63, 96, 132, 135, 137, 139, 141, 144, 145, 146, 191, 192, 193, 194, 218
annotation, 31
antagonism, 57
anthocyanin, ix, 155, 161, 176, 179
anthrax, vii, 3, 21, 28, 30, 37
antibacterial properties, 169
antibiotic, vii, 3, 14, 15, 16, 17, 18, 30, 36, 48, 192, 204, 207, 208, 210, 215
antibiotic resistance, vii, 3, 14, 15, 16, 17, 18, 30, 36, 204, 207, 208, 210
antibody, vii, 3, 18, 24, 25, 26, 27, 28, 29, 30, 31
antigen, 21, 24, 26, 28, 30, 32, 141, 144, 150, 196, 198, 206, 207, 208
antioxidant, 47, 51, 161, 179
antitumor, 45
aptitude, 65
Argentina, 63
arginine, 43, 136
argument, 108

Aristotle, 65
arithmetic, 99
aromatic compounds, 225
arthritis, 161
ascorbic acid, ix, 53, 155, 159, 171, 173, 176
Asia, ix, 55, 56, 60, 66, 82, 155, 156, 162, 174, 176, 184
assessment, 24, 32, 56, 86, 94, 99, 104, 107, 117, 124, 126, 127, 128, 146, 147, 202, 207, 209, 212, 221
assets, 47
assumptions, 1, 123, 124
asymptomatic, 141, 142, 193
atomic force, 214
ATP, 58, 202
attachment, 23, 151, 191, 203
attention, 41, 42, 49, 51, 54, 55, 68, 69, 146
Australia, 137
automation, 5, 11, 25
availability, 12, 31, 39, 55, 124, 170, 190
avoidance, 145

B

B vitamins, 46
Bacillus, vii, 3, 4, 6, 10, 11, 12, 16, 17, 21, 23, 30, 34, 35, 36, 37, 38, 43, 59, 63, 128, 163, 175, 178, 210, 242
Bacillus subtilis, 6, 17, 23, 38
bacteremia, 42, 88, 193
bacterial cells, 26, 127, 136, 202, 218, 239
bacterial infection, 21, 210
bacterial strains, 11, 170
bacteriocins, 42, 51, 59, 68, 75, 87, 159, 160, 167, 177, 181, 183, 186
bacteriophage, 54, 55, 82
bacteriostatic, 204
bacterium, 57, 61, 73, 80, 134, 196, 198
barley, 46
barriers, 211
BD, 85
beef, 29, 60, 105, 106, 125, 138, 140, 141, 142, 146, 148, 149, 150, 152
beer, 220
behavior, vii
beneficial effect, 48, 51
benzene, 47, 78
beverages, viii, 39, 40, 59, 72, 156, 171, 181, 182, 183, 220, 240, 242, 243
Bhutan, ix, 155, 163, 175
bias, 123, 222, 236, 237
bile, 64, 75, 160
bile acids, 160
binding, 24, 26, 28, 29, 43, 45, 58, 90, 135, 220
bioactive compounds, 48, 82
bioavailability, 44, 46, 58, 78, 84, 170
biochemistry, vii
biodegradation, 90
biodiversity, viii, 39, 71, 180
biofilm formation, 191, 211, 213
biological activity, 76, 81, 171
biological systems, viii, 35, 93
biomass, 76, 89, 168, 187
biomolecules, 31
biosensors, 28, 29
biosynthesis, 21, 84, 190, 196, 207, 208, 213
biotechnology, 181, 186, 188
biotin, 79
birds, 137
birth, 42, 132
bison, 137, 151
bladder, 47
blindness, 179
blood, 29, 45, 85, 135, 143
body composition, 79, 82
body fat, 67, 72, 89
body temperature, 118
body weight, 67, 74
bovine spongiform encephalopathy, 44
bowel, 143
brain, 44
breakfast, 60
breast cancer, 161
breast milk, 83
breeding, 42, 80, 193
brevis, 46, 47, 51, 52, 53, 54, 55, 57, 63, 67, 77, 88, 158, 163, 164, 165, 173, 174, 175, 177, 178
buffer, 28, 170, 223
Bulgaria, 66
Burkina Faso, 178, 185
by-products, 43, 194

C

Ca^{2+}, 198, 200
cabbage, ix, 17, 52, 53, 55, 56, 73, 77, 92, 155, 157, 162, 163, 166, 169, 171, 172, 173, 175, 176, 177, 179, 180, 181, 183, 188
calcium, 45, 46, 64, 67, 72, 77, 82, 89, 164, 167, 170, 172, 201, 227, 239
calibration, 235
campylobacter, 194
Campylobacter jejuni infection, 32
Canada, 3, 36, 137, 138, 139
cancer, 46, 71, 74, 84, 160, 161
cancer treatment, 74

capillary, 143, 223
capsule, 21, 34, 208
carbohydrate(s), 28, 29, 33, 44, 47, 48, 49, 51, 52, 53, 54, 55, 60, 79, 157, 167
carbon, 17, 42, 53, 57, 89, 187, 227
carbon dioxide, 42, 53
carcinogen, 75, 160
carcinogenesis, 81
cardiovascular disease, 46
cardiovascular risk, 81
carotene, ix, 155, 161, 176, 179
carotenoids, 162, 187
carrier, 82, 223
case study, 75
casein, 68
case-studies, 147
cast, 121
catalase, 61, 159
category a, 66
cattle, 68, 70, 137, 141, 145, 147, 148, 149, 150, 152, 153, 154, 193, 201, 205, 215
C-C, 73
cDNA, 5, 8, 17, 35, 36
CE, 84, 146
celiac sprue, 74
cell, ix, x, 9, 14, 17, 24, 27, 29, 33, 37, 45, 46, 62, 64, 67, 121, 134, 135, 144, 151, 159, 162, 178, 189, 190, 191, 192, 195, 196, 199, 200, 201, 202, 203, 204, 205, 206, 210, 211, 213, 223, 227, 241
cell adhesion, 134
cell cycle, 190
cell death, 135
cell line(s), 135
cell membranes, 159, 190, 192, 200
cell surface, 14, 196, 203, 211
cellulose, 167, 191, 205, 213, 215, 216
central nervous system, 143
ceramic, 220
certainty, 104
changing environment, 190
channels, 196
chemical composition, 179
chemical properties, 25
chemical structures, 196
chemiluminescence, 24
chemometrics, 241, 243
chicken, 13, 106, 141, 194, 208, 212, 213, 221
childhood, 153
children, 4, 67, 78, 89, 96, 137, 140, 142, 145, 152, 161, 213
China, 172, 176
chlorine, 241
cholera, vii, 3, 28, 30
cholesterol, 48, 77, 160, 161
chromatograms, 227
chromatography, 219, 239, 241, 242, 243
chromosome, 133, 134
chronic renal failure, 143
classes, 40, 159
classification, 5, 9, 26, 40, 65, 66, 88, 99, 108, 109, 119, 179, 190, 217, 220, 226, 229, 231, 238, 239, 241
cleaning, 145
cleavage, 12
clinical symptoms, 30
clone, 15
cluster analysis, 178
clustering, 73, 78
CO2, 110, 158, 159, 165, 168, 173, 174
coagulating enzyme, 66
coagulation, 65, 66
cobalt, 220
codes, 134
coding, 13, 21, 134
colitis, 149, 153
colon, 44, 47, 67, 161
colonisation, 206, 211
colonization, 52, 133, 160, 185, 198, 213
combined effect, 43, 128
commodities, 2, 43, 69, 180
commodity, 40
communication, 147, 181, 209
community, 16, 17, 81, 195
compatibility, 4, 30
competition, 43, 190
competitiveness, 57, 62, 167, 179
competitor, 63
compilation, 95
complement, 198
complex carbohydrates, 167
complexity, 1, 4, 7, 8, 31, 111, 115, 121
complications, 143
components, 21, 27, 28, 45, 58, 69, 87, 89, 99, 102, 104, 134, 191, 196, 200, 209, 221, 224
composition, 17, 47, 61, 66, 83, 85, 180, 185, 191, 203, 204, 214, 217, 238
compounds, 41, 46, 51, 58, 59, 64, 68, 69, 82, 159, 160, 164, 169, 170, 171, 177, 190, 191, 194, 195, 196, 199, 200, 201, 202, 203, 204, 205, 217, 218, 219, 220, 221, 223, 224, 225, 227, 228, 229, 230, 231, 232, 234, 235, 238, 239
comprehension, 104
computation, 222, 240
computing, 12
concentrates, 202, 219

concentration, 10, 17, 25, 26, 28, 43, 46, 49, 51, 52, 54, 64, 74, 77, 91, 96, 97, 100, 103, 104, 107, 108, 111, 115, 116, 117, 124, 128, 164, 166, 167, 168, 169, 171, 173, 174, 176, 178, 182, 185, 188, 219, 228, 237, 242
conceptual model, 123
conditioning, 116, 149
conductivity, 220, 231
confidence, 22, 103, 123, 124
confidence interval, 103, 124
Congress, iv
conservation, 20
constipation, 160
constraints, 12
construction, 8, 19, 31, 32
consulting, 147
consumers, 47, 124, 146, 147, 181, 218
consumption, viii, 1, 2, 13, 20, 41, 42, 47, 48, 50, 54, 66, 67, 68, 86, 94, 96, 103, 107, 108, 125, 128, 131, 132, 137, 153, 157, 163, 179, 192
consumption patterns, 94, 96
contaminant(s), 13, 71, 218
contaminated soils, 241
contamination, 4, 8, 9, 13, 18, 20, 23, 29, 30, 35, 42, 44, 49, 107, 108, 117, 118, 122, 123, 124, 129, 132, 136, 137, 141, 144, 145, 146, 148, 150, 154, 170, 192, 193, 195, 211, 212, 217, 218, 222, 227, 238, 240, 243
control, vii, viii, 1, 2, 3, 6, 7, 16, 17, 18, 19, 21, 23, 27, 28, 29, 30, 41, 43, 55, 66, 72, 131, 143, 144, 145, 146, 147, 148, 150, 152, 160, 165, 179, 181, 186, 193, 203, 215, 223, 224, 226, 227, 228, 230, 231, 232, 233, 235
convergence, 237
cooking, 43, 45, 52, 70, 89
cooling, 175
corn, 60, 78
correlation, 27, 195, 237
correlation coefficient, 237
cost benefits, 24
costs, 7, 8, 10, 18, 21, 22, 24, 94, 98
covering, 2, 6, 101
criticism, 147
crops, 186
cultivation, 17
culture, ix, 6, 10, 11, 17, 19, 27, 29, 47, 50, 56, 65, 66, 70, 71, 72, 73, 74, 76, 77, 78, 81, 82, 83, 84, 85, 87, 91, 144, 151, 155, 156, 164, 166, 167, 171, 177, 179, 181, 182, 185, 202
culture conditions, 17
curing, 60, 70, 142
cyclosporiasis, 20
cyst(s), 20, 44, 63
cytoplasm, 159, 190, 200
cytosine, 157
cytoskeleton, 134
cytotoxic action, 135, 153
cytotoxicity, 151
Czech Republic, 138

D

danger, 41
data analysis, 221, 228
data availability, 1
data mining, 31
data processing, 231
data set, 221, 226, 229, 232, 235, 237
database, 9, 12, 34, 109, 123
death(s), viii, 4, 36, 93, 113, 114, 117, 118, 127, 131, 135, 142, 149, 192, 218
death rate, viii, 113, 114, 131
decision-making process, 104
decisions, 2, 104, 124, 146
decomposition, 89
deficiency, 44
definition, viii, 59, 93
degradation, 25, 42, 46, 58, 68, 71, 84, 227
degradation process, 42
dehydration, 203, 243
demand, 69, 218, 222
denaturation, 25
Denmark, 57, 140
density, 5, 6, 9, 28, 35, 161
Department of Agriculture, 105, 126, 240
depolarization, 202
deposition, 5
derivatives, 28, 46, 78, 149, 181, 225
desiccation, 195, 212
desorption, 219, 223
destruction, 76, 117, 134, 199, 204, 218, 220
detection, vii, 4, 5, 6, 7, 8, 9, 10, 11, 12, 13, 14, 15, 18, 19, 20, 21, 22, 23, 24, 25, 26, 27, 28, 29, 30, 31, 32, 33, 34, 35, 36, 37, 38, 118, 141, 142, 143, 144, 147, 148, 150, 152, 217, 219, 220, 221, 223, 224, 231, 238, 239, 240, 241, 243
detection techniques, 18, 24
detergents, 196, 200, 202
developed countries, 4, 147
developing countries, ix, 4, 36, 84, 155, 170, 182, 192
dhamuoi, ix, 155
diarrhea, 78, 81, 132, 142, 151
diet, 45, 46, 48, 60, 66, 156, 173, 175
dietary fiber, 58
differentiation, 6, 12, 20, 26, 218

diffusion, 16, 28, 159, 202
digestibility, 45, 46, 171, 172
digestion, 74, 80, 83, 88, 159
digestive enzymes, 160, 171
diphtheria, vii, 3, 28, 30
direct observation, 118
discriminant analysis, 221, 241, 242, 243
discrimination, 8, 11, 12, 13, 20
disinfection, 145, 159, 227
dispersion, 229, 232, 234
disseminate, 29
distribution, 14, 81, 96, 100, 101, 102, 103, 104, 108, 121, 122, 218, 219
divergence, 13
diversification, vii, 3, 4, 31
diversity, viii, 12, 13, 14, 39, 55, 56, 57, 60, 61, 69, 75, 86, 121, 185, 197
division, 100, 106
DNA, v, vii, 3, 4, 5, 6, 7, 8, 9, 10, 11, 13, 14, 15, 16, 17, 18, 19, 20, 21, 22, 23, 24, 25, 26, 30, 31, 32, 34, 35, 36, 38, 70, 71, 144, 178
DNA damage, 70, 71
dogs, 137
doping, 220
down-regulation, 5, 15, 23
draft, 93, 147
drainage, 65
drinking water, 141, 145, 154, 201, 215, 239
drug resistance, 15, 198, 215
drug treatment, 16
drying, 52, 60, 62, 117, 149, 170, 175
DSM, 16
duodenum, 67
duration, 52, 54, 65, 96, 128
dyes, 5, 196, 242

E

E. coli, vii, viii, 3, 8, 9, 10, 14, 16, 24, 25, 26, 27, 29, 30, 33, 34, 42, 48, 51, 63, 65, 108, 121, 128, 131, 132, 133, 134, 135, 136, 137, 138, 139, 140, 141, 142, 143, 144, 145, 146, 147, 148, 149, 150, 151, 152, 153, 154, 170, 178, 191, 194, 196, 197, 198, 201, 210, 221, 222, 223, 224, 226, 227, 228, 230, 231, 234, 235, 237, 238, 239, 243
East Asia, 51
eating, 60, 82
ecology, vii, 54, 74, 83, 88, 147, 180, 206
economic losses, 4
ecosystem, 57, 59, 86, 195
edema, 21, 135
education, 146, 193
effluents, 192
egg, 13, 106, 208
Egypt, 162, 184
elderly, 4, 96, 142
electrical resistance, 231
electrodes, 202, 211, 215
electromagnetic, 239
electron, 17, 18, 57, 90
electron microscopy, 18
electrophoresis, 7, 8, 9, 10, 11, 13, 14, 34, 82, 178, 184
electrostatic interactions, 199
ELISA, 24, 26, 80, 144
elongation, 135
emergence, vii, 3, 11, 14, 15, 149, 192, 203
emergency preparedness, vii, 3
emerging issues, 32, 148
employment, 156
encapsulation, 35, 37
encoding, viii, 21, 131, 133
endocarditis, 13, 42
endothelial cells, 143
endothelium, 135
endotoxins, 197, 209, 212
energy, 23, 92, 114, 159, 167, 172, 224
England, 33, 126, 127, 129, 138, 139
enterovirus, 33, 63
entrepreneurs, 52
environment, 20, 40, 43, 64, 65, 97, 108, 136, 137, 141, 144, 159, 165, 191, 192, 193, 202, 207, 213, 218
environmental change, 195
environmental conditions, 25, 31, 36, 52, 115, 190, 197
environmental factors, 47, 110, 115, 123, 210
environmental stimuli, 205
enzymes, 25, 46, 47, 49, 58, 62, 65, 67, 72, 91, 160, 168, 171, 173, 190, 194, 197, 200
epidemic, 33, 137, 144
epidemiologic studies, 18
epidemiology, 2, 13, 26, 37, 136, 144, 147, 148, 150, 152, 207, 211
epithelial cells, 45, 92, 149, 150, 214
epithelium, 192, 194
equality, 229, 232
equilibrium, 43, 64, 162, 219, 227
equipment, 141, 144, 207, 219
erythrocytes, 136
Escherichia coli, v, viii, ix, x, 4, 6, 9, 10, 26, 28, 32, 33, 42, 85, 87, 89, 105, 106, 125, 126, 128, 131, 132, 147, 148, 149, 150, 151, 152, 153, 154, 155, 159, 187, 189, 191, 194, 205, 206, 207, 208, 210, 211, 213, 214, 215, 216, 218, 239, 240, 241, 243
ESN, 126

ester, 197
estimating, 1, 97, 105, 237
estimation process, 1
ethanol, 42, 43, 66, 156, 158, 227, 228, 229, 230
ethanolamine, 198
ethnic groups, 60
EU, 105, 148, 150
eukaryotic cell, 135
Europe, 56, 66, 128, 144, 162, 212
European Commission, 94, 107, 124, 126
European Community, 146
European Parliament, 152, 153
European Union, 142, 149, 153, 192, 193, 194, 207
evolution, 1, 4, 14, 31, 38, 87, 151
exclusion, 145, 160, 193, 211, 213
excretion, 68, 118
exercise, 82
exopolysaccharides, 59, 81
exotoxins, 12
experimental design, 25, 31, 117
expertise, 147
exposure, 2, 43, 44, 75, 87, 94, 95, 96, 97, 107, 108, 118, 127, 128, 195, 202
extracellular matrix, 191, 195, 216
extraction, 5, 7, 25, 31, 35, 52, 219, 223, 231, 241
extrapolation, 118, 119

F

factor analysis, 221, 240
failure, 24, 143
false negative, 7, 22
false positive, 8, 22, 24
family, 19, 46, 47, 191, 198
FAO (Food Agriculture Organization), viii, 93
farms, 141, 145, 210
fast food, 132
fasting, 45
fat, 60, 64, 67, 72, 76, 80, 84, 85, 87, 88, 90
fat reduction, 72
fatty acids, 59, 165, 197, 202, 218, 240, 241, 244
FDA, 22, 105, 106, 124, 125, 126, 172
feces, 83, 215, 218
females, 143, 153
fermentable carbohydrates, 49, 52, 54
fermentation, viii, ix, 15, 16, 17, 39, 40, 41, 42, 43, 44, 45, 46, 48, 49, 50, 51, 52, 53, 54, 55, 56, 58, 59, 60, 62, 63, 64, 65, 66, 69, 70, 71, 73, 74, 75, 76, 77, 78, 79, 80, 81, 82, 83, 84, 85, 86, 87, 88, 89, 90, 91, 92, 149, 155, 156, 157, 158, 159, 161, 162, 164, 165, 166, 167, 168, 169, 170, 171, 172, 173, 174, 175, 176, 177, 179, 180, 181, 182, 183, 184, 185, 186, 187, 188
fermentation technology, 84, 90
fertilizers, 193
fever, 27, 118, 143, 171, 192, 211, 215
fibrin, 143
fibroblasts, 135
film, 220
Finland, 57, 88, 139, 189, 193, 205, 211
first generation, 128
fish, 40, 68, 105, 179, 181, 185, 194
fitness, 48
flame, 202
flatulence, 44
flavonoids, 46, 71, 84, 161
flavor, 58, 60, 66, 184, 219, 243
flexibility, 4, 6, 30
flight, 223, 243
flora, 18, 47, 75, 83, 88, 115, 165, 167, 168, 169, 174
fluid, 56, 88, 135, 241
fluorescence, 24, 26, 202, 214, 242
fluorophores, 8, 28
focusing, vii, 3, 16, 67, 107
folate, 46, 58, 67, 74, 77, 79, 80
folded conformations, 25
folic acid, 46, 87
food additives, 172
food industry, 8, 10, 17, 24, 30, 31, 147, 156, 172, 207
food poisoning, viii, 21, 34, 35, 40, 131, 174, 240, 242
food production, ix, x, 16, 21, 80, 91, 180, 182, 189
food products, 7, 11, 12, 16, 20, 27, 97, 105, 159, 160, 179, 211, 217, 218, 220, 241
food safety, vii, viii, 2, 3, 4, 5, 10, 16, 20, 21, 22, 24, 25, 26, 29, 30, 32, 33, 36, 84, 85, 93, 98, 123, 127, 131, 136, 146, 147, 152, 170, 181, 185, 193, 209, 222, 243
forecasting, 240
foreign exchange, 174
France, 57, 140, 150, 152
free radicals, 161
freezing, 97, 170
fructose, 54, 57, 77, 167, 177, 184
fruits, viii, ix, 39, 40, 47, 49, 51, 52, 55, 71, 155, 156, 157, 161, 163, 164, 165, 166, 167, 168, 169, 170, 171, 174, 178, 179, 182, 183, 193, 218, 239, 242, 243
functional analysis, 26
fungi, 167

G

gamma radiation, 145
gas phase, 217, 238

gases, 175, 220
gasoline, 220
gastritis, 194
gastroenteritis, 9, 18, 33, 37, 44, 192
gastrointestinal tract, 45, 47, 67, 81, 145, 161
gel, 7, 8, 9, 10, 11, 13, 25, 34, 82, 178, 184
gender, 108
gene(s), vii, viii, 3, 5, 6, 7, 8, 9, 11, 12, 13, 14, 15, 16, 17, 18, 19, 20, 21, 22, 23, 25, 30, 31, 32, 34, 35, 36, 37, 38, 48, 81, 87, 131, 133, 134, 135, 144, 150, 154, 190, 195
gene expression, vii, 3, 5, 7, 14, 15, 17, 22, 23, 30, 31, 35, 38, 48, 195
gene transfer, 14, 15, 16
generalization, 1
generation, 53, 58, 193, 225
genetic diversity, 5, 19, 33, 36, 207
genetic marker, 7, 13, 26, 144
genetic screening, 11
genetics, vii, 14, 206, 214
genome, 4, 6, 8, 10, 11, 12, 13, 14, 15, 16, 17, 23, 30, 31, 32, 33, 36, 37, 185
genome sequences, 31
genomic instability, 12
genomics, 14, 33, 37
genotype, 12, 19, 20
Georgia, 139
Germany, 57, 60, 171, 208
germination, 227, 242
ginger, 173, 176
glass, 5, 17, 23, 26, 41, 175, 223
glucose, 15, 51, 54, 57, 60, 73, 74, 85, 157, 158, 167, 169, 177, 184, 197, 218
glucoside, 49
glycerol, 59
glycine, 64, 78
goat milk, 68, 78, 141
gold, 143
government, 93, 147
grains, 45, 46, 47, 56, 183, 239, 241
gram-negative bacteria, 201, 205, 208, 209, 211, 215
gram-positive bacteria, 36
graph, 228, 230, 233
Greece, 39, 57, 60, 62, 63, 75, 86, 87
groups, 14, 19, 29, 34, 65, 95, 96, 117, 142, 157, 159, 190, 197, 208, 221, 224, 229, 232, 233, 235, 238
growth, viii, ix, x, 15, 16, 17, 23, 27, 31, 42, 43, 45, 48, 49, 50, 52, 53, 54, 59, 60, 62, 66, 69, 77, 85, 86, 91, 93, 96, 98, 104, 105, 107, 108, 109, 110, 111, 112, 115, 116, 117, 123, 125, 127, 128, 129, 136, 149, 153, 155, 156, 158, 165, 166, 168, 169, 170, 174, 185, 186, 189, 194, 195, 197, 202, 204, 206, 219, 222, 227, 239, 241
growth rate, 54, 86, 109, 110, 111, 112, 115, 123, 128, 195
growth temperature, 115
guanine, 157
guidelines, 1, 93, 94, 104, 105, 125
gut, 44, 47, 80, 87, 133, 136

H

hands, 108, 122
harvesting, 4, 16, 165, 174, 179, 218
hazards, viii, 2, 42, 91, 93, 131, 243
headache, 143, 171
health, viii, ix, 2, 4, 10, 12, 21, 25, 30, 41, 42, 47, 48, 51, 54, 65, 67, 69, 78, 83, 84, 92, 93, 94, 95, 97, 98, 107, 131, 135, 146, 147, 155, 156, 157, 160, 161, 181, 183, 187, 192, 211, 215, 240
health care, 4
health care costs, 4
health education, 147
health effects, 83, 95
health problems, 41, 42
heart disease, 84
heat, 9, 10, 43, 45, 52, 55, 66, 96, 97, 99, 117, 132, 135, 136, 146, 159, 160, 164, 193, 220, 241
heating, 117, 146
heavy metals, 41, 42
heme, 151
hemoglobin, 151
hemolytic uremic syndrome, 149
hepatitis, 4, 18, 33
heredity, 48
heterogeneity, 99, 102, 117, 202, 209
histamine, 54, 56, 91, 171
HIV, 161
homeostasis, 62
homocysteine, 72
hospitals, 141
host, 14, 18, 47, 96, 118, 119, 120, 133, 134, 136, 149, 160, 192, 198, 210
host tissue, 198
housing, 208
HPC, 12
hua-chai po, ix, 155
human activity, 210
human brain, 221
human exposure, 44, 126
humidity, 61, 65, 220, 226
Hungary, 60, 63, 87
hybrid, 221, 242

hybridization, 5, 6, 7, 9, 11, 13, 14, 15, 16, 18, 19, 20, 21, 22, 30, 33, 38, 89, 144
hydrogen, 42, 74, 195
hydrogen peroxide, 42
hydrolysis, 41, 45, 46, 47, 58, 69, 75
hydrophilicity, 203
hydrophobicity, 203
hydroxyl, 197
hydroxyl groups, 197
hygiene, 2, 41, 108, 125, 145, 146, 153, 193, 208
hypertension, 143
hypothesis, 120, 121, 122

I

identification, vii, 3, 5, 6, 7, 8, 9, 10, 11, 12, 13, 14, 15, 19, 20, 21, 22, 26, 27, 31, 32, 35, 37, 38, 73, 76, 78, 92, 94, 95, 127, 132, 144, 147, 185, 190, 196, 218, 219, 220, 222, 240, 242, 244
identity, 11, 28, 226
ileum, 67
images, 203
immersion, 49
immobilization, 6, 25
immune response, 46, 48, 133, 192
immune system, 44, 161, 196, 198, 210
immunity, 159
immunization, 145
immunocompromised, 4
immunostimulatory, 197
implementation, 41, 111, 157
impregnation, 71
in situ, 6, 24, 43, 64, 97, 133, 206, 239, 241
in vitro, 16, 43, 64, 65, 85, 92, 96, 160, 161
in vivo, 69, 153, 160
incidence, 18, 37, 94, 124, 137, 142, 143, 148, 192, 203, 239
inclusion, 111, 117
income, 156
incubation period, 142
incubation time, 234
India, ix, 90, 155, 163, 174, 175, 176, 182, 186, 187
indicators, 56, 79, 213, 217, 218, 220, 238
indices, 123, 236
indigenous, 15, 55, 59, 70, 90, 144
indirect effect, 58, 69
Indonesia, 63, 176, 177, 208, 215
induction, 136, 204
industrial application, 48
industrial production, 36, 169
industry, 18, 33, 67, 68, 77, 124, 162, 181, 193, 243
infancy, 32
infants, 83, 161, 194
infection, vii, viii, 3, 4, 6, 14, 15, 18, 19, 27, 41, 42, 48, 91, 95, 96, 108, 117, 118, 119, 120, 121, 127, 131, 132, 133, 136, 137, 141, 142, 143, 144, 145, 146, 147, 148, 149, 152, 153, 154, 161, 192, 194, 197, 208, 210
infectious disease, 11, 32, 192
inflammation, 46, 75, 84
ingest, 120
ingestion, 45, 74, 82, 83, 95, 118, 142, 188
inhibition, 51, 59, 61, 111, 135, 160, 187, 200, 202, 206
inhibitor, 41, 199
inhibitory effect, 111, 158, 200
initiation, 54, 184
injections, 224
innate immunity, 205
inoculation, 54, 154, 223, 226, 230, 231, 234, 235, 238
inoculum, 166, 167, 223
input, 98, 221, 235, 236, 237
insight, 13, 17, 55
institutions, 104, 124
insulin, 74, 85, 86
insulin resistance, 86
integration, 1, 101
integrin, 153
intelligence, 21
intensity, 7, 26, 27, 28, 58, 178
interaction(s), vii, 3, 14, 23, 25, 29, 31, 33, 35, 45, 47, 75, 94, 119, 127, 134, 135, 190, 195, 198, 203, 208, 210, 211, 214, 215
interface, 210
interference, 27, 160, 231
international trade, 2
internet, 31
interpretation, 221, 238, 239, 241
interval, 117
intervention, 23, 48, 82
intervention strategies, 23
intestinal flora, 47, 149, 188
intestinal tract, 74, 86, 132, 218
intestine, 132, 133, 136, 137, 141
intoxication, 41, 42
intraocular, 21
invertebrates, 45
investment, 219
ionization, 224
iron, 17, 46, 133, 136, 200, 220
irradiation, 43, 55, 80, 117
irritable bowel disease, 67
irritable bowel syndrome, 80, 85
Islam, 144, 150
isoflavonoids, 71

isolation, 10, 73, 74, 132, 142, 143, 144, 192, 243
isomers, 45, 78, 172
Italy, 57, 60, 63, 70, 73, 74, 77, 78, 84, 86, 87, 124, 129, 131, 148, 150, 154, 163, 179
iteration, 101

J

Japan, 35, 137, 139, 140, 153, 163, 175, 176
justification, 119

K

K^+, 202
khalpi, ix, 155, 174, 178
kidney(s), 87, 134, 143, 172
killing, 198, 241
kinetic model, 109, 115, 116, 117
kinetics, 59, 77, 108, 109, 117
Korea, 35, 55, 162, 163, 172, 177
kuru, 44

L

labour, 25
lactase, 82, 89, 91, 160
lactic acid, ix, x, 16, 17, 23, 32, 40, 41, 42, 43, 45, 46, 48, 49, 50, 51, 52, 53, 54, 55, 56, 57, 59, 60, 62, 63, 64, 65, 66, 67, 68, 69, 70, 71, 73, 74, 75, 77, 78, 80, 81, 82, 85, 86, 87, 88, 89, 90, 91, 92, 156, 158, 160, 163, 178, 179, 180, 181, 182, 183, 184, 185, 186, 187, 188, 189, 195, 201, 206, 207, 208, 213, 242
Lactobacillus, ix, 16, 42, 50, 63, 64, 66, 68, 71, 72, 73, 74, 75, 76, 77, 78, 79, 80, 81, 82, 83, 85, 86, 87, 88, 89, 90, 91, 155, 156, 158, 160, 161, 162, 163, 165, 166, 170, 171, 173, 175, 177, 179, 181, 182, 183, 184, 185, 186, 187, 188, 201, 205, 206
lactoferrin, 207
lactose, 44, 48, 66, 67, 74, 77, 80, 82, 83, 84, 88, 91, 160, 161, 169, 184
lactose intolerance, 44, 48, 67, 160
lakes, 136
language, 105
large intestine, 134
Latin America, 60, 87, 156, 162, 174, 176, 180
Latin American countries, 87
laws, 99, 100
LDL, 51, 77, 161
leaching, 49
leakage, 202
learning, 222, 235, 237
learning process, 237
legislation, 146
legume, 40
lesions, 132, 133, 134, 142
lethal factor, 28, 30
lethargy, 143
Leuconostoc, ix, 45, 52, 65, 66, 71, 73, 80, 81, 89, 155, 156, 158, 162, 163, 165, 166, 170, 181, 184
liberation, 200
ligand(s), 24, 35
lignans, 51
likelihood, 147
limitation, 8, 207
linear function, 236
linear model, 115, 117, 121, 127
linkage, 197
links, 108
lipase(s), 62, 75
lipids, 67, 81, 159, 202, 203, 239
lipolysis, 61
lipopolysaccharides, 27, 134, 205, 207, 208
lipoproteins, 191, 196
liquid chromatography, 27
liquid phase, 169
liquids, 122
Listeria monocytogenes, 4, 10, 26, 28, 32, 34, 42, 72, 75, 77, 87, 98, 105, 106, 125, 126, 128, 129, 153, 159, 186, 218, 242
liver, 135, 157, 159, 179
liver cells, 135
liver disease, 159
livestock, vii, 3, 15, 44, 145, 153, 193
localization, 213
location, 192
locus, 12
longitudinal study, 150
low risk, 99
low temperatures, 51, 193, 197
low-density lipoprotein, 71
LPS, 134, 144, 192, 196, 197, 198, 199, 200, 202, 203
lycopene, 161
lysis, 68, 202
lysozyme, 199, 202

M

mAb, 25
macrophages, 192
magnesium, 46, 67, 77, 164, 172
magnet, 101
malabsorption, 74, 89
Malaysia, 82, 176

maltose, 57
mammal, 203
mammalian cells, 84, 151
management, 1, 2, 127, 143, 154
manganese, 60, 64, 220, 240
manipulation, 78, 231
mannitol, 58, 81
manufacturing, 4, 6, 17, 18, 30, 31, 48, 66, 70, 106, 146, 171
manure, 141, 145, 150
mapping, 6
marine environment, 210
market, 141, 174
marketing, 141, 211
mass spectrometry, 27, 217, 239, 241, 243
mastitis, 13
mathematics, 99
matrix, 25, 28, 48, 62, 65, 96, 119, 191, 195, 203, 219, 226, 228, 232, 235
maturation, 87, 141
maximum specific growth rate, 110
meals, 140
measurement, 202, 221, 224, 231, 238
measures, 2, 21, 44, 102, 108, 145, 186
meat, viii, 9, 16, 17, 18, 39, 40, 42, 47, 60, 62, 63, 64, 70, 71, 75, 80, 84, 89, 90, 106, 116, 137, 138, 140, 141, 142, 144, 145, 151, 154, 165, 179, 181, 192, 193, 194, 212, 243
media, 10, 35, 42, 70, 144, 183, 219, 238
median, 99, 145
medical care, 97
Mediterranean, 51
Mediterranean countries, 51
membrane permeability, 202, 205, 211
membranes, 191, 203, 206, 208, 210, 212, 214, 215
meningitis, 194
messengers, 28
meta-analysis, 37
metabolism, 17, 45, 59, 66, 75, 164, 167, 169, 172, 184, 194
metabolites, 43, 47, 59, 70, 159, 190, 194, 195, 205, 212, 218, 220, 222, 223, 227, 231, 235, 238, 239
metal oxides, 220
methane, 225
methionine, 227
Mexico, 60
Mg^{2+}, 198, 200
mice, 124
microarray, vii, 3, 4, 5, 6, 7, 8, 9, 10, 11, 12, 13, 14, 15, 16, 17, 18, 19, 20, 21, 22, 23, 24, 25, 26, 27, 29, 30, 31, 32, 33, 34, 35, 36, 37, 38
microarray detection, 10, 13, 19, 20, 33
microarray technology, vii, 3, 5, 11, 31, 32, 35, 37, 38
microbial cells, 77
microbial communities, 23, 36, 71, 204
microbial community, 23, 51
microenvironment(s), 42, 51, 57, 61, 69
micrometer, 29
microorganism(s), vii, viii, 4, 5, 9, 11, 12, 14, 15, 16, 17, 19, 22, 27, 30, 31, 38, 39, 40, 42, 43, 46, 47, 48, 53, 56, 63, 64, 65, 66, 68, 69, 71, 78, 87, 88, 89, 93, 94, 95, 96, 97, 98, 99, 105, 108, 107, 108, 109, 111, 115, 117, 118, 119, 120, 121, 123, 129, 132, 134, 136, 141, 159, 160, 162, 165, 167, 168, 171, 173, 175, 180, 181, 182, 184, 206, 217, 218, 220, 222, 236, 238, 241
microscope, 5
microscopy, 178, 210, 214
microstructure, 88
microwave heating, 76
Middle East, 66
milk, viii, 9, 17, 28, 39, 46, 65, 66, 67, 68, 69, 70, 71, 74, 76, 80, 82, 84, 85, 87, 88, 89, 92, 104, 105, 106, 125, 127, 128, 138, 140, 141, 142, 144, 145, 146, 149, 161, 179, 185, 193, 194, 211, 220, 240, 241
minerals, 46, 49, 67, 156, 157, 161, 165, 169, 175
mixing, 52, 56, 60, 107, 108, 128
mobility, 4
modeling, 126, 184, 240, 243
models, viii, 1, 63, 69, 93, 95, 96, 97, 98, 99, 102, 104, 105, 106, 107, 108, 109, 110, 111, 112, 113, 114, 115, 116, 117, 118, 119, 120, 121, 122, 123, 124, 127, 128, 129, 213, 243
modules, 231
moisture, 52, 61, 65, 66, 166, 175
moisture content, 65
mole, 158
molecular mass, 171
molecular weight, 25, 134, 196, 198, 202, 220
molecules, 23, 24, 28, 31, 45, 58, 159, 191, 196, 197, 198, 199, 201, 231
monolayer, 197
monomer, 135
monosaccharide, 36
Montana, 138
Montenegro, 137, 147, 152
Moon, 181
morbidity, viii, 4, 119, 131
morphogenesis, 206
morphology, 190, 198
mortality, viii, 119, 131, 132, 143, 144
mortality rate, 132, 143
moulding, 59

mRNA, 5
mucus, 172
multiple factors, 192
multiple regression, 221
multiplication, 100, 111, 118, 193
multivariate calibration, 221, 240
multivariate data analysis, 221, 222
muscle tissue, 44
muscles, 90, 193
mutant, 15, 29, 204
mutation(s), 15
mycobacteria, 190, 191, 206

N

NaCl, 17, 49, 52, 53, 54, 56, 62, 64, 77, 79, 92, 133, 166, 169, 176, 178, 185
nation, 147
neonates, 69
Nepal, ix, 155, 162, 163, 174, 175, 181
nervous system, 69
Netherlands, 37, 128, 242
network, 38, 58, 221, 226, 235, 237, 238, 239
neural network(s), 221, 235, 237, 238, 240, 242, 243
neurons, 221, 235, 236
next generation, 128
niacin, 67
nickel, 220
nitrate, 17, 60, 164
nitrogen, 223
NMR, 202
normal distribution, 100
North America, 161
Northern Ireland, 142
Norway, 140, 150
nucleic acid, 4, 7, 8, 12, 19, 29, 31, 32
nucleotides, 6, 24
nursing, 141
nursing home, 141
nutrient media, 221
nutrients, ix, 44, 45, 48, 52, 56, 72, 155, 159, 161, 165, 166, 167, 169, 190, 195, 196
nutrition, 37, 48, 70, 84, 156, 157

O

obesity, 86
observations, 123, 226, 237
occlusion, 143
oil(s), 43, 64, 168, 169, 174, 200, 209
Oklahoma, 139
oligonucleotide arrays, 6, 11, 14, 35
oligosaccharide, 196, 197, 207
olive oil, 51, 71, 72, 77, 78, 85
operon, 34
optimization, 7, 25, 31, 91, 239
organic compounds, 219, 220, 242
organic solvent(s), 219
organism, 35, 136, 173
organization, 153, 221
organochlorine compounds, 71
output, 98, 99, 101, 103, 217, 221, 226, 235, 236, 237
oxalate, 45
oxidation, 51, 53, 71, 83, 173
oxidative stress, 51, 92
oxides, 220
oxygen, 7, 23, 52, 61, 62, 159, 231
oyster, 55, 73, 106
ozone, 145, 206

P

packaging, vii, 56, 79, 179
pain, 45, 143
pairing, 7
Pakistan, ix, 155, 163, 175, 187
pantothenic acid, 79
paralysis, 135
parameter, 12, 110, 113, 115, 121, 122, 237
parasite(s), 18, 35, 44, 63, 97
passive, 51, 92
pasteurization, 17, 52, 85, 135, 193
pathogenesis, 13, 15, 21, 135, 136, 151, 152, 206, 210, 212, 214
pathogens, vii, ix, 3, 4, 5, 6, 7, 8, 10, 11, 12, 14, 15, 16, 19, 20, 21, 22, 24, 25, 26, 27, 29, 30, 31, 32, 33, 35, 38, 42, 47, 55, 62, 70, 80, 92, 95, 96, 98, 108, 115, 120, 121, 126, 127, 128, 142, 147, 148, 155, 159, 170, 179, 191, 193, 194, 195, 198, 203, 204, 207, 210, 211, 212, 214, 217, 218, 220, 239
pathways, 122, 158, 218, 220
pattern recognition, 217, 221, 222, 232, 238, 242
patterning, 34
PCA, 183, 221
PCM, 242
PCR, 5, 6, 7, 8, 9, 11, 12, 14, 15, 16, 18, 19, 20, 21, 22, 24, 26, 30, 31, 32, 35, 38, 70, 73, 76, 87, 144, 178, 181, 221
Pediococcus, ix, 49, 52, 63, 66, 155, 156, 158, 159, 162, 163, 165, 170, 172, 173, 175, 181
pepsin, 46, 65
peptic ulcer, 194
peptide chain, 135

peptides, 24, 46, 59, 69, 74, 81, 89, 159, 197, 198, 200, 203, 208, 210, 211, 215
percentile, 99
perception, 218
permeability, 191, 196, 198, 199, 201, 202, 211, 214, 215
permit, 109, 115, 116
peroxide, 92
personal hygiene, 141
pesticide, 42, 219
pH, ix, x, 15, 17, 22, 46, 49, 50, 51, 52, 53, 54, 55, 59, 60, 62, 63, 64, 66, 75, 77, 110, 114, 115, 116, 125, 133, 136, 141, 156, 159, 160, 162, 164, 165, 166, 169, 170, 173, 174, 175, 176, 177, 179, 181, 186, 189, 190, 195, 201, 207, 227
phage, 133, 145, 209
phenol, 46, 78, 92
phenotype, 13, 151, 195, 205
Philippines, 163
phosphates, 198, 215
phosphoenolpyruvate, 186
phospholipids, 199, 202
phosphorus, 46, 67, 164
physiology, 95, 179, 242
pigments, ix, 155, 161, 165, 176, 179
pigs, 87, 137, 148, 150, 193, 212
pilot study, 79
plants, 45, 46, 70, 80, 200
plasma, 74, 159, 202
plasma membrane, 159
plasmid, 12, 34, 36, 133, 135, 149, 150, 153
plasmolysis, 166, 169
PM, 80, 118
point mutation, 6
polarized light, 172
pollution, 24, 41, 42
polyacrylamide, 209
polymer(s), 48, 160, 190, 213, 219, 220, 227, 242
polymerase, 35
polymerase chain reaction, 35
polymorphism(s), 6, 11, 12, 20
polypeptides, 46
polyphenols, 45, 78, 88
pools, 117
poor, 8, 27, 51, 58, 63, 108, 135, 192
population, 1, 2, 4, 13, 14, 22, 43, 47, 49, 51, 52, 55, 60, 61, 64, 76, 87, 95, 96, 97, 102, 109, 112, 116, 117, 118, 136, 156, 165, 170, 194, 222
population density, 109, 112, 116
Portugal, 83
potassium, 60, 64, 73, 77, 78, 79, 180
potato, 157, 161, 170, 176, 185, 186
potential energy, 169
poultry, 9, 13, 106, 148, 149, 150, 152, 192, 193, 194, 212, 243
power, 12, 100, 106
prediction, 69, 111, 119, 129, 235, 236, 238, 243
preference, 57
premature death, 142
preschool, 72
preschool children, 72
preservative, 165, 201
pressure, 27, 52, 71, 117
prevention, viii, 48, 59, 67, 70, 71, 131, 161
prions, 44
probability, viii, 2, 22, 93, 94, 95, 96, 97, 99, 100, 101, 102, 108, 109, 115, 117, 118, 120, 121, 212
probability distribution, 100, 101, 102, 108, 115
probe, 6, 12, 31, 202, 210
probiotic, ix, 15, 16, 34, 47, 48, 64, 67, 70, 71, 73, 74, 75, 77, 78, 80, 81, 82, 83, 85, 86, 87, 88, 91, 92, 154, 155, 157, 161, 174, 178, 182, 184, 188, 201, 207
process control, 108
producers, 51, 177, 195
product design, 97
production, vii, 3, 10, 16, 17, 21, 30, 37, 39, 40, 41, 42, 43, 45, 46, 51, 52, 53, 54, 55, 56, 58, 59, 60, 63, 64, 65, 66, 68, 69, 70, 71, 73, 77, 78, 82, 83, 84, 86, 91, 97, 105, 133, 136, 142, 144, 145, 146, 147, 151, 154, 156, 157, 158, 162, 164, 165, 167, 168, 172, 177, 179, 180, 181, 185, 186, 187, 188, 193, 194, 205, 212, 215, 222, 239
production costs, 157
productivity, 4, 142
profit, 62
profit margin, 62
program, 223
programming, 105, 123
pro-inflammatory, 193
propagation, 222, 235, 238
propane, 225
prophylaxis, 44, 193
proteases, 62
protective role, 58
protein analysis, 35
protein arrays, vii, 3, 4, 24, 25, 26, 30, 31, 32
protein function, 23, 24
protein kinases, 23
protein synthesis, 25, 135
proteinase, 171, 177
proteins, vii, 3, 14, 21, 23, 24, 25, 26, 29, 31, 33, 40, 44, 45, 46, 58, 66, 67, 69, 74, 75, 84, 89, 133, 134, 159, 160, 161, 164, 190, 196, 198, 199, 200, 204, 209
proteinuria, 143

proteolysis, 61
proteolytic enzyme, 166, 169
proteome, 4, 25, 30, 32
proteomics, 35, 91
protocol, 34, 143
protons, 159
prototype, 132
protozoa, 20, 21, 36
protozoan parasites, 20
Pseudomonas aeruginosa, 10, 35, 205, 207, 208, 209, 210, 215
public health, 20, 36, 68, 94, 97, 132, 144, 193
pulse, 13, 34
pumps, 190, 196, 204, 211
purification, 7, 25, 26, 31, 35, 81, 187
pyrophosphate, 200

Q

quality assurance, 97, 239
quality control, 7, 8, 23, 24, 30, 32, 242
quartz, 220

R

radiation, 55
random weights, 237
range, viii, 6, 25, 39, 41, 42, 59, 60, 65, 67, 69, 100, 103, 104, 109, 119, 133, 166, 173, 220, 223, 231
raw materials, viii, 39, 40, 41, 43, 44, 45, 46, 52, 55, 56, 59, 60, 64, 69, 146, 156, 164, 193
reading, 5, 11, 19, 33
reagents, 25, 26, 144
real time, 219
reality, 98, 124
receptor sites, 45
receptors, 25, 28, 29, 153
recognition, 24, 29, 192, 217, 220, 221, 222
recovery, 7, 31, 161, 174, 220, 231
reduction, 15, 28, 45, 48, 60, 62, 64, 90, 98, 113, 114, 115, 124, 146, 148, 160, 170, 227
redundancy, 6, 7, 12, 13, 14
regression, 115, 116, 221, 222, 237, 240
regulation(s), 5, 4, 15, 37, 86, 146
rehydration, 78
relationship(s), 14, 48, 67, 91, 94, 95, 98, 106, 107, 115, 118, 119, 213
relevance, 6, 7, 24, 72
reliability, ix, 96, 104, 123, 155, 156
renal failure, 142
repetitions, 117
replacement, 73, 77, 79, 85, 180, 185
replication, 133, 142
residues, 42, 70, 197, 198
resistance, ix, x, 15, 16, 18, 25, 32, 34, 42, 44, 117, 136, 144, 151, 161, 189, 191, 195, 197, 198, 201, 204, 206, 207, 208, 209, 210, 211, 212, 214, 215, 231
resolution, 8, 11, 24, 99, 221, 239
resources, 94, 98, 99, 102, 104
respiratory, 19, 41
restriction enzyme, 11, 12
retail, 106, 141
retardation, 45
retention, 44, 49
rheological properties, 58
rheology, 59
riboflavin, 67
ribosomal RNA, 9, 34
rice, 45, 56, 60, 175, 176, 184
risk, 1, 2, 41, 42, 44, 75, 77, 94, 95, 96, 97, 98, 99, 100, 104, 105, 107, 108, 118, 122, 124, 125, 126, 127, 128, 129, 137, 142, 143, 144, 146, 149, 153, 154, 161, 174, 208, 218, 227
risk assessment, 42, 75, 94, 105, 107, 124, 125, 126, 127, 128, 129
risk factors, 41, 108, 122, 143, 208
risk management, 2
risk profile, 94, 99
RNA, 4, 5, 8, 10, 19, 25
robustness, 1, 31, 124
rods, 61, 157, 161
rotavirus, 44
routines, 226
rural population, ix, 155
Russia, 57

S

safety, vii, viii, 2, 3, 5, 22, 29, 30, 39, 40, 41, 42, 44, 47, 51, 56, 62, 64, 65, 66, 68, 69, 71, 87, 88, 131, 141, 146, 147, 159, 160, 180
salinity, 190
salmon, 221
Salmonella, vii, ix, x, 3, 4, 8, 10, 18, 24, 26, 27, 28, 30, 32, 35, 38, 42, 62, 98, 105, 106, 121, 124, 126, 127, 128, 129, 148, 151, 155, 159, 178, 189, 191, 192, 193, 195, 198, 201, 205, 206, 207, 208, 209, 210, 211, 212, 213, 214, 215, 216, 218, 227, 239, 241, 242, 243
salt(s), 17, 40, 43, 46, 49, 50, 51, 52, 54, 64, 66, 75, 82, 91, 92, 128, 162, 163, 165, 166, 167, 168, 169, 170, 172, 173, 174, 175, 176, 178, 185, 186, 188

sample, 5, 6, 7, 8, 11, 14, 24, 25, 27, 28, 31, 32, 203, 219, 220, 226, 227, 228, 231, 232, 235, 238, 241
sampling, 100, 101, 137, 142, 219, 223, 243
sampling error, 100
scarcity, 48
school, 137, 145
scientific community, 2, 41
scientific knowledge, 147
scores, 230, 233
seafood, 105
search(es), 208, 227
seasonality, 137
secretion, 14, 134
security, 156
sedimentation, 122
sediments, 87
seed, 45, 169, 175, 218, 227, 241
selectivity, 208, 220, 221
self-assembly, 205
semiconductors, 220
sensing, 231, 241
sensitivity, 6, 8, 10, 12, 14, 16, 18, 26, 27, 28, 29, 70, 104, 117, 122, 144, 179, 200, 204, 207, 220, 221
sensitization, 202
sensor technology, 220, 221
sensors, 29, 217, 220, 222, 224, 225, 226, 231, 232, 233, 234, 235, 239, 243
separation, 31, 102, 103, 119, 124, 144, 149, 154, 219, 229, 230, 233, 235, 238
sepsis, 81, 197
septic shock, 197
septum, 223
sequencing, 11, 12, 14, 15, 178, 190
series, 46, 47, 69, 98, 100, 231
serine, 22
serum, 29, 48, 160
severity, 1, 95, 97
sewage, 18, 192, 218
sex, 143
shape, 101, 110, 113, 118, 191, 203, 206, 215
sheep, 13, 68, 135, 137, 141, 148, 150
shellfish, 18, 35
shelter, 198
Shigella, vii, 3, 4, 8, 10, 30, 121, 134, 151
shock, 195
side effects, 44
signal transduction, 29
signalling, 192, 195
signals, 5, 28, 134, 224, 231, 236
signs, 137
silica, 219
silver, 209
similarity, 11, 21, 123, 134
simulation, 101, 124
single-nucleotide polymorphism, 37
sites, 12, 136, 148, 151, 198, 204, 210
skeletal muscle, 72
Slovakia, 141
sludge, 18
small intestine, 134
smoking, 51, 92
SNP, 12
society, 156
sodium, 43, 60, 64, 72, 73, 76, 77, 78, 79, 180, 201, 202
software, 5, 12, 24, 231
soil, 42, 78, 145, 150, 212
sol-gel, 25, 34
solid phase, 217, 240, 243
solubility, 46
solvents, 219, 225
South Dakota, 140
South Pacific, 52
Southern blot, 11
soy bean, 45
Spain, 1, 57, 63, 71, 93, 116, 139, 163
species, ix, x, 4, 8, 10, 11, 12, 13, 14, 15, 16, 17, 19, 20, 22, 36, 38, 42, 43, 44, 46, 50, 51, 54, 55, 60, 64, 65, 73, 74, 81, 85, 87, 121, 128, 132, 133, 135, 136, 137, 155, 160, 161, 166, 167, 174, 178, 181, 182, 189, 191, 192, 194, 195, 196, 199, 209, 218, 222, 231, 242
specificity, 2, 4, 5, 6, 7, 8, 9, 10, 13, 14, 16, 17, 19, 20, 25, 28, 30, 31, 220
spectroscopy, 240
spectrum, vii, 3, 5, 6, 18, 21, 28, 30, 51, 142, 227
speed, 5, 30, 239
spontaneous abortion, 77
spore, 191
spreadsheets, 110, 111
sprouting, 227, 241
SPSS, 224
Sri Lanka, 60
stability, 18, 25, 58, 62, 70, 78, 79, 85, 88, 89, 159, 167, 191, 199, 207, 237
stabilization, 50, 195
stages, viii, ix, x, 49, 54, 55, 61, 93, 94, 95, 97, 98, 99, 105, 189
standard error, 240
standardization, 50, 243
standards, 126, 179, 224
staphylococci, 61, 63, 64, 65, 75, 84, 86
Staphylococcus, ix, 4, 10, 11, 13, 15, 16, 18, 22, 26, 28, 30, 33, 34, 38, 43, 64, 71, 72, 77, 79, 83, 84, 87, 98, 106, 110, 116, 127, 128, 155, 159
starch, 40, 58, 82, 167, 176, 180, 181

starch granules, 58
starvation, 204
statistics, 99
steel, 65
sterile, 46, 108, 223
sterilisation, 96
stock, 223
stomach, 65, 136, 142
storage, 4, 16, 22, 23, 25, 27, 43, 47, 51, 54, 59, 62, 64, 66, 70, 72, 74, 76, 78, 81, 84, 85, 87, 89, 92, 96, 99, 115, 116, 117, 141, 145, 149, 171, 173, 174, 177, 181, 182, 185, 193, 206, 223, 224, 228, 231, 241
strain, vii, 3, 4, 5, 6, 9, 12, 13, 15, 16, 22, 31, 42, 45, 46, 47, 48, 51, 59, 64, 66, 68, 73, 77, 82, 83, 84, 87, 108, 144, 153, 164, 176, 188, 192
strategies, 4, 21, 27, 29, 30, 43, 44, 48, 55, 98, 143, 145, 193, 210, 215
strength, 243
stress, 17, 108, 117, 192, 195, 204, 208, 209, 210, 212
stressors, vii, 3, 15, 22, 30
structure formation, 195
subgroups, 46, 224, 226, 228, 230, 232, 234, 238
substitutes, 40, 78
substrates, 23, 26, 28, 40, 50, 69, 73, 80, 173, 179
subtraction, 100, 133
sucrose, 51, 59, 73, 169
Sudan, 181
sugar, 55, 78, 83, 156, 157, 160, 169, 170
sulfur, 220, 225, 227
sulphur, 47, 169, 171
summer, 137, 162, 174
suppliers, 165
supply, vii, 3, 4, 15, 16, 21, 22, 27, 67, 147
suppression, 54
surface area, 203
surface chemistry, 23
surface component, 196
surface layer, 191
surface properties, 25
surface structure, 198
surveillance, 11, 16, 21, 38, 136, 147
survival, viii, ix, x, 7, 23, 36, 42, 44, 47, 63, 74, 81, 82, 83, 86, 93, 107, 109, 113, 116, 117, 127, 133, 145, 149, 153, 174, 184, 189, 195, 197, 198, 201, 206, 208, 210, 214, 215, 239, 241
survival rate, 174
survivors, 114, 143
susceptibility, 18, 29, 34, 119, 199, 203, 215
Sweden, 140, 193
swelling, 143
Switzerland, 150, 212
symbiosis, 86
symptoms, 4, 80, 132, 134, 137, 143
syndrome, 86, 143, 150, 154
synergistic effect, 42, 160
synthesis, 5, 6, 8, 31, 37, 133, 134, 197, 213
systems, vii, 3, 14, 21, 22, 23, 30, 31, 38, 147, 160, 170, 183, 191, 192, 204, 243

T

Taiwan, 91, 163, 178, 180
tanks, 54, 173
tannins, 45
target population, 1
target response, 12
targets, 4, 6, 9, 16, 20, 23, 30, 134, 211, 237
taxonomy, 187, 214
technology, 4, 5, 28, 30, 32, 34, 56, 60, 74, 80, 85, 156, 240, 241, 242
telephone, 124
TEM, 203
temperature, 7, 23, 43, 49, 52, 56, 61, 64, 65, 75, 76, 77, 81, 87, 96, 99, 110, 114, 115, 116, 125, 128, 136, 146, 162, 164, 165, 166, 168, 173, 174, 175, 181, 185, 190, 197, 212, 213, 219, 223, 224, 225, 226, 227, 231
tension, 7, 23
tetanus, vii, 3, 28, 30
Thailand, ix, 55, 155, 163, 176, 177, 187, 217
theory, 121
therapeutics, 215
therapy, 44, 81, 143, 145
thermodynamics, 108
thermophiles, 65, 166
thiamin, 58, 67, 90
threat(s), 4, 12, 25, 26, 192, 204
threshold, 12, 120, 121, 122
thrombocytopenia, 143
time, 6, 8, 9, 11, 13, 15, 16, 21, 22, 24, 25, 26, 27, 28, 30, 31, 38, 44, 49, 53, 62, 65, 81, 96, 98, 99, 101, 103, 104, 109, 110, 111, 112, 113, 114, 115, 116, 117, 123, 142, 143, 162, 175, 217, 218, 219, 221, 223, 224, 225, 227, 237, 238, 239, 243
timing, 145
tin, 220
tissue, 19, 52, 60, 151
TLR, 197
total product, 164
toxic effect, 43, 45
toxic products, 40
toxic shock syndrome, 13, 33
toxicity, 135, 176

toxin, 4, 6, 9, 21, 24, 25, 27, 28, 30, 36, 37, 42, 43, 128, 132, 134, 135, 136, 147, 149, 150, 151, 152, 153, 154, 158, 214
tracking, 16
trade, 4, 66, 218
tradition, 60
training, 226, 235, 237
traits, 12, 17
transcription, 5
transducer, 29
transferrin, 200, 207
transformation(s), 111, 236
transition, 31, 214
translation, 108
transmission, viii, 9, 15, 18, 21, 99, 131, 148, 194
Transmission Electron Microscopy (TEM), 203
transport, 17, 47, 141, 153, 159
transportation, 106, 179, 193
trial, 27, 78, 80
trypsin, 46
TTGE, 178
tungsten, 220
Turkey, 106
typhoid, 27, 192, 208, 215
typhoid fever, 192, 208
tyramine, 43, 51, 54, 56, 64, 84, 88, 171
tyrosine, 171, 172

U

UK, 83, 126, 127, 137, 148, 149, 153, 180, 181, 185, 211
uncertainty, 100, 102, 103, 104, 108, 124, 128
United States, 4, 32, 36, 38, 98, 105, 149, 151, 192, 211, 214, 218, 240
universal gas constant, 114
updating, 1
urine, 29
USDA, 105, 106, 109, 124, 126, 240
USSR, 80
UV, 43, 195
UV light, 43
UV radiation, 195

V

vacuole, 166
vacuum, 47, 71
Valencia, 37
validation, 1, 25, 31, 38, 123, 124, 226, 240
validity, 121, 124
values, 12, 28, 60, 62, 96, 97, 99, 100, 101, 104, 105, 108, 114, 121, 123, 124, 142, 201, 204, 207, 237
variability, viii, 19, 20, 37, 43, 55, 64, 93, 99, 102, 103, 104, 115, 119, 122, 124, 128, 137
variable(s), 6, 22, 25, 42, 95, 98, 99, 100, 102, 103, 104, 107, 109, 114, 115, 121, 122, 137, 190, 191, 196, 197, 198, 209, 221, 228, 229, 230, 232, 233, 237
variance, 100, 221, 224, 229, 232
variance-covariance matrix, 224
variation, 7, 11, 22, 52, 54, 56, 104, 115, 121, 231, 237, 243
vector, 224, 226, 230, 233, 237
vegetable oil, 220, 241
vegetables, viii, ix, 10, 17, 20, 23, 34, 39, 40, 46, 47, 52, 53, 55, 56, 79, 80, 84, 89, 138, 139, 141, 142, 144, 146, 155, 156, 157, 158, 159, 161, 162, 163, 164, 165, 166, 167, 168, 169, 170, 171, 172, 173, 175, 176, 177, 178, 179, 180, 181, 183, 185, 187, 193, 218, 239, 242
vehicles, 42, 136, 141, 218
vero, 152
vertebrates, 45
Vietnam, ix, 155, 163, 176
viral infection, 18
viruses, 10, 18, 19, 20, 25, 37, 44, 45, 63, 88, 97, 128
viscosity, 28, 67
vitamin B1, 67, 87
vitamin B12, 67, 87
vitamin B2, 67
vitamin C, 171, 172
vitamin E, 92
vitamins, ix, 49, 67, 155, 156, 157, 161, 165, 169
vomiting, 143, 171

W

Wales, 33, 138, 140
warrants, 18, 21
wastewater, 8, 35
water supplies, 4, 20, 192
wavelet, 239
wealth, 6
web, 214
well-being, 69
West Africa, 60
West Indies, 176
wheat, 46, 56, 57, 58, 73, 74, 78, 80, 81, 83, 88, 91, 92
wildlife, 210
wine, 17, 40, 43, 156, 219, 221, 242, 243
winter, 175
women, 81, 96

workers, 141, 218

Y

yeast, ix, 17, 40, 42, 43, 57, 74, 82, 86, 155, 166, 169, 176, 186, 187
yield, 9, 19, 67, 105
young adults, 74, 86
young women, 82

Z

zinc, 46, 67, 220
zinc oxide, 220